THE CHLORINE REVOLUTION

THE CHLORINE REVOLUTION

Water Disinfection and the Fight to Save Lives

Michael J. McGuire

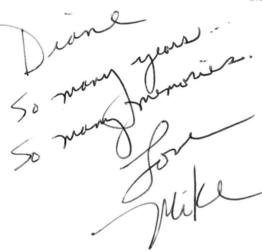

American Water Works Association

The Chlorine Revolution: Water Disinfection and the Fight to Save Lives

Disclaimer

AWWA Sr. Project Manager: Gay Porter De Nileon
Technical Editor: Nancy Zeilig
Cover Art: Daniel Feldman
Interior Design: Cheryl Armstrong
Production: TIPS Technical Publishing, Inc.
Front cover photo: Summer scene of drinking water from street pump in New York City, circa 1910–1915. Published by Bains News Service, obtained from the George Grantham Bain Collection, Library of Congress.
Back cover photos, clockwise from top right: A bacteriologist working in the Louisville Water Company laboratory, circa 1896. Courtesy of Louisville Water Company. Laying wooden stave pipe, around the turn of the 19th century. Courtesy of Denver Water, February 2013. Film still from the Everett Collection/Shutterstock.

Library of Congress Cataloging-in-Publication Data

McGuire, Michael J., 1947-
  The chlorine revolution : water disinfection and the fight to save lives / Michael J. McGuire.
    pages cm
  Includes bibliographical references and index.
  ISBN-13: 978-1-58321-920-1 (hbk.)
  ISBN-10: 1-58321-920-X (hbk.)
  ISBN-13: 978-1-58321-913-3 (pbk.)
  ISBN-10: 1-58321-913-7 (pbk.)
  1. Water—Purification--Chlorination—United States—History. 2. Water—Purification—Chlorination—Health aspects—United States—History. 3. Water quality—United States—History. I. American Water Works Association. II. Title.
TD462.M35 2013
628.1'662--dc23

2012049592

Printed in the United States of America

American Water Works Association

6666 West Quincy Avenue
Denver, CO 80235-3098
303.794.7711
www.awwa.org

To my wife, Deborah Marrow, and our children,
David M. McGuire and Anna M. McGuire.

Contents

Acknowledgments

I especially want to thank Laura Cummings and Joe Bella of the Passaic Valley Water Commission in Clifton, New Jersey for allowing me access to their archives and museum. Their cooperation and assistance made this project possible. Several public and private library research staff members provided invaluable help, including John Beekman, Jersey City Public Library; Bruce Bardarik and Mary Wilson, Paterson Public Library; Myles P. Crowley at the Massachusetts Institute of Technology Archives; and Maureen O'Rourke, New Jersey Historical Society. Kurt Keeley, Heidi Riedel, and the library staff at AWWA provided an astonishing amount of valuable material. Many thanks to Gary Schaeffer of Siemens Water Technologies Corp. for giving me access to the old Wallace and Tiernan files. I am grateful to my niece, Eliza Bemis, for doing some important research for me at MIT's library.

I owe a big debt of gratitude to my friends and colleagues who read drafts of this book and provided helpful comments: Tom Yohe, Joe Bernosky, Dr. David L. Rabin, Julius C. Calhoun, Esq., Laura Cummings, Dr. Peter Vinten-Johansen, and Kurt Keeley. I want to particularly thank the great grandson of John L. Leal, Hank Morehouse, for reviewing Leal's biographical chapter. I am also grateful to the Leal and Fuller families for giving me access to photographs and documents related to these two extraordinary men. Thanks also to Dr. William Gurfield for his medical insights.

Many thanks to my wife, Deborah Marrow, and our children, David and Anna, for putting up with the many hours that I was away from them and for my distracted consciousness while I was in their presence. Friends and family have observed that for days at a time, I was lost in the late 1800s instead of paying attention to the early 21st century. Deborah provided a very helpful final read of the manuscript at a critical juncture—many thanks.

I am very grateful to Gay Porter DeNileon and the publications staff of the American Water Works Association. They went out of their way to make the publication process enjoyable.

I have known my editor, Nancy Zeilig, for over 25 years. She was the editor of the *Journal - American Water Works Association* for many years. It was a special treat to work with her again on this project. She is quite simply the best editor I have ever worked with. Couple that expertise with the nicest personality and tons of patience and you have a good idea of what she is like. Words fail me. Thanks.

Abbreviations

AMA	American Medical Association
APHA	American Public Health Association
AWWA	American Water Works Association
EJWC	East Jersey Water Company
gpm	gallons per minute
JCWSC	Jersey City Water Supply Company
mgd	million gallons per day
MIT	Massachusetts Institute of Technology
ppm	parts per million
PVWC	Passaic Valley Water Commission
USEPA	U.S. Environmental Protection Agency

Foreword

Many years ago, an engineer colleague told me that if I ever wanted to thank one person for the excellent state of public health in the United States, make sure I thanked an engineer. When I presented a puzzled mien, he proceeded to explain the role of the engineering profession in developing methods for treating wastewater and disinfecting drinking water. I certainly understood the difficulty in, and importance of, treating wastewater, but I scoffed at the disinfection aspect, especially the difficulty. "What's the big deal with disinfecting the drinking water? Just add some chlorine," I replied. He responded with a glare that I suspect I'd given to some of my students over the years, and proceeded to lecture me on drinking water disinfection with chlorine. Little did I realize that, lo these many years later, I would be penning the foreword to a book titled, *The Chlorine Revolution: Water Disinfection and the Fight to Save Lives,* and being amazed at and enthralled by the wealth of information in Dr. Michael J. McGuire's book. Suffice it to say that in writing this book, Mike added two more hats to his engineering one—those of historian and detective. The book is meticulously researched and documented, and well-written.

Even I, as a non-engineer (hydrogeologist), have known Mike by reputation for a number of years. After all, he *is* a fellow water wonk and (unlike me) a member of the National Academy of Engineering, an organization that does not suffer fools. I first met him several years ago when we both served on a National Research Council committee, and that in-person experience did nothing but enhance his reputation and my perception of his expertise.

Whence the origins of this book? It is an outgrowth of a paper Mike published in the *Journal - American Water Works Association* in 2006, "Eight Revolutions in the History of U.S. Drinking Water Disinfection." The title reminded me of my aforementioned comment disparaging the difficulty of drinking water disinfection. If a person like Mike sees eight revolutions, then I must be missing something. For once, my hunch was right. I had missed something about disinfection.

Foreword

As Mike states in his Preface, the article was well-received, but he was not satisfied:

The first revolution fascinated me. What was the decision-making process behind the first continuous use of chlorine in the United States? Why was chlorination of a water supply first accomplished in Jersey City, New Jersey? Who were the people primarily responsible for implementing chlorination? What in their experience led them to take such extraordinary steps? What were the consequences of their actions? These questions led me to begin research for this book and to explore in depth this historic era in drinking water disinfection.

After reading the above, I was hooked. I wanted to learn what the big deal was. After all, I knew enough about drinking water chlorination to know that it entails adding a *poison* to something about to be ingested by humans. You'd better have the biology, chemistry, and engineering right.

Now comes the surprising part. Despite the first sentence of my foreword, it was two men not trained as engineers who played the key roles in the struggle to chlorinate drinking water supplies: Dr. John L. Leal, a medical doctor turned public health and water expert, and George Warren Fuller, an MIT-trained chemist who became the foremost U.S. sanitary engineer before his 40th birthday. Leal was the one who speculated upon the health benefits of chlorinating public water supplies on a continuous basis. His research into bacteriology and the development of germ theory of disease led him to the works of Pasteur, Koch, and others to posit that chlorine dose to drinking water would be the key to drastically reducing or eliminating health threats from typhoid and other waterborne diseases. Fuller's prodigious engineering skills would implement Leal's theories; all he had to do was develop a continuous feed system that would *safely* chlorinate 40 million gallons of water per day!

Mike's book goes into incredible detail. The court trials. The chemophobia that was prevalent at the turn of the last century. The pioneering bacteriologists. John Snow and his cholera map. The understanding of what caused disease (no, not miasmas). Even the improvements to the microscope that led to advances in bacteriology do not escape Mike's sleuthing.

Recalling the first sentence of this foreword, I must now confess that for years I have been telling my students whom to thank. But now I know a lot more: the *whole* story and *exactly* whom to thank. And I know exactly what I will tell the students in my *US Water Management* class tomorrow morning.

Foreword

The Chlorine Revolution: Water Disinfection and the Fight to Save Live is an extraordinary book telling a remarkable story. Engineer, water wonk, or not, you'll be glad you read it.

I am.

Michael E. Campana, PhD
Past President, American Water Resources Association
Professor of Hydrogeology and Water Resources Management
Oregon State University
Corvallis, OR
21 January 2013

Preface

On November 30, 2005, the editor of the *Journal - American Water Works Association* invited me to write an article on the history of drinking water disinfection, including advances in technology and research. The article would be included in a special March 2006 *Journal* issue commemorating AWWA's 125th anniversary. The assignment captured my imagination except for the unfortunate fact that the manuscript was due in only 37 days. In that short time, I was able to gather documents, shape an outline, and write the article. I was physically with my family between Christmas and New Year's Eve 2005, but my mind was occupied with early twentieth century events as I wrote the 12,000-word article.[1] Rather than recount a tedious series of technological advances, I decided to present the most important disinfection breakthroughs by designating eight "revolutions" that I considered critical to the progress of public health protection. I defined revolutions in the context of the article as "events or series of events that so changed the practice of disinfection in the United States that it was never the same again."

The published article was well received, but I was not satisfied. The first revolution fascinated me. What was the decision-making process behind the first continuous use of chlorine in the United States? Why was chlorination of a water supply first accomplished in Jersey City, New Jersey? Who were the people primarily responsible for implementing chlorination? What in their experience led them to take such extraordinary steps? What were the consequences of their actions? These questions led me to begin research for this book and to explore in depth this historic era in drinking water disinfection.

The resulting book focuses on a 20-year period, 1890 to 1910, and a partnership between two innovative figures in the field of drinking water disinfection—Dr. John L. Leal and George Warren Fuller. These two men came from different backgrounds and

---

1 McGuire, M.J. "Eight Revolutions in the History of U.S. Drinking Water Disinfection." *Journal - American Water Works Association*, 98:3 123–49.

initially chose different career paths, but they both ended up as experts in drinking water treatment and public health protection. Their collaboration resulted from their work together in professional societies and, most importantly, from a critical meeting on June 19, 1908, in New York City. This book explains the significance of that meeting, the events leading up to it, and the impact on U.S. public health resulting from the eradication of typhoid fever and other waterborne diseases.

As an engineer accustomed to writing and editing technical books and articles, I had to learn how to work as historians do. Secondary sources were available in libraries or via the Internet, but primary sources required more sleuthing. I knew that neither Leal nor Fuller had left journals or diaries, and, to my knowledge, there are no archives of their private papers. I also checked with many of their descendants, and no one had a treasure trove of documents buried in a trunk in an attic. In early 2006, I made a short deviation from a business trip to investigate the files at Boonton Reservoir, site of the first chlorination system; unfortunately, nothing of interest showed up. I traveled to New Jersey in January 2008 to see if I could find out more about the two trials that seemed, from my secondary research, to be important drivers in the widespread adoption of chlorination.

In the Jersey City Library, the librarian guided me to a set of the first six volumes of the twelve-volume transcript of the two trials. I skimmed through the volumes while I was in Jersey City, but they contained nothing about using chlorine for disinfection. The library also had some documents that gave credit for adding chlorine to the water supply to someone other than John L. Leal. I certainly needed more than the library's collections could supply.

During the trip, I remembered that Laura Cummings, an old friend from California, worked for the Passaic Valley Water Commission (PVWC), located somewhere in northern New Jersey. I got her on the phone and asked if she knew of any old documents that might help in my research. She urged me to come over to the treatment plant in Little Falls and look at some of its records. In the middle of our conversation, it dawned on me that she worked at the treatment plant that was built on the original site of George Warren Fuller's groundbreaking mechanical filtration plant. I told her I would be right over.

After I had checked out a disappointing pile of old administrative reports, Laura suggested that we go over to the "museum" and look for documents there. The museum is about the size of a large garage, and when we opened the door, a musty smell wafted out from the unheated, darkened interior. Inside, the dim light revealed old chlorinators, valves, and wooden pipes scattered around the floor. In the distance, I could see a large bookcase that covered most of the back wall. As I walked toward it, I became more excited with every step. The shelves were stacked with old books and papers.

Incredibly, the bookcase contained the library of the law firm Collins & Corbin. William H. Corbin was the attorney who defended the private water company in the trials I was investigating. As I note at the end of the fifth chapter of this book, the PVWC is the public water agency that was formed from the remnants of several private water companies, including two that John L. Leal worked for. Somehow, the law library from Corbin's firm survived and ended up in the PVWC museum. After spending several hours looking through the books and papers, I identified three complete transcripts of the two trials. I knew then that I had the essential archival information I needed for the book.

In this book, I cover the two trials in depth. Transcripts of testimony can be as dry as dust, but there were flashes of history-defining insights during these trials:

- The witnesses were under oath. Although that did not prevent them from stretching the truth, the transcripts are the best history we have of the events surrounding the first use of chlorine. Because of the transcripts, it was possible to construct a clear timeline of how the decision to add chlorine was made, who made it, and under what conditions.
- The voices of the witnesses can be heard in the transcripts. The testimony gives us a fascinating glimpse into the professional demeanor and personality of the many expert witnesses. I am not aware of any surviving films or recordings of the voices of these superstars of sanitary science, but through their printed testimony, we hear them speak.
- The witnesses described significant technological advances during their testimony. The very first mention of the orthotolidine reagent to determine low concentrations of chlorine in water occurred during this trial. For decades, this test (after many modifications) was the best tool water utilities had to ensure that a chlorine residual was maintained in the water delivered to their customers.

- "Pure and wholesome" was defined. The court provided a definition that I had not seen anywhere else. The term "pure and wholesome," when applied to water, was a nineteenth-century (and possibly earlier) doctrine of English common law. Its echoes survive today in several national laws and in the California Safe Drinking Water Act, which requires all drinking water in the state to be pure, wholesome, and potable.
- Chemophobia haunted the testimony. Chemophobia, the fear of chemicals, was prevalent at the turn of the twentieth century and continues to be an issue. Chemophobia almost derailed the use of chemical disinfectants then, and it often dominates discussions of drinking water safety today.
- The progress of water bacteriology was recorded. The witnesses' testimony tracked the development, pros and cons, and interpretation of microbiological analyses in the early twentieth century. Bacteriological science was changing rapidly during this period, and the trial transcripts offer us a window into those changes.

As readers may gather from the content of this Preface, I am a drinking water guy. I was trained as a civil engineer at the University of Pennsylvania and earned M.S. and Ph.D. degrees in environmental engineering at Drexel University. I have always worked for water utilities or as a consultant helping water utilities achieve their mission of providing safe drinking water to consumers. I have brought all of the knowledge and perspective gained during my education, training, and experience to the development of this book. I hope both technical and lay readers enjoy it as much as I enjoyed learning about this period and writing about the chlorine revolution.

Michael J. McGuire
Santa Monica, California
October 26, 2012

Journey to Launch a Revolution

"**revolution** . . . a : *sudden, radical, or complete change* . . . e : *a changeover in use or preference especially in technology*"

—Merriam-Webster, "Revolution"

On a fair Jersey City day in mid-June 1908, Dr. John L. Leal, a physician turned water expert, sat in the law offices of Collins & Corbin to discuss a matter of great importance with William H. Corbin. Leal worked for the Jersey City Water Supply Company, which had hired Corbin to protect its contract interests. Corbin had recently completed a months-long trial defending his client's actions under a contract signed in 1899. The judge's decision on May 1, 1908, had largely supported the company's contractual right to be paid $7.6 million (more than $175 million in current dollars) for building one of the largest new water supplies on the East Coast. The judge agreed that the Jersey City Water Supply Company should be paid for constructing a large dam, creating Boonton Reservoir, and for laying a 23-mile pipeline to supply Jersey City.

But there was a problem. The judge found that the company was not at all times providing the city with water that was "pure and wholesome." Because of sewage contamination in the watershed, water from the Rockaway River that contained high concentrations of bacteria would short-circuit the natural purifying processes of the reservoir during rainstorms two or three times a year. In his May 1 opinion, the judge made it clear that sewers could capture sanitary wastes and divert the contaminated materials away from Jersey City's water supply. The cost of the sewers, which was significant, would be deducted from the $7.6 million contract price. However, Dr. Leal was convinced that because of contaminated runoff from agricultural and urban lands, the sewers would not solve the problem identified by the judge. He also saw an opportunity to end waterborne disease in Jersey City, but he knew that would require a revolution in drinking water treatment.

Corbin and Leal knew they had only one chance to overturn the negative part of the judge's decision. The final decree was issued by Chancery Court Vice Chancellor Frederic William Stevens on June 4, 1908. Between May 1 and June 4, Corbin and Leal had convinced Stevens that they should be given a chance to install "other plans or devices" that would reduce bacterial concentrations in the delivered water, which was the intent of the judge's requirement to build sewers. Thus, in his final decree, Stevens referred the assessment of sewer costs and the suitability of the "other plans or devices" to one of the special masters of the New Jersey Chancery Court, the Honorable William J. Magie. Leal was given only three months from June 4 to develop his alternative plan, and the urgent meeting in Corbin's law offices on June 19 was to chart the next steps in developing this alternative.

Dr. Leal was uniquely qualified to develop and implement his radical idea. He was a physician educated at Princeton and trained at one of the foremost medical schools in the country—Columbia College of Physicians and Surgeons. Fascinated by the new science of bacteriology, he had taken courses and performed experiments that had furthered his knowledge in this emerging field. He was also an expert in public health. For 10 years, he had been the health officer for Paterson, New Jersey (a significant manufacturing center in the nineteenth and early twentieth centuries), and he had dealt with all of the public health issues of the day—smallpox outbreaks, diphtheria, typhoid fever epidemics, high rates of infant mortality, contaminated milk supplies, collection and disposal of sewage, and protection of public water supplies.

Leal's revolutionary idea was nothing less than introducing a "poison" into the Jersey City water supply to kill the bacteria that were present in high concentrations two or three times per year. As a physician, he was well aware of the risks of putting a poison into a water supply. However, he had specialized knowledge that this poison, chlorine, could be used safely and, in small concentrations, would be highly effective in controlling bacteria. Leal was aware of the historical uses of the poison—chloride of lime (a convenient form of chlorine)—as a disinfectant to purify contaminated households after an infectious disease had struck a family. He also knew about some limited uses of chlorine in European water supplies over the previous 10 years. During the period

1898–1901, his interest in chlorine had prompted him to conduct laboratory experiments on its effectiveness in killing bacteria.

Leal's proposal was not to have the Jersey City Water Supply Company conduct more laboratory experiments with chloride of lime in the hope of convincing the special master that this approach would be judged a suitable alternative plan or device to replace the use of sewers in the Rockaway River watershed. Leal knew he had to recruit the best talent in the sanitary field of water supply and have these experts test the efficacy of chloride of lime in their laboratories. Although Leal could have done the tests himself, he understood that outside experts would carry more weight with the special master.

In addition, Corbin and Leal had great respect for Jersey City's attorney, James B. Vredenburgh, who had made short work of partially formed arguments in the original trial. Leal also recalled that Vredenburgh had made effective use of his own expert witnesses. There was no doubt that Vredenburgh would recruit top professionals in the water sanitary field to attack the safety, effectiveness, and reliability of using chlorine as a disinfectant. Leal needed to supply overwhelming evidence from top researchers so the plaintiffs' experts would not cancel out the defendants' experts.

Leal's biggest challenge was that no water company or city in the United States was using chlorine as a disinfectant as of June 19, 1908. Leal was convinced that the only course open to the private water company he worked for was to install an actual chlorine feed system, treat the entire Jersey City water supply, and present the improved bacteriological evidence at the hearing to be conducted by the special master. But his course of action involved a major problem: no one had designed and constructed a reliable chlorine feed system capable of treating 40 million gallons per day—a huge flow of water. Leal had experimented with an electrochemical generation method for producing chlorine, but he knew it was not ready for full-scale application. He did know from his years as a municipal health officer that reliable sources of chloride of lime existed. He also knew that only one man could design a reliable feed system for chloride of lime in the insanely short time period available.

Leal presented Corbin with a two-pronged plan. First, he had to recruit world-class experts in the field of chemistry and bacteriology to perform the laboratory-scale disinfection testing. Second, he had to build a full-scale operating plant and show how effective chloride of lime could be.

Leal realized he would have a golden opportunity to recruit the experts he needed in August, when the annual conference of the American Public Health Association (APHA) would be held in Winnipeg, Manitoba, Canada. He knew that several former presidents of APHA were active researchers in this area, and he believed they would likely be attending the conference. The first part of his plan would have to wait a few weeks, but the second part of his plan needed to be started right away; the only thing in his way was the Hudson River.

Leal left the Collins and Corbin law offices at 243 Washington Street in Jersey City and walked three blocks to the ferry terminal on the Jersey City waterfront. If he bought a *New York Times* or a local newspaper on the docks, he would have seen that the news was dominated by the completion of the Republican presidential ticket with the nomination of William Howard Taft and James S. Sherman. At the terminal, he boarded a ferry to take him across the Hudson to a rendezvous with his co-conspirator in this revolutionary plan.

As Leal stood on the deck of the ferry waiting to depart, he might have taken a look at the water of New York Harbor, which he was about to traverse. He knew from his work that it was polluted. However, it did not take a medical degree or years as a health officer to know the harbor was filled with contamination.

A glance toward the shore would have shown garbage and unidentified organic matter piled up and sloshing around from the propulsion effects of the ferries. Leal would have been able to see the swarms of flies feasting on the refuse. The water would likely have had an oily sheen that streaked and blackened the hulls of commercial and pleasure boats in the area. Because the day was fairly warm, he would have been able to see bubbling gases rising to the surface from the huge sediments created by unrestricted discharges of sewage from both New York City and the cities of New Jersey. He might have read about the sea of floating garbage that had been spotted offshore.

Still, it would have been the noise of the train sheds and the transfer of cargo to ships on the waterfront and the rotten egg smell from putrefaction of the floating waste and sewage sediments that would have dominated his attention.

The thoughts that went through Leal's mind during that short ferry ride to the Manhattan ferry terminal must have been dramatic. After today, there was no turning back. He was about to set in motion a series of events that would brand him as a visionary or a criminal. He knew that jumping ahead of the entire sanitary engineering discipline and the water treatment practices of decades was fraught with risks. In the field of public health, in which human lives were at stake, there was no room for error on this scale.

From the ferry terminal on the New York City side, it was only a half-mile walk to Leal's destination. As he walked through the crowded streets, he would have dodged horse-drawn wagons and carriages and the new automobiles that had begun to take over the streets of New York City.

There is no record that Leal made an appointment with his co-conspirator, but we know he arrived at the 170 Broadway office of the consulting engineering firm of Hering and Fuller sometime during that day in June. There he sat down with two individuals to discuss an assignment and to map out the execution of his radical idea. Present at the meeting was one principal of the firm, George Warren Fuller. The other principal, Rudolph Hering, was traveling in Europe at the time of the meeting. Also present was a junior member of the firm, George A. Johnson.

Rudolph Hering was 61 years old and the senior member of the firm. He was famous in the engineering field for his work in the areas of water supply and waste disposal; his clients included the cities of Philadelphia, Chicago, and New York. However, Leal had come to see the junior partner of the firm, George Warren Fuller. At the time of the meeting, Fuller was only 39 years old, but he was at the height of his intellectual power and the foremost sanitary engineer in the early twentieth century. He was also an expert chemist and bacteriologist. He certainly understood what Leal wanted to accomplish and was aware of up-to-date investigations of the use of chlorine to disinfect water supplies. In fact, he had investigated a small-scale application of chlorine to the water at Louisville, Kentucky, in his landmark filtration studies.

Everyone at the meeting knew they had only one chance to get this right. If the engineering part of the plan failed, none of the theoretical benefits of water disinfection developed by Dr. Leal would be worth much. A failure in this high-profile lawsuit might set the idea of drinking water disinfection back several decades.

The good news was that George Warren Fuller had designed and installed a feed system for a critical chemical six years earlier at the Little Falls, New Jersey, water treatment plant, and this feed system had been operating successfully. Leal was familiar with Fuller's feed system for sulfate of alumina (now known as aluminum sulfate or, more simply, alum) at the Little Falls plant because Leal had been responsible for the plant's performance for the East Jersey Water Company since its startup in 1902. Leal asked Fuller to repeat his successful design and adapt it to feed chloride of lime in very small doses (fractions of a part per million) to a very large flow—40 million gallons per day.

There is no record that a contract was signed on that day; it is more likely that the work was agreed to on the basis of a handshake. We know, however, that the work was completed, and successful chlorination of water from Boonton Reservoir began on September 26, 1908—99 days after that fateful meeting at 170 Broadway in Manhattan. The purification of water supplies has never been the same since. Though Leal's journey from Jersey City to New York City was short in distance, it was long in its impact on public health. Millions of lives have been saved since Dr. John L. Leal made that journey to launch a revolution. This partnership between a physician and an engineer brought about a permanent change in drinking water treatment, and a court case brought the revolution to the attention of the world.

2

Demons, Miasma, and the Death Spiral

"A person with typhoid in such a place soon fills the air with the specific miasmas . . . "
— Barnes, *Great Stink of Paris*, 194–5.

Disease and death in the U.S. during the late 1800s and early 1900s was commonplace. In the preface of his 1908 book on typhoid fever, civil engineer and bacteriologist George C. Whipple stated the risks succinctly: "Few people, according to vital statistics, die of old age; almost every one dies of disease; and when your turn and mine shall come to shuffle off this mortal coil, we shall have the unwelcome and involuntary lot of more than two hundred different diseases."[1] Many of these diseases were epidemic and waterborne, and their death tolls helped to keep the average life expectancy in 1900 at 47 years.[2]

Before French chemist and microbiologist Louis Pasteur established the germ theory of disease in the 1860s and 1870s, several primitive concepts were believed to explain the causes of disease. Chapter 2 of William T. Sedgwick's celebrated book on public health, published in 1902, contained a concise history of early disease theories. Sedgwick started with the Demonic Theory of Disease, which held that sick people were possessed by an evil spirit or demon.[3] Alongside this theory was the belief that epidemics were God's punishment for sinful souls.[4]

The Four Humors Theory of Disease was developed in ancient Greece during the time of Hippocrates.[5] Hippocrates lived from 470 to 360 BCE[6] and was known even in ancient times as the Father of Medicine. As described by Sedgwick:

"The dominating theory of disease was the humoral, which has never ceased to influence medical thought and practice. According to this celebrated theory, the body contains four humors—blood, phlegm, yellow bile, and black bile—a right proportion and mixture of which constituted health; improper proportion and irregular distribution, disease."[7]

Prior to the nineteenth century, the humoral theory underwent several modifications and maturations, but what is relevant to the chlorine revolution is nineteenth and early twentieth century epidemic disease, which was spread through the not-so-delicately-named fecal–oral route.

IN RESPONSE TO EPIDEMIC DISEASE, the miasma theory became prevalent. Many modern books covering the history of infectious diseases give only passing notice to the miasma theory, but the tenets of this theory are important because the widespread belief in it had horrendous implications for U.S. city dwellers in the late nineteenth and early twentieth centuries. The miasma theory, old and well established, drove municipal public works decisions that led to high death rates from waterborne disease.

The miasma theory held that the cycle of epidemic diseases was due to weather changes and seasonal characteristics (e.g., wet spring, cold and damp winter). Public health historian George Rosen called the atmospheric state that produced a particular disease an "epidemic constitution." So long as a specific epidemic constitution was dominant, the epidemic would continue. When this one-to-one relationship did not hold, alternative explanations were given.[8] For example, if the weather changed from damp and cold to dry and warm but the epidemic of remittent fever had not abated, the cause was said to be the phase of the moon or an unusual alignment of the stars.

Further refinements of the miasma theory included the belief that miasmas rose out of the earth and that astrological conditions might be responsible for epidemics.[9] The association of miasma with soil led to bizarre worries that the earth should not be disturbed, lest miasmas be released.

> ". . . in the spring of 1885 the New York City Board of Health resolved that the 'laying of all telegraph wires underground in one season . . . would prove highly detrimental to the health of the City in that portion densely populated through the exposure to the atmosphere of so much subsoil, saturated, as most is, with noxious gases.'"[10]

In the late nineteenth century, the belief that sewer gas caused disease was strong enough that a book devoted entirely to the subject was published in 1898.[11] The author, Herman A. Roechling, admitted that there were two sides to the argument. Roechling was aware of the role of both aerobic and anaerobic bacteria in the degradation (or "oxydation," as he termed it) of fecal material to its ultimate mineralized state. He was also aware that certain gases of degradation could cause instant death by acting as poisons, carbon dioxide and hydrogen sulfide being the major culprits. He contended that the gases produced during the process of putrefaction were possibly pathogenic but, more importantly, that they acted on the body to make a person more susceptible to contracting disease—especially typhoid fever. The primary evidence for sewer gases weakening the body's defenses against disease came from an 1895 study by Giuseppe Alessi, with the Hygienic Institute of the University of Rome, who claimed that the odor of putrefaction increased the susceptibility to death of 408 animals when they were injected with an attenuated typhoid bacillus.[12] An attempt in 1918 to replicate Alessi's results was not successful.[13]

Three horrific citywide odor invasions in the nineteenth century affected beliefs in the miasma theory. The Great Stink in London in 1858 was ghastly by all accounts,[14] but because no link between the smell and any epidemic was evident, the miasmatists had to perform some linguistic jujitsu to explain the connection.[15] The first Great Stink in Paris in the late summer of 1880 was purportedly linked to increases in disease, and this encouraged the miasmatists. The second Great Stink in Paris in the summer of 1895 was generally believed to not cause disease even though the smell was putrid, disgusting, and nauseating.[16]

In the age of miasmas, filth was the source of all evil.

". . . filth was dangerous not merely because it was a vehicle of disease, or an unfavorable condition, but also because it was a *source* of disease, the supposition being either that specific disease germs could be generated *de novo* [anew, starting from the beginning] from other germs in filth, under favorable circumstances, or that at least germs capable of producing disease found in filth the conditions of their more perfect development, some even requiring residence for a time in filth in order to reach their full maturity. In regard to typhoid fever, for example, it was held that the micro-organisms of the disease required a stay, longer or shorter, in the earth or heaps of filth, and only after such a period attained their natural and dangerous development."[17] (italics in original)

With some perspective gathered from his training and experience, Dr. John Leal later defined filth and its roots as part of the process of putrefaction. He was also aware that, in some sense, filth was in the eye of the beholder.

"From a general point of view anything dirty—any matter which fouls, soils, defiles or pollutes, waste matter, nastiness—is filth. From a sanitary point of view filth is decomposing or decaying vegetable or animal matter and other matter involved with it."[18]

In the years between the two great stinks in France, Louis Pasteur and his germ theory had a huge effect on the collective understanding of disease, filth, and miasmas. A commission appointed to report on the second Great Stink did not brand odors as causing disease.[19] In many ways, the miasma theory lost its attraction more quickly in France as a result of Pasteur's influence.

Nevertheless, progress from belief in miasmas to belief in the germ theory was not linear, and aspects of a compromise theory popped up throughout the timeline. Belief in the miasma and germ theories overlapped for a significant period of time. Precursor concepts of the germ theory showed up in the early nineteenth century, and vestiges of the miasma theory lasted well into the twentieth century.

Chloride of lime: disinfectant, deodorizer, and miasma destroyer. Prior to acceptance of the germ theory of disease, chemicals were used to "disinfect" and deodorize filth and the emanations of miasmas.

"From setting fires in the streets to burning incense or sulfur indoors, communities have attempted to neutralize disease-causing influences by chemical and other means since ancient times. By the mid-nineteenth century, the favored forms of disinfection—defined by Larousse dictionary in 1870 as the destruction of 'certain gases or certain exhalations produced by living matter, and called miasmas'—included the application of liquid chemicals, the burning of materials such as sulfur, and mechanical devices producing artificial ventilation through the forced circulation of air."[20]

The need for an effective and economical laundry bleaching agent led to the development of the most effective miasma and microbe disinfectant—chloride of lime. James Watt, of steam engine fame, is often credited with first using chlorine as a bleaching agent,[21] but the dose must have been too high and the pH too low because the early experiments destroyed the linen fibers that were supposed to be bleached. It was not until chlorine gas was passed over slaked lime[22] that a stable bleaching product could be made. Chlorine gas was injected into a gas-tight room where slaked lime was laid out in ridges on the floor. The slaked lime absorbed the chlorine gas, and chloride of lime was born.[23] Commercial chloride of lime contained 20 to 35 percent available chlorine, and full-scale manufacturing of chloride of lime began about 1800.[24] Other terms and chemical names for the same material were hypochlorite of lime, calcium hypochlorite, chlorine of lime (or just "lime"), bleaching powder, and bleach.

In 1832, an early treatise on "chlorine" by Chester Averill, professor of chemistry at Union College, laid out in some detail the "disinfecting" powers of chloride of lime. The occasion of his essay was a cholera epidemic in Schenectady, New York. The treatise primarily described how chloride of lime protected against disease transmission by eliminating miasma-induced causes. However, Averill also speculated that chlorine destroyed the "virus" responsible for several contagious diseases.[25]

Even when the miasma theory of epidemic disease was the only explanation available, one physician, Dr. Ignaz Semmelweis, used his powers of observation and chloride of lime as effective weapons to stop needless deaths among women giving birth.

> "In 1846, Semmelweis of Vienna was struck with the different rates of mortality from this fever in those women who were attended in child birth by medical students who, naturally, were often working in the dissecting room and brought into connection with other sources of contamination, and those who were attended by midwives who were not so exposed. In the former, the deaths from childbed fever were 11.4 per cent. In the latter, only 2.7 per cent!

> "He made experiments on rabbits and became convinced that the students carried the poison to the mothers. He, therefore, compelled every student [under his authority] to cleanse his hands first with *chlorine water* and later

with *chloride of lime*. Observe that both of these were antiseptics or germicides, as now we know. Of course, he knew nothing of germs. Nobody then did. The result was that very shortly the deaths from childbed fever in his wards fell to 1.27 per cent!"[26] (emphasis added)

In 1854, a British Royal Commission used chloride of lime as a deodorizer for London sewage.[27] During this period, chloride of lime was also added to foul mixtures to stop putrefaction, a process sometimes erroneously labeled "disinfection." During the cholera epidemic in 1854, chloride of lime was used liberally on the streets of London.

The Great Stink invaded London in 1858, and chloride of lime was used to mitigate the olfactory assault. "For many weeks the atmosphere of Parliamentary committee-rooms was only rendered barely tolerable by the suspension before every window of blinds saturated with chloride of lime, and by the lavish use of this and other disinfectants."[28] In this case, the chloride of lime was being used as a deodorizer. The heavy smell of chlorine emanating from the blinds saturated with chloride of lime apparently had some effect on the foul smell coming from the River Thames.

At the beginning of New York City's cholera epidemic in 1866, the New York Metropolitan Board of Health swung into action after one of the first cases was identified. "At the order of Elisha Harris, the bedding, pillows, old clothing, and utensils—anything that might 'retain or transmit evacuations of the patient'—were piled in an open area and burned. Chloride of lime was generously strewn through the house, and five barrels of coal tar and other disinfectants distributed so as to cover the surrounding area."[29]

It was news in Trenton, New Jersey, when the APHA Committee on Disinfectants prepared an 1885 report comparing available disinfectants, deodorizers, and antiseptics for use by public health professionals.[30] The full report, which comprised three progress reports, was a comprehensive treatise on the ability of various disinfectants to kill bacteria and other agents of disease. The report concluded that the best chemical disinfectant for all types of bacteria was a 4 percent solution of chloride of lime.[31]

WITH PERFECT HINDSIGHT, we now see that deadly outbreaks of waterborne disease in the United States were exacerbated by concentrations of population in cities and industrial centers.

In 1800, only 6.1 percent of the U.S. population lived in urban centers. By 1900, 40 percent of the population lived in cities.[32]

By 1900, the country stretched from ocean to ocean, and waves of immigrants fed the insatiable manufacturing machine that was building a new economy. Before the Civil War, when the country's economy was primarily agrarian and the population was more dispersed, death rates from cholera, typhoid, and dysentery were not epidemic in rural areas. But wherever there were concentrations of people, such as in the big cities, epidemics occurred and people died as a result.

A well-known "urban penalty" was associated with living in a crowded city.

> "The increasing use of mortality statistics provided new support for a traditional faith in the healthfulness of rural, as opposed to urban life. An infant born in the pure air of a farm or village could expect to live years longer than the child forced to breathe the city's polluted atmosphere."[33]

Michael R. Haines, noted economist at Colgate University, observed that before 1920, the mortality of urban dwellers was far higher than the mortality of people living in rural areas. In 1900, for example, the life expectancy of rural white males was 10 years higher than that of urban white males. By 1910, the excess life expectancy had dropped to 7.7 years and to 5.4 years in 1930. By 1940, the excess life expectancy for rural white males was only 2.6 years.[34] For infants, the penalty for urban living was far worse. In 1890, excess mortality for urban infants (compared with the mortality of rural infants) was 88 percent. By 1900, it had dropped to 48 percent, and it declined further through the early twentieth century. By the 1930s, infants had a better chance of survival past the age of one year in cities because of improvements in drinking water and milk supplies.[35]

Waterborne Diseases

Three types of waterborne illness—cholera, typhoid fever, and diarrheal diseases—are the most commonly described in the early literature on waterborne disease. Today, the origins and impacts of cholera and typhoid fever are the best understood. The fatal importance of diarrheal diseases, including dysentery, on the survival of young children before 1920 is not widely comprehended in the twenty-first century.

NEWS THAT CHOLERA was invading a city caused terror among the populace. Those who could afford to flee fled. A review of the disease's symptoms and its progress to a rapid death would cause anyone to run screaming for the countryside.

> "In the first, or premonitory, stage the sufferer might experience nothing but a vague unease and perhaps a mild diarrhea, as if having eaten spoiled food. The second stage was characterized by vomiting, muscular spasms, and pains in the lower chest and upper abdomen, accompanied by a profuse diarrhea. The diarrhea, widely accepted as the signature symptom of the disease, was of a peculiar type—there was virtually no fecal color or smell to the stool, which instead appeared watery with small white particles suspended in it. Because it looked like water in which rice had been boiled, it was dubbed 'rice-water stool' and considered a hallmark indication of second-stage cholera. The third stage was one of profound collapse. Victims retained mental function until near the end, but the body became cold, a pulse could scarcely be felt, and the face and extremities often turned dusky. Blue, corrugated skin made even young patients seem aged. . . ."[36]

The delay between ingestion of contaminated water and the onset of cholera symptoms was somewhat variable but generally was only a few days. Once infection took hold, progression of the disease could be rapid. Many accounts of cholera epidemics described people who were feeling well in the morning and were dead when the sun went down. Cholera could be fatal in up to 50 percent of cases. Even during epidemics, cholera did not kill as many people as tuberculosis, diarrheal diseases, or diphtheria did in a normal year, but cholera was spectacularly fatal, and the velocity of its spread seemed demonic.

Treatment of cholera. Medical therapies for treating cholera from 1832 to 1890 were ineffective. Ineffective is not merely a judgment from the twenty-first century; the public felt that way during the 1800s. Physicians still held on to the humoral theory when it came to treating patients with disease, and anything a physician could justify as capable of restoring the internal balance of the four humors was fair game. The list of accepted and quackery-based treatments for cholera was long: sulfur, opium, drinking large quantities of water, sulfuric acid, tannic acid, silver nitrate, creosote, charcoal, camphor cigars, and lime water. Some doctors favored saline injections with the saline water introduced rectally. In 1832, Irish physician William Brooke O'Shaughnessy proposed the novel (and later proven correct) therapy—venous injections of saline water to replace the huge amounts of water lost from choleric diarrhea. Dr. Thomas Latta, a Scottish physician, implemented the suggested therapy and achieved some success. Unfortunately, Latta died a year after effectively treating patients with this method, and it was not until the 1890s that hydrating patients with intravenous injections of saline water found favor again.[37]

During the 1831–32 worldwide pandemic, cholera was thought to be a disease of the blood because the blood of patients prior to death became dark and viscous as a result of severe dehydration. Bloodletting from the venous system was practiced to restore the balance of the blood with the three other humors. It is hard to imagine a more devastating treatment for a cholera sufferer than extracting blood from his or her failing circulatory system. In 1848, articles began appearing in British medical journals promoting the use of chloroform (as an inhalant) for the treatment of cholera.[38]

Worldwide cholera epidemics. It is generally accepted that up to the present, seven cholera pandemics have occurred. For centuries prior to 1817, cholera was endemic in the Ganges River Valley and other parts of South Asia. However, lack of transportation prevented its spread much beyond the local areas of infection.[39]

Chloride of lime was used during the 1854 cholera epidemic in London to "disinfect" the streets. An article from 1854 described its prolific use in the neighborhood surrounding the Broad Street pump.

> "An oil-shop puts forth a large cask at its door, labeled in gigantic capitals 'Chloride of lime.' The most remarkable evidence of all [of the severity of the cholera attack], however, and the most important, consists in the continual presence of [chloride of] lime in the roadways. The puddles are white and milky with it, the stones are smeared with it; great splashes of it lie about in the gutters, and the air is redolent with its strong and not very agreeable odour. . . . The fact is that the parish authorities

have very wisely determined to wash all the streets in the tainted district with this powerful disinfectant; and, accordingly, the purification takes place regularly every evening."[40]

Of course, cleaning the streets did nothing to stem the cascade of deaths caused by cholera bacteria in water from the Broad Street pump.

THE STORY OF DR. JOHN SNOW and how he discovered the cause of a cholera epidemic in the Golden Square neighborhood of London in 1854 has reached almost mythical proportions in public health literature. Three excellent books and a Web site describe Snow's life (see sidebar for a brief overview) and the details of the cholera outbreak he linked to the Broad Street pump.[41] However, some aspects of Snow's career and personality are relevant to the story of chlorination because they were reflected in another physician, Dr. John L. Leal, half a century later.

With emphasis on the Broad Street pump episode in most historical accounts, Snow's pioneering work in epidemiology, based on the occurrence of cholera in a London district served by two water supplies, usually gets lost. Snow was able to demonstrate conclusively that households in the areas of London that were being served contaminated water from the tidal portion of the Thames Estuary were far more likely to experience cholera deaths than households served water from an unpolluted upland source.

Dr. John Snow died June 16, 1858, forty-two days after the birth of John L. Leal, who eventually carried on Snow's concern about the ability of contaminated water to spread disease. If Snow's discoveries had been accepted by engineers, sewer planners, and drinking water providers beginning in 1854, millions of deaths would have been avoided. But Snow was only one person trying to overcome the juggernaut of the miasma theory. He was far ahead of his time.

After Snow died, his findings proving that cholera epidemics were caused by infected drinking water supplies were rarely quoted by others. One exception was a short article summarizing a talk by chemistry professor Edward Frankland at the Royal Institution of Great Britain in 1896. "The first effect of Dr. Snowe's [sic] cardinal discovery was the removal of the intakes of the [Thames] river water companies to positions beyond the reach of the tide and of the drainage of London. The second was the greater attention paid to the efficiency of filtration."[42] In the period 1890–1910,

Dr. John Snow

Snow was born March 15, 1813, in the city of York. He qualified as a licensed apothecary in 1838 and became a surgeon with a London practice in October 1838. With an office in the parish of Saint Anne's Church in Soho, Snow would have a medical career for only twenty years before he died of a stroke at the age of 47.

At 17, Snow became a vegetarian and soon thereafter committed to drinking only boiled or, preferably, distilled water on the basis of the writings of vegetarianism promoter John Frank Newton. He embraced abstinence from alcohol around 1836. Snow was known to be quiet, frugal, and energetic, a man of integrity, and a surgeon with an indifferent bedside manner. He refused to dispense pills and other medicines just because his patients wanted them. He was able to make a living and acquire some success as a physician when he perfected the administration of chloroform as an anesthetic for use during surgeries and infant deliveries. He even delivered two of Queen Victoria's babies.

Snow never married. He made all of the knowledge he developed available for free to any doctor who wanted it, and he made no attempt to patent his many devices for dispensing chloroform and ether.

One overriding personal characteristic of this ascetic doctor of the Victorian era was courage. He worked hard to develop his ideas and used the scientific method and laboratory investigations to establish his case in whatever area he was working. Once he became convinced of the rightness of his position, nothing could deter him. Only his courage made it possible for him to go up against the establishment and argue that something other than foul air was causing the deadly disease of cholera.

a second remembrance of Snow's accomplishments and lessons appeared in William T. Sedgwick's book, first published in 1902 and reprinted four times up to 1914.[43]

Although Snow's work did not lead directly to the successful chlorination of the Boonton Reservoir water supply, his findings (and the findings of others) that sewage-contaminated water caused disease were accepted by some sanitary engineers and many public health experts long before the medical community and the public fully embraced the germ theory of disease. Snow's thesis was crucial to building a foundation for the educations in public health and bacteriology that Dr. John L. Leal and George Warren Fuller received on their way to their singular accomplishment at Boonton Reservoir.

TYPHOID FEVER and the disease typhus (spread by lice) were considered the same disease until the nineteenth century. Dr. William Jenner (not to be confused with Edward Jenner, who discovered the smallpox vaccination through the use of cowpox) published a paper in 1849 that clearly distinguished between the two diseases.[44]

In the 1800s, typhoid fever struck the rich and the poor, the famous and the unknown. Abraham Lincoln's son Willie died from the disease during the Civil War years. Queen Victoria's husband and consort, Prince Albert, died from typhoid fever in 1861, and her eldest son almost succumbed to the disease as well. Theodore Roosevelt's mother died of typhoid fever on February 14, 1884.[45]

In the nineteenth century, typhoid fever was typically diagnosed by the "remittent" nature of the fever. The fever would come and go during the course of the disease, with maximum body temperatures usually occurring at night and minimum temperatures in the morning hours. In 1896, the Widal Reaction test became available and provided certain diagnosis of typhoid fever (about 95 percent effective) if applied after the first week of infection.

Many publications in the sanitary engineering literature dealing with typhoid fever contain dry recitations of statistics and death rates. However, the typical progression of a typhoid case is instructive in understanding the devastation of the disease (see sidebar).

The ratio of typhoid fever cases to deaths from the disease varied significantly during the late 1800s. When the bacterial strain was particularly virulent or the population was particularly susceptible, 10 to 25 percent of those infected died. As nursing care and medical therapies improved, less than 10 percent of the cases were fatal.[46] Although people afflicted with the disease might wish they were dead after the fifth week of fever, delirium, and diarrhea, the good news was that once a person contracted typhoid fever, he or she acquired immunity from any further infection.

Treatment of typhoid fever. In the nineteenth century, medical treatment for typhoid fever included changes in diet, cold baths, copious amounts of imbibed water, and subjection of the patient to an assault from "medicines" and chemicals. Medicines in particular favor were "antiseptics," which were intended to kill the organism in the intestines. It is hard to believe nowadays, but purgatives, considered part of an antiseptic cure, were given to typhoid fever patients—as if they did not have enough problems in that region of the body. Like many medicines given to patients during this era, the antiseptic cure was many times worse than the disease. Some of the "antiseptics" prescribed were calomel (mercurous chloride), naphthalin, beta-naphthol, thymol, carbolic acid (phenol) with tincture of iodine, salol (a mixture of salicylic acid and carbolic acid), and urotropin (a derivative of formaldehyde).[47] Some of these chemicals are recognizable as acute poisons!

A concentrated solution of chlorine and water was a well-known antiseptic treatment for people afflicted by typhoid fever in the late 1800s.[48] In his 1868 article detailing the "chlorine water" treatment, J. Burney Yeo, a physician associated with Kings College, London, described, rather colorfully, the reason why he thought the treatment was necessary.

> "During an attack of typhoid fever, we must regard the whole of the intestinal canal as a long sewer, into which the morbid offending material [not yet identified as a bacillus] present in the circulating fluids is constantly being poured. . . . if we shut up these morbid products in the intestines, we must strive to render them innocuous. In short, we must disinfect our sewer."[49]

Yeo's method of producing chlorine water created a medicine that would have been unpleasant in the extreme for any typhoid fever patient to imbibe.

> "A solution of chlorine should be made by the action of strong hydrochloric acid on finely powdered chlorate of potash [potassium chlorate]. This is easily done by putting about one drachm [1.77 grams] of the powdered salt into a pint bottle, and pouring upon it about two drachms [7.4 milliliters] of the strong acid. The gas

Typical Progression of
a Case of Typhoid Fever

"Between the time when the typhoid bacillus enters the body
and the time when the patient realizes that he is seriously ill,
there is a so-called incubation period, or prodromal period,
which usually lasts about two weeks, but which may vary from
one week to three. During this time the health may be apparently unaffected, but more often the patient feels played out,
loses appetite, and 'aches all over.' The true onset is generally
accompanied by symptoms which compel the patient to take
[to] his bed and call his physician. There may be shivering, or
perhaps a chill, headache, a coated tongue; perhaps nose-bleed
or a bronchial cough; fever, restlessness and insomnia, muscular
weariness, thirst, nausea; there may be either diarrhea or con-
stipation. These symptoms continue during the first week, the
temperature gradually rising to 103 or 104 degrees at night, with
a corresponding [decrease] in the morning temperature, which
is lower than at night, the pulse also showing a slight elevation.

"During the second week most of the symptoms become worse,
but headache and nausea disappear. The temperature rises
to 104 or 105 degrees, with morning remissions of one or two
degrees; the pulse rises to 100 or 110 and becomes weaker.
Prostration and apathy become great, the voice feeble, the
tongue dry and brown. The bowel discharges are frequent and
loose, pale yellowish brown in color, and more or less lumpy.
About the eighth or tenth day rose-spots about one-eighth
of an inch in diameter appear on the abdomen, coming and
going in successive crops, lasting only a few days each, and
disappearing altogether during the third week. Nervous trem-
ors become conspicuous, and there may be some delirium.

"During the third week the night temperatures continue
high, but the morning remissions may be somewhat lower.
The patient becomes emaciated, semi-conscious, and perhaps
delirious. The stools may become tinged with blood, the urine
lessened in amount. During this week, pulmonary complications are most likely to develop—sometimes pneumonia.

"During the fourth week night temperatures slowly fall to 102 or 101 degrees, while the morning temperatures begin to approach normal, even becoming sub-normal in some instances. With lower temperatures all the symptoms tend to improve, the pulse grows stronger, the mind becomes clear, the tongue becomes moist, and the patient begins to desire food.

"The fever sometimes persists for a few weeks longer, but usually the fifth week begins the period of convalescence, which may last anywhere from two or three weeks to as many months, and if no complication sets in the patient recovers. Indiscretions in eating, in exercise, or exposure, however, may cause a dangerous relapse, during which there is a repetition of the original symptoms. The second attack is seldom as severe as the original one; but, on the other hand, the reserve strength of the patient is correspondingly less, so that relapses are always to be dreaded."[1]

Then, of course, there were the complications.

"It is said that not over a third of the deaths from typhoid fever are due directly to the effects of the disease, i.e., to the effects of the typhotoxin. Two-thirds of the deaths are due to the numerous complications, among which [are] pneumonia and tuberculosis. This is a most important matter to sanitarians in connection with death certificates."[2]

---

1   Whipple, Typhoid Fever, 2–5.

2   Ibid., 6.

[yellowish-green] will at once be given off in abundance. The mouth of the bottle [12 imperial ounces] should be stopped by the finger, and, after a short time, water added slowly, in small quantities, thoroughly shaking the bottle on each addition [until bottle is full]. By this process a solution of chlorine will be obtained, acidulated with hydrochloric acid."[50]

The dosage was one to two tablespoons of the mixture every half hour or more often if considered necessary by the physician. It is clear from the preparation instructions that this "medicine" was a solution of chlorine dioxide—about 3,000 parts per million—possibly mixed with chlorine.

Although the results for the patient were undoubtedly not pleasant, patients during these times were accustomed to being given disgusting concoctions and in many cases would not trust a physician who didn't prescribe a ghastly "physic." Needless to say, dosing patients with chlorine water was not effective and was abandoned as a treatment for typhoid fever in the early 1900s.

Although the details of chlorine water treatment and the sometimes bizarre medical procedures it represented can be interesting, what's important here is that the addition of chlorine to the Jersey City water supply exposed the population to a minimal amount of chlorine (a fraction of a part per million) compared with the chlorine water treatment for typhoid fever (thousands of parts per million). This comparison would be used during the second Jersey City trial.

Typhoid fever epidemics. In the nineteenth and early twentieth centuries, typhoid was endemic in the United States and England. W.H. Corfield, professor of hygiene at University College, London, published more than a hundred case histories of typhoid fever outbreaks that occurred in England in the 1800s.[51] Endemic typhoid fever meant that the disease was always present in the population, particularly in urban settings. Unlike cholera which involved dramatic symptoms and a high probability of death, a variety of gastrointestinal diseases could be confused with typhoid fever. Only when the number of cases rose above the background of endemic typhoid cases and normal diarrheal disease was an epidemic declared.

Hundreds of descriptions of typhoid epidemics were published in the medical literature and popular press from the 1890s to early 1900s. Two typhoid outbreaks that were important as the sites of some of the first large-scale uses of chlorine as a water disinfectant occurred in Maidstone, England, in 1897 and Lincoln, England, in 1905.

Dysentery and Diarrheal Diseases

Unlike those who studied cholera and typhoid fever, not a single historical person is credited with determining where diarrheal diseases came from and how they were spread. Part of the problem with diarrheal diseases is that they are caused by many different organisms and involve varying symptoms. Also, contaminated food, including milk, was a problem during the nineteenth and the early twentieth centuries. In many cases, it was impossible to separate a foodborne from a waterborne outbreak of diarrheal disease. However, one early epidemiology expert, William T. Sedgwick, was certain that contaminated water played a major role in epidemics of diarrheal diseases.[52]

Table 2-1 lists a number of waterborne diarrheal diseases, causative organisms, symptoms, and treatment methods, along with historical references. Many more organisms cause diarrhea, but the ones shown in Table 2-1 include the bacteria, viruses, and protozoa most often connected to diarrheal diseases that can cause death, especially in infants less than one year of age.

As shown in Table 2-1, the symptoms of diarrheal diseases ranged from a stomachache to severe diarrhea leading to dehydration and death. Both typhoid fever and cholera are included in this list of diarrheal diseases because diagnoses of all of these diseases were intertwined and confused throughout this period.

Essentially no effective treatment for diarrheal diseases existed at the turn of the twentieth century. The diseases were self-limiting, chronic, or fatal. Dr. John Leal's father died from a seventeen-year infection that appeared to be a chronic case of amoebic dysentery.

Undoubtedly, many outbreaks of diarrheal disease occurred during the period 1890–1910, but because diarrhea typically was endemic in the United States, epidemics were rarely noted, and there were few statistics on the rate at which diarrheal diseases occurred in the total population.

Fuller described an epidemic of diarrheal disease that occurred in Lowell, Massachusetts, in 1903. The water supply from the Merrimack River was grossly contaminated with sewage. By 1896, Lowell had changed its water source from the river to a series of wells that tapped an uncontaminated supply. The 1903 epidemic occurred when river water from a system supplying industrial water was used to fight a fire, and a valve separating the potable supply from the fire-fighting supply failed. Numerous cases of diarrhea were followed by an outbreak of typhoid fever.[53]

More than fifteen years later in Peabody, Massachusetts, 1,500 people were sickened during a virulent waterborne outbreak of gastroenteritis. No water treatment of any kind was provided by that city.[54]

Table 2-1 Diarrheal diseases: causes, symptoms, and treatment

Disease	Causative Organism	Symptoms	Treatment
Historical: dysentery (bloody flux)	Many	Severe diarrhea with bloody discharge	Leeches, opium, purgatives, blood-letting
Historical: cholera infantuum	Many	Severe diarrhea leading to death resulting from dehydration for infants <1 year	No effective treatment
E. coli infection	*E. coli 0157:H7* and other Shiga toxin-producing *E. coli*	Severe stomach cramps with diarrhea (sometimes bloody), lasting about one week; severe complications	Historical: none. Current: oral rehydration; antibiotics are NOT recommended
Travelers' diarrhea	Enterotoxigenic *E. coli*, or *ETEC*	Diarrhea, abdominal cramps, nausea, vomiting, lasting one to two days	Historical: none. Current: disease is usually self-limiting, oral hydration, antibiotics
Viral gastroenteritis (stomach flu)	Norwalk-like viruses, adenoviruses, rotavirus, coxsackievirus	Acute illness for a few days with diarrhea and vomiting	Historical: none. Current: disease is usually self-limiting, oral rehydration; prevention with vaccine for rotavirus
Amoebic dysentery	*Entamoeba histolytica* and other amoeba species	Fulminating dysentery, bloody diarrhea, weight loss, fatigue, and severe abdominal pain	Historical: none, often fatal. Current: several antibiotics are effective
Shigellosis	*Shigella sonnei, S. flexneri, S. dysenteriae*	Diarrhea (sometimes bloody), stomach cramps, fever, lasting about one week; severe disease in children <2 years old	Historical: none. Current: fluid replacement, antibiotics
Giardiasis	*Giardia lamblia*	Diarrhea, gas, stomach or abdominal cramps, nausea, vomiting, lasting about two weeks	Historical: none. Current: antibiotics including Flagyl
Cryptosporidiosis	*Cryptosporidium parvuum* and other species	Stomach cramps, nausea, vomiting, fever, watery diarrhea over a two-week period	Historical: none. Current: none—disease is self-limiting except for immunocompromised individuals for whom the disease can be fatal

(continued)

Table 2-1   Diarrheal diseases: causes, symptoms, and treatment   (continued)

Disease	Causative Organism	Symptoms	Treatment
Cholera	*Vibrio cholerae*	See text for detailed description	Historical: see text. Current: venous rehydration
Typhoid fever	*Salmonella typhi*	See text for detailed description	Historical: see text; typhoid fever is fairly uncommon in infants <1 year. Current: antibiotics; prevention with typhoid vaccine
Salmonella poisoning	*Salmonella enteritidis*	Acute diarrhea and vomiting; usually associated with food poisoning	Historical: none. Current: oral rehydration and antibiotics

Diarrheal diseases and death rates for young children. Infant mortality was a serious problem for centuries. During the Middle Ages and afterwards, if a child survived the first year, people deemed it God's will, and the child had a decent chance of making it to adulthood or at least young adulthood. If a child died a few months after birth, well, that was also considered God's will.

One author ascribed the frailty of the first year to several factors:

> "For centuries the world has taken it for granted that a large percentage of infants born must die during the first year of life as a matter of course. . . . In the human race the death of infants during the first year of life was generally considered a matter of constitutional weakness and inherited disability to endure the climate or to properly digest nourishment."[55]

Some of the earliest statistics on infant mortality come from the eighteenth century. In France, infant deaths from all causes were more than 200 per 1,000 live births, and a statistic from Geneva, Switzerland, recorded 296 deaths per 1,000 live births.[56]

In the latter part of the nineteenth century, infant death rates were a major problem.

> "The percentage of infantile lethality during those few months [July, August, September] is so high that

we would fain apply to our own country [Canada] the remarks written in 1878 by Mr. Bergeron, perpetual secretary to the Academy of Medicine of Paris: 'No doubt,' said he, 'the first months and notably the first weeks of life, which conspire to bring about so many causes of diseases and of death, will always, against all odds, furnish a greater proportion of deaths than any other period of life, excepting old age. But is it not distressing to learn that, in our times and in our own country, in spite of the general degree of comfort reached, and the progress of public and private hygiene, the mortality of the new-born should be so great as to justify the assertion supported by figures that a new-born baby offers less chances than a man 90 years old to survive one week, and less chances also than an octogenarian to outlive one year?'"[57]

A 1921 summary of infant death data noted that in some large cities in the world infant death rates were as high as 400 per 1,000 births. The U.S. census in 1900 showed eight cites with annual infant death rates per 1,000 births ranging between 300 and 419. A large percentage of these deaths resulted from infant diarrhea, also called "cholera infantum" and "summer diarrhea."[58]

The number of deaths from diarrheal diseases among children less than one year old in Paterson, New Jersey, was astounding. This cause of death was more prevalent than all others published for children less than one year of age in annual Board of Health reports to the city of Paterson in the 1890s. Death rates of young children from diarrheal disease were among the highest for all causes and all ages in the city.[59]

One aspect of diarrheal deaths among infants less than one year old was that the number of deaths peaked dramatically in the summer months—usually July and August in temperate climates. Peaks in May and June were noted in the southern part of the United States, including New Orleans.[60]

The high death rate from diarrhea in children under one year old was recognized, but in 1902 the relation between diarrheal diseases and drinking water was considered somewhat of a mystery by experts, including William T. Sedgwick, who was one of the foremost epidemiologists of his time.

"Moreover, there are certain members even of the group of diseases known as 'diarrhoeal' which do not seem to be as readily conveyed by drinking water as are others [typhoid and cholera] of the same class. Cholera infantum, for example, is a common, severe and often fatal diarrhoeal disease of children. But it seems seldom, if ever traceable to polluted drinking water, with which typhoid fever and Asiatic cholera can very often be directly connected."[61]

A few years later, Sedgwick and public health officer J. Scott MacNutt published a lengthy article summarizing vital statistics from translated reports written by Dr. J. J. Reincke, the health officer from Hamburg, Germany. Reincke had presented a significant amount of data showing that the quality of drinking water served to a community was directly related to infant death rates.

"One of the surprising results of our study will probably be the disclosure of the remarkable relation subsisting between polluted water and infant mortality. This subject has been more fully elucidated by Dr. Reincke than by anyone else. . . . Students of preventive medicine will do well to extend these studies, which promise to shed much light upon the solution of one of the most serious problems of the time."[62]

Sewer Pipe–Water Pipe Death Spiral

Rampant waterborne disease during the latter half of the nineteenth century was directly caused by the construction of sewer pipes to remove human wastes as quickly as possible to sources of drinking water: rivers, lakes, and reservoirs. Building a centralized water-pipe distribution system solved water supply problems but created an efficient system for delivering pathogens.

It was a common misconception among chroniclers of the period 1850 to 1900 that the act of installing sewers, in and of itself, was an effective public health protection strategy. One of the major proponents of this misconception was Englishman Edwin Chadwick, who in the 1840s became one of the leaders of the European sanitary movement. [63]

Chadwick envisioned vast sewer systems collecting human waste and transporting it to rural areas, where it would be put to beneficial use as fertilizer for farms. A piped water system would supply cities with water from protected sources that were not affected by any locale's sewage. Unfortunately, only one of the three parts of Chadwick's vision was implemented in London and elsewhere. Sewers were built, but the crucial disposal of human waste on farmland was not carried out. Instead, sewage was discharged into rivers and lakes, after which no surface supply of drinking water was safe because filtration and chlorination were not immediately implemented.

After the Great Stink attacked the personal aesthetics of Londoners in 1858, Parliament was moved to divert sewers that were dumping directly into the Thames River in and around London. Joseph Bazalgette, chief engineer of London's Metropolitan Board of Works, was given the Herculean task of building intercepting sewers to carry sewage away from London and discharge it downstream. Steven Johnson, author of the 2006 book *The Ghost Map*, hailed this as a major milestone in public health protection.

> "But Bazalgette's sewers were a turning point nonetheless: they demonstrated that a city could respond to a profound citywide environmental and health crisis with a massive public works project that genuinely solved the problem it set out to address."[64]

Unfortunately, the intercepting sewers solved only the aesthetic problem. The public health problem was transported downstream. Tidal action of the Thames River brought the discharged wastes to drinking water extraction points for the city. Bathing beaches and shellfish beds received the full brunt of London's sewage. It is hard to imagine any municipal agency being so reckless. What Bazalgette's sewers actually accomplished was simply transporting the human waste some distance away. The virulence of its potential to cause disease was not changed at all. The example provided by Bazalgette was repeated in England and the United States, where water supply intakes were located downstream of sewage discharges.

In his book *The Great Stink of London*, contemporary social historian Stephen Halliday claimed the cholera epidemic that hit Hamburg in 1892 did not visit London because of Bazalgette's sewer system.[65] His claims were unsubstantiated. He presented no data and no proof; he simply based his claim on the coincidence of dates. He also ignored the extraordinary efforts of British officials to quarantine any ships or travelers from Hamburg during the epidemic. The United States also avoided a cholera epidemic in 1892, but this was not because of any sewer systems built in the previous years. Like their counterparts in Britain, U.S. quarantine officials isolated immigrants from Germany and other places that were suspected as sources of disease at the borders and in the harbors.

In the late nineteenth century, others agreed with Halliday's assessment.

> "How far this reduction in the typhoid rates is due to the execution of works for the supply of good water and how far to the carrying out of proper sewerage works had been decided by [sewer inspector Sir George] Buchanan in favour of sewerage works, and from his and further careful investigations by Continental observers it may safely be concluded that sewerage works contribute to it in a more prominent degree than waterworks."[66]

Dr. John Snow proved that a water supply contaminated by human sewage produced cholera in large numbers of people who consumed water from that supply. He had strong opinions about sewers and drinking water systems.

> "Snow, who distilled his own drinking water, agreed that London water should be improved, but he considered the abolition of cesspools and the increasing preference for water closets a sanitary disaster . . . water closets connected to sewer lines that emptied into rivers also used for metropolitan drinking water were, in his mind, primarily an efficient means of recycling the cholera agent through the intestines of victims as rapidly as possible. Sanitary reforms were needed, but flushing the waste of a town into the same river by which one quenched ones' thirst seemed sheer stupidity."[67]

Snow published his observations in the mid-1800s, but engineers did not apply his lessons. Instead, engineers continued to find better ways to provide piped water to every house in crowded European cities and to efficiently gather sewage and deposit it in the nearest stream.

George Warren Fuller and James R. McClintock provided one of the clearest summaries of the progression of waste disposal practices in the United States, as rural and village life gave way to urban congestion.

> "In early days, each household disposed of its filth as best it could. Cesspools for household wastes were officially recognized by sanitary authorities and there was little or no supervision of the disposal of the contents.

In large towns polluting matters accumulated near the dwellings and leached into the soil, unless carting to land was readily negotiable.

"The use of waterclosets, effectively established in England, in 1810, rapidly increased in favor, with the introduction of pressure water supplies, making tight cesspools less practicable. Cesspool overflows led to street drains, frequently uncovered, and were discharged into the nearest water course. Under these conditions the cesspool, with its expense and difficulty of cleaning, passed from favor and underground sewers were gradually developed to take the place of the poorly arranged and badly built street drains. Household wastes in increasing proportions reached the watercourses. The resulting nuisances became a serious problem about the middle of the last century [1850s]. . . . The sewage problem thus took its modern form of pollution of waterways. . . .[68]

U.S. cities needed systems to move the miasma out of town, and sanitary engineers followed the British model of building sewers.

"'Sanitary Engineering' . . . consists [of] removing dangerous impurities from water—supplying an abundance of both of these life-giving elements, and *removing as speedily as possible, before decomposition commences, all refuse matter*, whether of animal or vegetable origin."[69] (emphasis added)

Pitifully few sewage treatment systems existed in the United States in 1893, when Moses N. Baker conducted a census of U.S. sewage treatment systems from his vantage point as editor of *Engineering News*. Baker's 1893 publication is an extraordinary historical document. He identified only 30 U.S. towns that had any type of sewage treatment system in 1893.[70]

Water Pipes for Cities

Similar to the progression of waste disposal described by Fuller and McClintock, domestic water systems grew in size and complexity along with cities. In the beginning, water supplies were shallow wells, springs, and buckets dipped in streams. Individuals carted their own water or hired a specialized group of workers—water carriers.

As a town grew in size, additional wells were hand-dug, and springs located farther away were tapped. If the river appeared to be clean, some kind of water wheel or pumping mechanism was usually installed to provide a more reliable source of water. Eventually, someone organized a water supply by running pipe from a more distant source or enlarging the pumping station on the river to provide water to fountains or central distribution points. Both private water companies and municipal governments began to provide this function. In some cases, the need for a piped water supply was dictated by the need for fire protection. In 1801, Philadelphia's Schuylkill River water system supplied "fire plugs" in a network of wooden pipes in the high-value real estate of the central city.[71]

As towns grew into cities, the difficulty became twofold: find a supply that was big enough to provide water even during droughts and that was, at least, thought to be pure and wholesome. Demand for water in cities increased further when the water closet became popular because piped water was necessary for flush toilets to operate.

All of the water supply treatises from the late 1800s stressed the need for cities and water companies to find upland, pure water supplies. In many cases, however, these supplies were not readily available or were too expensive to develop, and growing cities turned to the abundant supplies available in local rivers and lakes. Philadelphia had the Schuylkill and Delaware rivers; Chicago had Lake Michigan; Paterson, New Jersey, had the upper reaches of the Passaic River; and Jersey City, New Jersey, had the lower portion of the Passaic River. New York City probably would have tapped the Hudson River early in its history, but the estuary was too salty.

Describing the state of water treatment systems in the United States in 1896, Moses N. Baker noted that ". . . at the beginning of this century [1800] there were practically no water-works in the United States. . . ." He determined that in 1896, there were 3,341 water works serving almost 4,000 U.S. cities and towns.[72] When the U.S. Census Bureau began publishing statistics for public water supplies in 1900,[73] approximately 33 million people (about 44 percent of the U.S. population) were served by public water supplies. Few of these water supplies were filtered, and none of them was disinfected.

Completion of the Death Spiral

In the 1890s, some authors tried to prove that the construction of sewers saved lives.

> "The beneficial effects of sewerage are plainly seen in the statistics of towns where an efficient [sewerage] system has been carried out. By sewering certain towns in England, the death rate from pulmonary diseases alone was reduced 50 per cent. A marked decrease in the amount of sickness and a prolongation of life has always followed proper sanitary works."[74]

The authors of this text presented a table with data on death rates from six towns in England. The data showed a marked decrease in overall deaths, especially those caused by pulmonary diseases. In Salisbury, England, the reduction in pulmonary disease rates purportedly resulting from the addition of sewers was 49 percent.[75]

It is unlikely that installing sewers would prolong life for people suffering from pneumonia or tuberculosis. In fact, deaths caused by tuberculosis were decreasing dramatically during the latter part of the nineteenth century as a result of improved care and isolation of patients suffering from what then was called consumption. The improvement in typhoid death rates is explainable only if moving sewage more quickly out of these cities was accompanied by abandoning sewage-contaminated wells and connecting the town to a pure, upland water supply.

When Steven Johnson examined the 1866 cholera outbreak caused by the East London Water Company, he seemed to imply that all that would have been needed to prevent this outbreak was for one of the last segments of Bazalgette's sewer system to be put into operation. In fact, the problem was due to cholera-contaminated water being introduced into the company's water system at the Old Ford Reservoir without being treated by the company's slow sand filters.

Sewers were not the solution to waterborne disease epidemics. In many cases, sewers were actually the problem. Sewers concentrated the wastes from cities and dumped their contents into nearby streams, rivers, and lakes. Centralized drinking water distribution systems in cities downstream of a community with sewers could then efficiently distribute the human waste–contaminated water to their customers. Thus, the sewer pipe–water pipe death spiral was complete (Figure 2-1).

Lawrence, Massachusetts—Poster child for the death spiral. Lawrence, Massachusetts, was a manufacturing center during the mid- to late 1800s. Located on the banks of the Merrimack River, Lawrence and the other industrial cities in the Merrimack Valley used available water power to run textile and shoe manufacturing industries. The valley was developed both for its abundant water power and its access to markets in Boston. Lawrence was less than 25 miles from that important trading and population center.

The sources and transmission of typhoid fever in Lawrence were under intense scrutiny in the 1890s. The Lawrence Experiment Station was located in that city, and William T. Sedgwick led research investigations at the station.

The Merrimack River watershed is shown in Figure 2-2. The river rises in the White Mountains of New Hampshire and flows

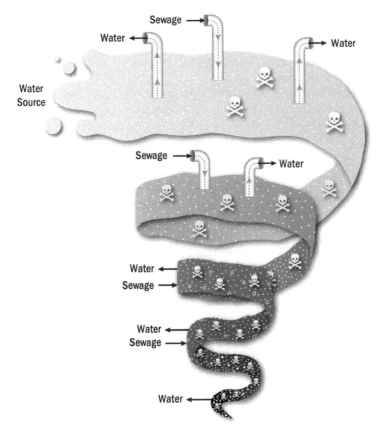

Figure 2-1   The sewer pipe–water pipe death spiral

south through Concord, Manchester, and Nashua, New Hampshire. After it crosses the state line and meets up with the Concord River in Lowell, Massachusetts, the river takes a hard left turn and heads for the Atlantic Ocean. Only nine miles downstream of Lowell is Lawrence, followed by Haverhill and Newburyport.

In 1890, Lowell had the largest population in the Merrimack Valley at about 78,000—almost exactly the same population as Paterson, New Jersey, another prominent industrial town founded on water power.

Sewage from cities in the watershed was collected in sewer systems and discharged without any treatment into the Merrimack. Sewers installed in Lawrence as early as 1848 did not help the incidence of typhoid fever in that city by one jot.[76]

In general, Lowell's raw water supply received sewage from the three major upstream cities—Concord, Manchester, and Nashua. The untreated water supply for Lawrence, in turn, was overwhelmingly influenced by sewage discharges from Lowell a few miles upstream. Only Lowell and Lawrence used the Merrimack River for their water supply. Other cities on the river used alternative sources.

According to an analysis of typhoid death rates in these cities from 1889 to 1893, Concord, Manchester, Nashua, Haverhill, and Newburyport, which were supposedly not using the Merrimack River for water supply, had relatively high death rates from typhoid fever. Sedgwick explained that even though these city water systems were not hooked up to the river, many of the factories in these mill towns provided their workers with water taken directly from the river or from the canals and mill races connected to the river. [77]

Lowell and Lawrence, which relied on the river for their water supplies, had horrendously high typhoid fever death rates. During 1890–91, they were 195 and 187 per 100,000 for Lowell and Lawrence, respectively. The death rates during this year made it obvious that the two cities were experiencing an epidemic.[78]

An analysis of the monthly death rates for the cites along the Merrimack—April 1890 to March 1891—showed that a typhoid fever "wave" appeared to travel down the river, with the peak monthly death rate hitting Lowell first. The discharge of Lowell's human waste containing stupefying amounts of typhoid bacilli traveled downstream, and one to two months later, the death rate peaked in Lawrence.

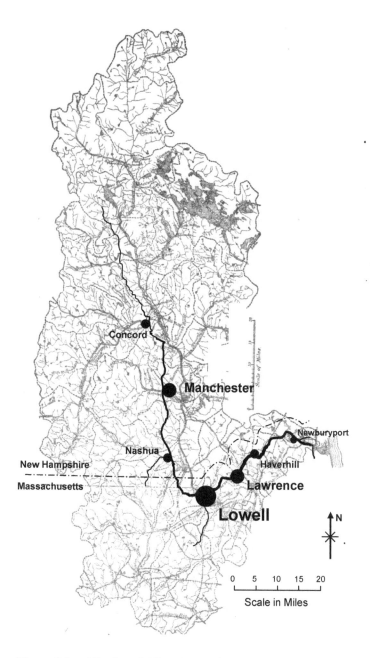

Figure 2-2   Merrimack River Watershed

Installation and operation of the intermittent slow sand filter in Lawrence in 1893 disrupted the Sewer Pipe–Water Pipe Death Spiral on the Merrimack River. The slow sand filter at Lawrence was called the first filter in the country installed to remove bacteria.[79] In 1901, the filter's bacteria removal efficiency was reported to be 99.14 percent.[80] The annual death rate from typhoid fever in Lawrence had dropped from 107.4 per 100,000 in 1892 to 18 in 1901.[81]

By 1919, the Death Spiral in Lawrence was interrupted further as a result of the addition of chlorine to the water supply. No details are available on the dosage or whether the feed was continuous. But the typhoid death rate in Lawrence was 8.6 per 100,000 in 1919,[82] still relatively high and indicative of noncontinuous operation or the use of other contaminated sources by the public.[83]

Chicago—The death spiral, turned on itself. Over the period of time when sewers were constructed and drinking water was not properly protected, Chicago collected some of the most reliable health statistics and had some of the worst waterborne disease and death rates in the country.

Chicago sits on the shore of Lake Michigan, and the Chicago River has been an integral part of the city's landscape since its founding in 1833. The river is a flat, sluggish stream that served as a commercial–industrial transportation corridor and a destination for the city's sewage.

The city grew from little more than a village of a few hundred souls in 1833 to the booming metropolis of well over a million people at the time of the 1893 World's Fair and Columbian Exposition. Providing sufficient water for the exploding population was no problem; Lake Michigan provided an inexhaustible supply. However, severe water quality problems were apparent early on. Drinking water withdrawal points for the city were initially on the shore of the lake. As sewage and industrial activities polluted the water near shore, the water system intakes were moved farther out into the lake in "crib" structures connected to the shore by pumping plants, tunnels, and pipelines.

So long as nothing unusual happened, the water system functioned fairly well—accompanied by the usual typhoid deaths, cholera epidemics, and diarrheal disease problems.

"A great population upsurge in 1846–1847 laid the groundwork for an upsurge in disease. In 1848 both cholera and smallpox broke out in epidemic proportions. In the early 1850s Chicago experienced outbreaks of dysentery and cholera."[84]

After a devastating cholera epidemic in 1854, when 6 percent of the population died, the Illinois legislature authorized creation of the Chicago Board of Sewage Commissioners.[85] The 1854 epidemic was part of the same worldwide Asiatic cholera pandemic that Dr. John Snow confronted in London. In Chicago, the sewage commission hired engineer Ellis Sylvester Chesbrough to prepare an overall plan for the sewer system and to recommend how to dispose of the city's wastes. The plan was submitted to the mayor and city council in 1855. According to one source, no other American city at that time had developed a comprehensive plan for a sewer system.[86] It was not until two years later that consulting engineer Julius W. Adams was given the task of planning a sewer system for Brooklyn, New York.[87]

As part of his plan, Chesbrough considered four options for sewage disposal. Sewage could be drained

- into the Chicago River and its tributaries and then allowed to flow into Lake Michigan,
- directly into Lake Michigan,
- into reservoirs that would store the waste and then transport it to farms for fertilizer (the Chadwickian vision), or
- into the Chicago River, which would be diverted into the Illinois River and then to the Mississippi River through a "steamboat canal."[88]

The first option was deemed the best and least costly. However, the fourth option was eventually implemented when the flow of the Chicago River was reversed in 1900. The choice not to implement the third option was based on the miasma theory of disease. "There would be danger to the health of the city during the prevalence of winds from the quarter in which the sewerage might be used as a manure, especially, if only a few miles distant and spread over a wide surface."[89]

To implement the first option, water for flushing purposes was designed to be transported to the Chicago River by a canal, ". . . especially during the warm and dry season of the year, when there would be danger of sickness from putrid exhalations."[90]

Sewers were dutifully constructed according to Chesbrough's plan, and most of the foul wastes were dumped into the Chicago River. But even with the redirection of some sewers to the river, the Chicago River still emptied into Lake Michigan, and some major sewers still discharged directly into the lake, contaminating the water supply.

A plot of the death rate from typhoid fever in Chicago from 1857 to 1917 (Figure 2-3) indicates that significant typhoid epidemics occurred in Chicago in 1857, 1863–66, 1872, 1881, and 1891.[91] Another interpretation of the data in Figure 2-3 is that Chicago experienced a continuous typhoid epidemic from 1857 to 1911.

In 1885, an unusual weather event happened—torrential rainfall and a flood of biblical proportions. The city was lucky that cholera and typhoid epidemics did not result from the storm's flushing raw sewage into the lake. Instead, a myth was created. It has been reported by a number of otherwise knowledgeable authors and journalists that as a result of the 1885 deluge, a cholera epidemic wiped out 90,000 Chicagoans—about 12 percent of the population. In fact, no case of cholera had been detected in the city from 1873 onward. Libby Hill's book on the Chicago River published in 2000 took on the myth and demonstrated that the epidemic never happened.[92]

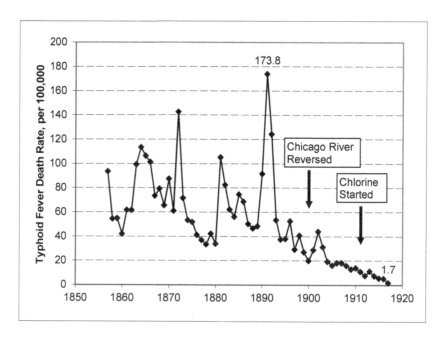

Figure 2-3   Typhoid fever death rate in Chicago, 1857–1917

Even though 90,000 people did not die from cholera in 1895, about 2,000 people died in 1891 from typhoid fever. Figure 2-3 shows that the death rate from typhoid fever in 1891 was 173.8 per 100,000—astonishingly high by anyone's reckoning.[93] An account in a New York newspaper took Chicagoans to task.

"Greater folly has never been revealed in the recent history of any great city's attempts to procure a clean and healthful supply of water. While Chicago has been drawing water from the lake by devices to which the attention of the world has been directed, it has been deliberately poisoning the waters of the lake by the sewers of a population constantly and rapidly increasing. It is not difficult to find the causes of the continuing epidemic of typhoid fever in that city."[94]

One can imagine the panic among the city leaders in 1891 with the 1893 Chicago World's Fair only months away.[95] Over the next few years, most of the sewers were directed to dump into the Chicago River. This set the stage for implementing one of Chesbrough's alternatives for ultimate sewage disposal—connecting the Chicago River through the Illinois and Michigan Canal and the Illinois River to the Mississippi River through a "sanitary and ship canal." Much to the dismay of the city of St. Louis (43 miles downriver of the connection to the Mississippi), the connection was completed on January 2, 1900.

Peculiar to the history of Chicago's sewage and drinking water systems was the overpowering presence of the Union Stock Yards. The Chicago River was the destination not only for raw sewage but also for the putrid wastes from the enormous slaughterhouses and meatpacking plants that were in large part responsible for the city's growth. The South Fork of the river was the main recipient of blood, guts, and animal waste. Particularly poisonous was the South Branch, which emptied into the South Fork. Conditions were so reprehensible in this stretch of the river that it was given its own name—Bubbly Creek. It was not bubbly because it was an effervescent stream from a pure source. It was bubbly because of the gases of anaerobic and aerobic waste decomposition that bubbled up to the surface—methane, mercaptans (sulfur-containing organic chemicals), volatile organic acids, carbon dioxide, and hydrogen sulfide.

There are hundreds of colorful descriptions of the offensive nature of this putrid tributary of the Chicago River, but the best one had a literary source. Upton Sinclair, in his acclaimed 1904 novel *The Jungle*, topped everyone with his view of the visual vulgarity of Bubbly Creek.

> "The grease and chemicals that are poured into it undergo all sorts of strange transformations, which are the cause of its name; it is constantly in motion, as if huge fish were feeding in it, or great leviathans were disporting themselves in its depths. Bubbles of carbonic acid gas will rise to the surface and burst, and make rings two or three feet wide. Here and there the grease and filth have caked solid, and the creek looks like a bed of lava; chickens walk about on it, feeding, and many times an unwary stranger has started to stroll across, and vanished temporarily. The packers used to leave the creek that way, till every now and then the surface would catch on fire and burn furiously, and the fire department would have to come and put it out."[96]

There were also creative descriptions of the stench from Bubbly Creek: " . . . the filthy layer of green scum on top of a stagnant pool, emitting the most abominable stenches and nauseating odors. . . . "[97]

Chicago's normally high incidence of typhoid fever and the 1891 epidemic that killed 2,000 people were clear indications that something was terribly wrong with Chicago's sewer and water systems. The data in Figure 2-3 show that the typhoid fever death rate decreased after the Chicago River flow was reversed in 1900, but it was still high prior to 1910.

Beginning in 1911 and extending to 1916, chlorine was installed in the nine pumping stations that were withdrawing water from the lake. The death rate from typhoid dropped from about 14 per 100,000 to approximately 5 per 100,000 during those years. In 1917, after the installation of chlorination at all nine pumping plants was complete, the typhoid death rate dropped to 1.7 per 100,000.[98] In 1918, the average chlorine dosage was 0.15 mg/L, enough to achieve adequate disinfection because of the general good quality of Lake Michigan water.[99] It is a good thing for Chicagoans that the city got on the chlorine bandwagon, even though it took years to complete the chlorination systems. A plant to filter Lake Michigan water was not operational until 1947.

Chicago's history of sewer construction and drinking water contamination is a perfect example of the Sewer Pipe–Water Pipe

Death Spiral, albeit one turned on itself. A fully executed sewer plan did not solve the city's high typhoid death rate. Reversing the Chicago River was helpful but did not unwind the death spiral. It was only after chlorination was installed that the Sewer Pipe–Water Pipe Death Spiral was shattered in Chicago.

It's the Drinking Water, Stupid[100]

Building sewers in a city was a necessary but insufficient solution for protecting public health. Sewers were ultimately needed to transport human wastes to sewage treatment plants that would be built many years in the future, but the solution required for protecting public health was a safe water supply. Just getting the smelly stuff out of the city as quickly as possible was not enough.

Authors have tried to shoehorn the history of construction of sewer and water distribution systems into a Chadwickian ideal that saved lives. What actually happened in the United States in the late 1800s and early 1900s was a Chadwickian nightmare that turned into the Sewer Pipe–Water Pipe Death Spiral. It was not until cities across the country began disinfecting drinking water with chlorine that the Sewer Pipe–Water Pipe Death Spiral was broken and the typhoid death rate in the United States declined until it was ultimately eradicated.

1   Whipple, *Typhoid Fever*, ix.

2   U.S. Bureau of the Census, *Statistical History*, 25.

3   Sedgwick, *Principles of Sanitary Science*, 21.

4   Wishnow and Steinfeld, "The Conquest of the Major Infectious Diseases," 428.

5   Sedgwick, *Principles of Sanitary Science*, 21–7.

6   The dates are approximate and BCE refers to "Before the Common Era."

7   Sedgwick, *Principles of Sanitary Science*, 26.

8   Rosen, *History of Public Health*, 79–80.

9   Ibid., 80–1.

10 Duffy, *The Sanitarians*, 129.

11 Roechling, *Sewer Gas and its Influence Upon Health*.

12 Ibid., 62–9.

13 Winslow and Greenberg, "Effect of Putrefactive Odors."

14 Halliday, *Great Stink of London*, ix–xi.

15 Budd, "Observations on Typhoid or Intestinal Fever," 486.

16 Barnes, *Great Stink of Paris*, 290.

17 Sedgwick, *Principles of Sanitary Science*, 114–5.

18 Leal, "Facts vs. Fallacies," 131.

19 Barnes, *Great Stink of Paris*, 19.

20 Ibid., 141.

21 Race, *Chlorination of Water*, 1-2; Hooker, *Chloride of Lime*, 1-2; although White claimed that Charles Tennant developed chloride of lime in 1790, White, *Handbook of Chlorination*, 3.

22 Slaked lime or calcium hydroxide is a dry, granular solid substance made by combining just the right amount of water and lime (calcium oxide).

23 Kienle, "Relation of the Chemical Industry to Water Works," 501.

24 Race, *Chlorination of Water*, 2.

25 Averill, *Disinfecting Powers of Chlorine*, 12–3.

26 Keen, "A Surgeon's Answer," 1.

27 Race, *Chlorination of Water*, 3.

28 Sedgwick, *Principles of Sanitary Science*, 354—quoting from William Budd, *Typhoid Fever: Its Nature, Mode of Spreading and Prevention*, London, 610–2.

29 Rosenberg, *Cholera Years*, 205.

30 *Trenton Evening Times*, June 25, 1885.

31 Sternberg et al., *Disinfection and Disinfectants*.

32 Haines, "The Urban Mortality Transition," 2.

33 Rosenberg, *Cholera Years*, 177–8.

34 Haines, "The Urban Mortality Transition," 14.

35 Ibid., 3, 8.

36 Vinten-Johansen, et al., *Cholera*, 168.

37 Ibid., 186–7.

38 Ibid., 165.

39 Colwell, "Global Climate and Infectious Disease," 2025–6; Pollitzer, *Cholera*, 11–50; Macnamara, *History of Asiatic Cholera*; Dziejman et al., "Comparative Genomic Analysis of *Vibrio cholerae*," 1556.

40 Vinten-Johansen, et al., *Cholera*, 296; quoted from *The Times*, "The Cholera Near Golden Square," September 15, 1854).

41 Vinten-Johansen, et al., *Cholera*; Johnson, *Ghost Map*; Hemphill, *Strange Case of the Broad Street Pump*; Frerichs, "John Snow," http://www.ph.ucla.edu/epi/snow.html.

42 Frankland, "Water Supply of London," 620.

43 Sedgwick, *Principles of Sanitary Science*, 170–81.

44 Corfield, *Etiology of Typhoid Fever*, 21.

45 Tomes, *Gospel of Germs*, 5.

46 Brannan, "Typhoid Fever," 713.

47 Ibid., 733.

48 Stewart, *Treatment of Typhoid Fever*, 61–4.

49 Yeo, "On the Treatment of Typhoid Fever," 117.

50 Ibid.

51 Corfield, *Etiology of Typhoid Fever.*

52 Sedgwick, *Principles of Sanitary Science*, 217.

53 Fuller and McClintock, *Solving Sewage Problems*, 108.

54 Weston, "Epidemic of Gastro-Enteritis," 193.

55 North, "Milk and Public Health," 253.

56 Kennedy, *Brief History of Disease*, 105.

57 Simard and Fortier, "Nourishment of Children," 367.

58 Ravenel, *Half Century of Public Health*, 253.

59 Leal, "Report of Board of Health 1892;" and annual reports up to 1898.

60 Why did more infants die in the summer than in the other seasons? Dr. Bill Gurfield, a pediatrician in Santa Monica, California, says that it is all about dehydration. "While adults have a good deal of fluid reserve to cope with acute dehydration, infants have little reserve for this. In the summer, fluid loss (via evaporation) through the skin to maintain body temperature is increased. If there is fever with the disease, then this type of fluid loss is accentuated. Infants have a higher surface area relative to body weight than adults and older children; hence their loss of fluid by this mechanism is greater." (e-mail, September 19, 2011) With the additional fluid loss caused by diarrhea, an infant less than one year old did not have a chance to survive unless its fluids were replaced.

61 Sedgwick, *Principles of Sanitary Science*, 166.

62 Sedgwick and MacNutt, "Mills-Reincke Phenomenon," 562.

63 Halliday, *Great Stink of London*, 37–8.

64 Johnson, *Ghost Map*, 207.

65 Halliday, *Great Stink of London*, 125.

66 Roechling, *Sewer Gas and its Influence Upon Health*, 75.

67 Vinten-Johansen, et al., *Cholera*, 256.

68 Fuller and McClintock, *Solving Sewage Problems*, 2–3.

69 Adams, *Sewers and Drains*, 11.

70 Baker, *Sewage Purification in America*, 1.

71 Kyriakodis, "Not Your Ordinary Fire Plugs."

72 Baker, *Manual of American Water-Works*, G.

73 U.S. Bureau of the Census, *Statistical History*, 240.

74 Staley and Pierson, *Separate System of Sewerage*, 29.

75 Ibid.

76 Todd and Sanborn, *Report of the Lawrence Survey*, 226.

77 Sedgwick, "Typhoid Fever in Lowell and Lawrence," 676–7.

78 Ibid., 669.

79 Todd and Sanborn, *Report of the Lawrence Survey*, 220.

80 Clark, "Purification of Water," 314.

81 Sedgwick, "Typhoid Fever in Lowell and Lawrence," 697; Clark, "Purification of Water," xviii.

82 Clark, "Study of Massachusetts Water Supplies," 203–16.

83 In 2007, Lawrence built a modern water treatment plant but, upon the start of operations in April, was cited for violations of the drinking water regulations by the state

83   In 2007, Lawrence built a modern water treatment plant but, upon the start of operations in April, was cited for violations of the drinking water regulations by the state of Massachusetts. After a series of problems with the plant staff and mounting fines, operation of the facility was turned over to an outside contractor. The Massachusetts Department of Environmental Protection found a number of treatment violations including inadequate treatment of turbidity in the influent water, which constituted a warning that the microbiological barriers in the plant were not satisfactory. The last thing the state or the city of Lawrence needed was an outbreak of typhoid fever that brought back the bad old history of the late nineteenth century. *Eagle-Tribune*, May 4, 2007; *Eagle-Tribune*, November 18, 2007.

84   Hill, *Chicago River*, 98.

85   Contaminated surface wells were the primary cause of the cholera epidemic in 1854; Chicago Department of Health, "Bulletin of the Department of Health," 2.

86   Hill, *Chicago River*, 99.

87   Adams, *Sewers and Drains*, v.

88   "Public Works of Chicago," 42.

89   Ibid., 43.

90   Ibid., 55.

91   Whipple, *Typhoid Fever*, 391-2, for data 1857 to 1906; Jennings, "Chlorine Compounds in Water Purification," 251, for data 1907 to 1917.

92   Johnson, *Ghost Map*, 214-5 repeated the myth; Hill, *Chicago River*, 116-9 disproved the myth.

93   Hill, *Chicago River*, 118.

94   *New York Times*, February 14, 1892.

95   The city had been notified on February 24, 1890, that its bid to hold the 1893 World's Fair was successful.

96   Sinclair, *The Jungle*, 97.

97   "Progress in 1906," *Engineering-Contracting*, 172.

98   Jennings, "Chlorine Compounds in Water Purification," 251.

99   Spalding and Bundesen. "Control of Typhoid Fever in Chicago," 360.

100 With apologies to James Carville

3

Germs, Disease, and Bacteriology

"... from 1865 to 1895 Western medicine underwent
a virtual civil war over the truth of the germ theory."
— Tomes, *Gospel of Germs*, 28

The germ theory of disease gave rise to the tools to identify contamination in water supplies, to figure out ways to eliminate that contamination, and, ultimately, to protect human health.

For the better part of the nineteenth century, three competing theories about the cause of epidemic disease existed simultaneously:[1]

- Miasma theory (sometimes called the pythogenic theory), which held that epidemic outbreaks of infectious disease were caused by atmospheric conditions.[2]
- Specific contagium or germ theory, which held that epidemic infectious diseases were caused by specific "contagia" that late in the nineteenth century were identified as various bacteria.
- Compromise theory, which held that contagia caused infectious disease but had to act in conjunction with the atmosphere, condition of the soil, or social factors.

Before there was a comprehensive articulation of the germ theory, many authors described aspects of the theory.[3] An early form of the germ theory posited in the sixteenth century captured some of its precepts.

In 1546, Italian physician Girolamo Fracastoro published his theories on disease contagion, which he believed was facilitated by "minute infective agents" called seminaria or "seeds." A half century later, George Rosen believed that the seeds were not the equivalent of microorganisms but that Fracastoro's construct of another agent, fomites, could be. In a complex mix of miasma and contagion, the fomite concept could help to explain part of disease transmission.[4]

"... Fracastoro recognized three modes of contagion: by direct contact from person to person; through intermediate agents such as fomites; and at a distance, for example, through the air. He postulated that under unusual conditions the general atmosphere becomes infected producing pandemics, and that such conditions might occur in association with abnormal atmospheric and *astrological conditions*."[5] (emphasis added; yes, astrological)

ANY DISCUSSION OF THE GERM THEORY of disease has to begin with an exploration of the discoveries of Louis Pasteur. Many full-length books have described his life and the extent of his influence,[6] but one excellent paragraph presents an overview of Pasteur's contributions.

"To some extent, Pasteur's interest in practical problems evolved naturally from his basic research, especially that on fermentation. The germ theory of fermentation carried quite obvious implications for industry and medical doctrine. By insisting that each fermentative process could be traced to a specific living microorganism, Pasteur drew attention to the purity and special nutritional and oxygen needs of the microbes employed in industrial processes. He also suggested that the primary industrial product could be preserved by appropriate sterilizing procedures, labeled 'pasteurization' almost from the outset. Furthermore, the old and widely accepted analogy between fermentation and disease made any theory of the former immediately relevant to the latter. The germ theory of fermentation virtually implied a germ theory of disease as well. This implication was more rapidly exploited by others, particularly Joseph Lister and Robert Koch, but Pasteur also perceived it from the first and devoted the last twenty years almost exclusively to working out some of the practical consequences of the germ theory of disease."[7]

Solving practical problems was key for this chemist, self-invented bacteriologist, and invader of the medical field. Much controversy existed on both sides of mid-nineteenth century understanding of the causes and transmission of disease.

Translations of three of Pasteur's later papers (1878–1880) are included in a slim volume, along with Lister's seminal contribution to antiseptic surgery originally published in 1867.[8] Reading the original words of Pasteur and Lister has far more impact than reading secondary and tertiary sources, which included the authors' own spin on events of the last half of the nineteenth century.

In the middle of various controversies and a professional war with his nemesis, French journalist Jules Guérin, Pasteur published a paper extending his germ theory to explain the etiology of several diseases. Though Pasteur's scientific observations and conclusions are important, it is his plea for a fair hearing of his ideas that is most compelling.

> "I have detailed the facts as they have appeared to me and I have mentioned interpretations of them: but I do not conceal from myself that, in medical territory, it is difficult to support one's self wholly on subjective foundations. I do not forget that Medicine and Veterinary practice are foreign to me. I desire judgment and criticism upon all my contributions. Little tolerant of frivolous or prejudiced contradiction, contemptuous of that ignorant criticism which doubts on principle, I welcome with open arms the militant attack which has a method in doubting and whose rule of conduct has the motto 'More light.'"[9]

American historian Thomas S. Kuhn's analysis of other scientific revolutions is applicable to the analysis of the problems with acceptance of the germ theory of disease.[10] Kuhn's thesis— "normal science" leading ultimately to a "paradigm shift"—was controversial, but he remains one of the most cited authors in the English language. His description of parallel beliefs and the eventual ascendance of "truth" are appropriate for following the early struggles and ultimate acceptance of the germ theory of disease. Pasteur was a revolutionary, even though his work was originally incremental ("normal science") in design. It is the genius of Pasteur's interpretation of his findings and the integration of his observations with the broader world of public health that led to the paradigm shift that made his contributions so groundbreaking.

JOSEPH LISTER'S LIFESPAN ENCOMPASSED the harshest debates over the germ theory of disease and its general acceptance.[11] Lister found that the use of carbolic acid (phenol) before, during, and after surgery virtually eliminated infections, especially the dreaded gangrene infections, which killed many people who survived the physical shock of surgery. Lister's seminal paper on antiseptic principles in surgery, curiously, did not mention Pasteur's influence on his research.[12] However, he acknowledged in other writings his debt to the French bacteriologist.

> "'Permit me,' wrote Lister, 'to thank you cordially for having shown me the truth of the theory of germs of putrefaction by your brilliant researches, and for having given me the single principle which has made the antiseptic system a success.'"[13]

Lister helped Louis Pasteur by supporting Pasteur's findings in France with practical examples in Scotland. His confirmation of Pasteur's theory was crucial because it gave other physicians simple tools to determine whether germs were causing infections in their patients. A well-equipped laboratory and training in scientific methods were needed to confirm that spontaneous generation was a fraud or to demonstrate that fermentation was caused by yeast. All physicians had to do was wash their hands between patients. A physician who washed his hands would notice immediately that his patients who were giving birth stopped dying in droves. If he removed his bloody apron, applied an antiseptic, wore clean clothes and gloves, and sterilized his instruments, most surgical patients avoided death from infections.

Lister and his work had a direct influence on the historic march toward disinfection of drinking water. Harriette Chick was a member of the Lister Institute staff for fifty years. She worked there in 1908 when she did her celebrated research on the mode of action of disinfectants on bacteria. Her research was the basis for labeling the equation she developed as Chick's Law.[14]

HUNDREDS OF AUTHORS have written about the germ theory of disease and the brilliant (and not so brilliant) men and women who were involved in its expression, refinement, and proof. Many accounts during 1860–1880 contained confusing chronologies of who said what first, who influenced whom, and who was responsible for the major advances that led science to the "Golden Age of Bacteriology."

Synthesis of the many publications on this topic is useful, but synthesis is also a tricky process because it involves the risk of oversimplification. In her wonderful book *Gospel of Germs*, history professor Nancy Tomes took the synthesis approach in one notable paragraph:

"The phrase 'germ theory of disease,' which came into common use in the English-language medical literature around 1870, was scientific shorthand for propositions associated with the work not only of Pasteur, but also of Koch, Tyndall, Lister and other investigators. Put simply, the germ theory consisted of two related propositions: first, that animal and human diseases were caused by distinctive species of microorganisms, which were widely present in the air and water; and second, that these germs could not generate spontaneously, but rather always came from a previous case of exactly the same disease."[15]

Tomes's clear statements of these two propositions are helpful, but lumping together Pasteur, Koch, Tyndall, and Lister implied that they were all somehow working together in 1870 to put forward this theory. In fact, in 1870 the German physician Robert Koch was a country doctor with scientific curiosity and not much else.[16] It was not until December 1875 that he did his famous experiment with anthrax by injecting a rabbit with material from a diseased source and infecting the rabbit with the disease. He did not publish the paper describing his groundbreaking anthrax research until December 1876.[17]

Pasteur initiated his important work on fermentation in 1856. His first publication on the fermentation of lactic acid came out in April 1857. It has been called the "birth certificate of microbiology."[18]

At a higher level of synthesis and integration, Nancy Tomes described the conflict over the competing theories about disease propagation. ". . . from 1865 to 1895 Western medicine underwent a virtual civil war over the truth of the germ theory"[19]. That a war existed is not controversial. When it ended is not agreed to by all. Tomes believed the war for doctors was over in the 1890s. As she put it, "In the end, advocates of the germ theory triumphed: by the 1890s medical students were being educated to revere the germ theory as scientific orthodoxy and to regard Pasteur and Koch as heroes."[20]

Despite Tomes's claim, the war for doctors was not over in the 1890s. It would take the passing of the older generation of physicians and the ascendance of forward-looking scientists and physicians trained in the 1890s to solidify the victory of the germ theory of disease.

The Microscope—Window into the Microbial World

Throughout the history of scientific advances, the development of tools helped scientists achieve incremental increases in knowledge as well as allowing them to break barriers and make discoveries that would otherwise not have been possible. Such is the case with the invention and improvement of the microscope.

Lenses that magnified things were around for hundreds of years. Then multiple lenses were assembled in tubes, creating the compound microscope. But it was not until the seventeenth century that a big leap was made, thanks to Anton van Leeuwenhoek.

Others improved the microscope, including Joseph Lister's father, Joseph Jackson Lister. In 1832, two hundred years after van Leeuwenhoek's birth, the elder Lister was able, through manipulation of the lenses in the tube, to eliminate the "chromatic effect" or light halos around the objects being observed.[21] Thus, a relatively sophisticated tool was available for Pasteur to view his yeasts, bacteria, and other microbes.

BUT ROBERT KOCH NEEDED even better microscopes to make his discoveries. In the early 1870s, the Zeiss Optical Works obliged him with microscopes designed by mathematical equation instead of trial and error. In parallel with the advances in microscope design and fabrication were the discoveries of stains that could highlight certain subfeatures of bacterial structure. In the mid-nineteenth century, several advances came together, including perfection of the oil immersion objective, which dramatically improved resolution of the bacteria under study.[22]

Adding to the scientific toolbox was a condenser that was developed by the Zeiss Company and that effectively focused light on the object being viewed. Now scientists could grow the bugs, fix them to a glass slide, stain cellular structures, and look at their private parts with the highest magnifications available—just in time for Robert Koch. In fact, the relationship between Koch and microscope advances was interactive. He used the advances to the fullest and, in turn, improved on them.[23]

Pasteur and Lister were critical to developing the foundation of the germ theory of disease. They took enormous personal risks to advance the theory. However, they did not develop efficient tools that others could use to advance the infant science of bacteriology. Pasteur and his followers used liquid media for all of their cultures. Because liquid cultures of this type were inevitably

Anton van Leeuwenhock

Anton van Leeuwenhoek was born in 1632 in Delft, in what is now the Netherlands.[1] In the same year, Galileo published his famous work *Dialogue*, in which he argued that Copernicus was right—the sun was the center of our solar system. To put it mildly, science was in its infancy. The Catholic Church rewarded Galileo for his insight by declaring him a heretic and holding him under house arrest for the rest of his life.

Of the many descriptions of van Leeuwenhoek's life, the most entertaining is the lyrical narrative in American microbiologist Paul de Kruif's classic 1926 book *Microbe Hunters*. De Kruif described van Leeuwenhoek as a janitor and shopkeeper, and, indeed, he was. However, van Leeuwenhoek was also obsessed with grinding lenses, making better microscopes, and viewing the as yet unviewed microbial world.[2]

While looking around his house for common items to study with his inventions, he decided to look at drops of water and discovered "beasties" swimming around. After a significant amount of time, during which he perfected his tool and honed his descriptions of the microbial world, van Leeuwenhoek began corresponding with the Royal Society in London. Despite initial skepticism, the Royal Society elected him to its august body. Van Leeuwenhoek did not share well with others, preferring to keep his improvements to the microscope to himself, but his hundreds of letters to the Royal Society include many descriptions of bacteria. He was the first person to make these observations.[3]

---

1   De Kruif, *Microbe Hunters*, 2.

2   Ibid., 3.

3   Ibid., 8–9.

mixed bacterial cultures, they were not helpful for finding specific bacteria responsible for individual diseases.

Robert Koch developed the tools that spawned the next generation of advances in bacteriology, and these advances provide a direct link to the two Jersey City trials. Without his breakthroughs, there would not have been any bacteriological data to determine if the Boonton Reservoir was providing pure and wholesome water to Jersey City.

In 1881, Koch published his seminal paper on bacterial growth on a solid medium. Called the "Bible of Bacteriology," the paper (in German) described in some detail how Koch combined the liquid medium in which pathogens would grow with a solidifying agent—gelatin.[24] The transparent nutrient gelatin could be fixed onto a transparent glass plate, and the use of a magnifying lens made counting the bacterial colonies that grew on the nutrient medium quite easy.[25] The incubation temperature was typically

ROBERT KOCH

Robert Koch was born December 11, 1843, in the small city of Clausthal in what was then called Lower Saxony. The city is 120 miles south and a little east of Hamburg and about the same distance west and a little south of Berlin. American microbiologist Thomas D. Brock's excellent 1999 biography of Koch chronicled his life, triumphs, and tragedies.[1]

Koch studied many diseases besides those that were waterborne. In addition to his innovative work in water bacteriology, he became world-famous for isolating and accurately describing the tubercle bacillus, the cause of anthrax disease (*Bacillus anthracis*), the cholera germ, and the genus of *Staphylococcus* organisms that cause many infections in humans.

It was Robert Koch who revolutionized our understanding of microscopic organisms in water and their relation to specific diseases. Once again, tools were crucial to progress. Although Koch had basic microscopes, not everything could be described or investigated under a microscope. He needed methods to examine what made microorganisms grow and die. So he and the scientists in his laboratory developed the tools that advanced the science of bacteriology, many of which are still in use today.

In 1880, Koch changed from a German country doctor performing clever experiments in a spare bedroom to a professional researcher at the Imperial Health Office in Berlin. In Berlin, Koch realized that the key to advances in bacteriology was development of pure cultures of the organisms causing disease. He was aware of early work in which a limited number of bacteria were grown on the solid surface of potato slices.[2] However, the human pathogens he was interested in studying did not grow very well on a potato substrate.[3]

In 1908, Koch and his wife visited the United States as part of a world tour. In many ways, this trip was Koch's victory lap. But the trip was the beginning of the end for Koch; he died two years later in Baden-Baden on May 27, 1910, at the age of 67.

---

1   Brock, *Robert Koch*, 6.

2   Ibid., 96–7.

3   In one of his first publications in 1891, George W. Fuller was able to show that the typhoid bacillus did grow on potato slices; Fuller, "The Specific Organism of Typhoid Fever," 140–1.

20–25 degrees Celsius, and the incubation period was usually three to five days.

In a paper published five years later, in 1886, German chemist Gustav Bischof examined a number of modifications to Koch's method and the changes made by British chemist Percy F. Frankland, son of Edward Frankland. Bischof also applied his own method for total bacteria counts to water supplies in the London area to show quantitative comparisons between contaminated and acceptable water supplies. In his comparison, Bischof used Koch's value of 100 total colonies per milliliter as the bright line below which a water supply could be classified as "very good."[26]

Two critical advances in bacteriology were made by researchers who built on Koch's achievements. German physician Walther Hesse worked in Koch's laboratory for about six months during 1881–1882 and was familiar with Koch's gelatin plate technique.

However, the inability of gelatin to be incubated at body temperature led Hesse to use nutrient agar as a growth medium.[27] Although Hesse never published the details of his discovery of agar's efficacy as a growth medium, Koch mentioned it in a paper on tuberculosis published in 1882.[28] Nutrient agar is used throughout the world today as a solid medium for bacteria growth.

Richard J. Petri, another assistant in Koch's laboratory, developed the circular glass dish that largely replaced the rectangular flat glass plates originally used by Koch. Petri's 1887 paper gave rise to dramatically improved bacteriological methods. Petri dishes with their covers were an elegant improvement to the complicated and expensive apparatus Koch used.[29]

Koch studied the disinfecting power of a number of substances, including chlorine.[30] His findings were published in 1881, at about the same time as his publication on the gelatin plate technique. He found that carbolic acid was not an effective disinfectant. An APHA committee subsequently conducted a major study of bacterial disinfectants; their results, which agreed with Koch's carbolic acid findings, were published in a full report in 1888.[31]

The Rise of Water Bacteriology: 1880–1905

Harold W. Wolf, who was a professor of civil engineering at Texas A & M University, defined the science of water bacteriology as beginning in 1880 with the identification of *Klebsiella pneumonia* and *K. rhinoscleromatis* as bacteria related to human fecal contamination.[32] The earliest use of the word "bacteriology" did not occur until 1884,[33] and it was the advances in Koch's laboratory that made possible the Golden Age of Bacteriology, usually set between 1880 and 1905. The growth of water bacteriology occurred during the same period, with many improvements added in later decades.

Table 3-1 lists the microbes identified during this period as causing the diseases transmitted primarily by contaminated water.

Improvements in the microscope led to a better understanding of the microbial world. Higher magnifications and sharper resolution made accurate descriptions of microorganisms possible. On the other hand, changes and "improvements" in methods for culturing bacteria in water produced chaos and argument.

Germs, Disease, and Bacteriology

Table 3-1 Timeline for discovery of waterborne disease organisms[34]

Year	Disease	Organism	Discoverer
1880	Typhoid fever	*Salmonella typhi*	C. J. Eberth
1883	Cholera	*Vibrio cholerae*	Robert Koch
1885	Diarrheal disease	*Escherichia coli*	Theodor Escherich
1898	Dysentery	*Shigella dysenteriae*	Kiyoshi Shiga
1900	Paratyphoid	*Salmonella paratyphi*	H. Schottmüller

During the late 1800s, bacteriologists realized that the task of isolating individual pathogens in contaminated water was fraught with difficulties. As early as Koch's work in the 1880s, sewage contamination of a water supply was demonstrated by analyzing for bacteria shed in large numbers by humans. Total bacteria counts and analysis for Bacillus coli (B. coli), now known as total coliforms, were two surrogate measurements for pathogen contamination.

In the 1890s, anyone who had access to a simple laboratory could develop his or her own test to identify and enumerate bacterial characteristics of human waste. Unfortunately, this is exactly what happened.[35] It took George Warren Fuller and an APHA committee several years to standardize the total bacteria count and B. coli tests.

"Fuller's interest in standard methods of analysis dated back to a meeting of the American Public Health Association (APHA) in Montreal in 1894. It was suggested at that time that a cooperative investigation should be made in bacteriological testing to bring some order out of the prevailing chaotic state. Fuller became a member of the subcommittee assigned with the task. It was chaired by Charles Smart; other members were J. George Adami and Wyatt Johnson. They concluded that laboratory techniques were so diverse that a much more detailed study was essential. This assignment was given to a new committee chaired by Adami. Fuller was also a member of this group and in the company of a galaxy of stars: A. C. Abbott, T. M. Cheesman, W. T. Sedgwick, Charles Smart, Theobald Smith, and W. H. Welch. Their deliberations culminated in the 1897 report of the Bacteriological Committee. It was the forerunner of the remarkably successful series of *Standard Methods of Water and Sewage Analysis* still produced in 1976 [and continuing through 2012 with the twenty-second edition]."[36]

The date when water bacteriology began to mature can be set at 1905, when the first edition of *Standard Methods* was published under the auspices of the APHA. However, during the two trials involving the Jersey City Water Supply Company and Dr. John Leal, plaintiffs and defendants used different methods for counting total bacteria and *Bacillus coli*. Comparing results from the different methods of identifying and enumerating bacteria was not unlike comparing apples, oranges, pears, and bananas.

AN 1885 PAPER by Percy F. Frankland gave details of the Koch method of determining bacteria counts in water and then used the method to determine the efficiency of slow sand filters for removing bacteria. Frankland noted that he increased the amount of gelatin in the nutrient gelatin medium by 50 grams per liter to decrease the tendency for the gelatin to liquefy at 20 degrees Celsius.[37] A subsequent discussion of Frankland's paper raised the concern that adding eggs with their shells to the medium to improve clarification of the substrate would result in a gelatin plate technique that differed from the original method published by Koch just four years earlier.

Franklin replied that "eggs were used by a number of bacteriologists; although this was Dr. Koch's process, it was now the property of the scientific world, and anyone was at perfect liberty to introduce an improvement."[38] Apparently, many newly minted water bacteriologists believed improvements were necessary. Changes to the Koch gelatin plate method were fast in coming, and this made interpreting the results difficult.

In the first edition of *Standard Methods*, the preparation of nutrient gelatin and agar were spelled out clearly so that researchers in different laboratories could obtain comparable results.[39] The method stated that adhering to the specified parameters was important because of the many differences in methods used for total bacteria count at the time.

> "In the present state of bacteriology there is no method known by which the absolute number of living bacteria in a sample of water can be determined, and all

quantitative determinations of bacteria are necessarily of a relative character. This being the case, strict adherence to a standard procedure is of especial importance."[40]

In the second edition of *Standard Methods*, the growth medium for total bacteria was specified exclusively as nutrient agar. Gelatin was no longer acceptable as a solid growth medium. In the same edition, the incubation temperature was fixed at 37 degrees Celsius (blood temperature) to more closely resemble bacteria that might be human pathogens. The incubation period was specified as 24 hours so the results could be determined more quickly than with an extended incubation period.[41]

From the second edition to the twentieth edition of Standard Methods, what was generally called the "pour plate" method for heterotrophic plate count bacteria did not change much. Incubation temperature was lowered to 35 degrees Celsius, and the incubation time was set at 48 hours.[42] However, numerous observations have indicated that bacteria transferred from water into melted agar (44–46 degrees Celsius) experienced thermal shock and significantly decreased in number. In the twentieth edition, two improved nutrient agars were recommended for enumerating total bacteria—R2A and m-HPC (previously called m-SPC). These low-nutrient agars yielded higher counts, especially when incubated at room temperature for several days.[43] In some ways, the total bacteria count method came full circle back to a lower incubation temperature and a longer period of incubation.

In 1886, THEODOR ESCHERICH, a pediatrician practicing in what we now know as Munich, Germany, published a treatise on the physiology of infant digestion, which identified the important role of intestinal bacteria. Though Escherich was not a bacteriologist, he was interested in the roles bacteria play in the human body. Using the available tools of the day (microscope and culture techniques), he identified rod-shaped organisms that were prevalent in the gut of infants.[44] He called the organism Bacterium colon commune, a name later changed to Bacillus colon communis. Over time, the organisms were called by a number of other names, including B. coli and, eventually, total coliforms. Years later, a particular bacterial species found in the gut of warm-blooded animals was named *Escherichia coli* in his honor.[45] A 1971 AWWA publication recounted some of the early history of the development of fecal indicators of water contamination and their use in regulations and control of drinking water quality.[46]

After untold debates and publications, the methods for determining total bacteria and concentrations of B. coli in water were based on compromise decisions of committees that published their findings in *Standard Methods*. The target organisms were defined by the methods.

To understand the early differences in defining and detecting coliforms, it is helpful to look at the destination—the current definition of the coliform group. The coliform group is now defined as aerobic, facultative anaerobic, gram-negative, non–spore-forming, rod-shaped bacteria that ferment lactose in 24 to 48 hours at 37 degrees Celsius. In the late 1800s and early 1900s, bacteria that fermented dextrose and other sugars were included in the B. coli group.

In 2006, Paul A. Rochelle and Jennifer L. Clancy published an excellent update of water microbiology, which chronicled advances in analytical methods from the pour plate to the latest molecular methods.[47]

View of the Germ Theory from 1902

In his 1902 landmark book, William T. Sedgwick presented an interesting perspective on the germ theory after it was well established but before it was embraced by all.

"The principal objection to the germ theory was and is that already referred to as met and overcome by Koch, viz., that germs may be seemingly the consequence, not the cause, of disease. Another objection is that in certain diseases the most careful search has failed thus far to reveal causative micro-organisms. The answer to this latter is simply that in the absence of all positive evidence of the true cause of disease, we are at liberty to choose the most likely working hypothesis, and no hypothesis has yet been found for any infectious disease more reasonable or more probable than the germ theory.

"A great scientific theory has never been accepted without opposition. The theory of gravitation, the theory of undulation [wave theory of light], the theory of evolution, the dynamical theory of heat—all had to push their way through conflict to victory. And so it has been with the germ theory of communicable diseases."[48]

Some writers would have us believe that a switch was thrown after the discoveries of John Snow, Louis Pasteur, Joseph Lister, and Robert Koch and that sometime in the 1890s, everyone embraced the germ theory of disease without reservation and without pining for the comfort of the old, well-worn slipper of the miasma theory. The following quotes illustrate the "switch" problem.

"The concept of miasma persisted, however, until Pasteur."[49]

"By the last decades of the nineteenth century, the germ theory of disease was everywhere ascendant, and the miasmatists had been replaced by a new generation of microbe hunters charting the invisible realm of bacterial and viral life."[50]

"Although many physicians continued to have reservations about the germ theory of disease, the general principle that microorganisms played a central role in causing communicable diseases had by 1900 achieved widespread acceptance in both Europe and America."[51]

Parts of the miasma theory persisted long after Pasteur. It would take decades of hard work and constant repetition of the germ theory in the scientific and popular press to displace the long-held beliefs of miasma theorists. Vestiges of the miasma theory hung on long after the germ theory of disease became mainstream. The compromise theory, which espoused aspects of both the germ and miasma theories, explains why some parts of the miasma theory persisted.

An amazing example of the compromise theory was expressed by T.E. Hayward, a public health officer in England, as late as 1899. He showed that portions of the miasma theory were still deeply entrenched.

"But it must not be forgotten that, although the typhoid germs are diffused by water, they are bred in polluted earth."[52]

"As to the causation of this epidemic, seeing that careful inquiry failed to show that milk-supply had anything to do with it, and that there was no evidence at all to trace it to general pollution of the water supply, I feel convinced that it was due to:

(a) Soil polluted with excremental filth.

(b) The existence of the special germs of typhoid in this soil.

(c) The multiplication and diffusion of these germs being favoured by the special climatic conditions of the year, especially with regard to the hot and dry autumnal season."[53]

At least Hayward did not believe in miasma alone. "It must be remarked *that bad smells do not* cause typhoid."[54] (emphasis in original)

Ignorance was not confined to the other side of the Atlantic. A few years before Hayward's paper was published, John W. Hill, a civil engineer from Cincinnati, Ohio, made an extraordinary presentation at the 1893 AWWA conference. He began with an incorrect statement: "The connection of drinking water with disease is not so well established now as it will be in due time, and much of the problem is still within the realm of speculation."[55]

In the face of proof, Hill liked to raise questions: "It is not settled among the medical fraternity whether the coma bacillus of Koch is the cause or effect of cholera. . . ."[56] Of course, just such a link had been proven by Koch and others ten years earlier.[57]

Hill then went on to confuse the presence of organic matter in water and diseases caused by the ingestion of pathogens. Perhaps the penultimate confusion of miasma and germs was his statement about the cause of malaria. "And I have reason to believe that malarial fever can be traced to water charged with vegetable matter in a state of decay."[58] Hill's presentation illustrated just how far some drinking water professionals had to go to embrace the germ theory of disease and get on board with preventing waterborne disease by filtering water supplies and installing proper chemical treatments.

The fraternity of scientists and engineers could react only so fast and accept only so much. Although the new explanation of disease causation eventually became the paradigm, scientists and the public held on to some parts of previous theories until time and further evidence eroded even these remnants. Ptomaines and bad air were examples of such remnants. It is only because of continuing beliefs in vestiges of the miasma theory that we can begin to explain turn-of-the-century efforts to build sewers to remove sewage swiftly and dump the contents into the nearest water supply.

Ptomaines. Putrefaction is now accepted as a mostly biological process that requires a mass of microorganisms (and higher animals) acting on organic material (composed of carbon, nitrogen, oxygen, and trace elements) and that produces gaseous products (carbon dioxide, methane, and hydrogen sulfide), additional masses of microorganisms, and a soup of degradation by-products that are generally simpler molecules compared with the ones that have been putrefied. In the 1890s, ptomaines were vaguely defined as products of putrefaction.

At the APHA annual meeting in Minneapolis in 1894, one of the discussants of a series of papers, Gardner T. Swarts, raised the issue of ptomaines specifically in regard to bacterial contamination of water supplies. His discussion began with compliments to George Warren Fuller on his presentation summarizing filtration studies at the Lawrence Experiment Station.[59] Swarts then moved on to explaining *why filtration may not be warranted* if ptomaines are really causing disease instead of bacteria. What is extraordinary about Swarts's statements is that he had personally conducted water filtration studies using the then-novel mechanical filters in Providence, Rhode Island.[60] A few years after Swarts's address, Leal made it clear that ptomaines were produced from decomposition of organic matter.[61]

As late as 1912, a summary of then-current beliefs about ptomaines was published. "Ptomaines are chemical compounds of an alkaloidal nature formed in protein substances during the process of putrefaction. . . . So far about sixty ptomaines have been isolated and studied and of these about one half are more or less poisonous.[62]

> "The symptoms of ptomaine poisoning vary in kind and severity, depending on the nature and quantity of the poison consumed. . . . As a result we may have vomiting, abdominal pain, diarrhea or constipation, usually attended with great prostration."[63]

We know now that ptomaines are amines produced by microbial degradation of proteins and that they are not responsible for causing disease. Products of putrefaction smell so disgusting that it is not surprising people thought they caused disease. Of course, the bacteria ubiquitous on spoiled meat—*Streptococcus, Staphylococcus,* and *Campylobacter* (or, if you are really unlucky, *Salmonella* or *E. coli 0157:H7*)—are the causes of gastrointestinal distress.

Bacteria were eventually shown to produce toxins that cause food poisoning, but that is not what drove the belief in ptomaines. The deadly toxin produced by *Clostridium botulinum* in an anaerobic environment is the classic example of a bacterial toxin.

Bad Air. In the 1890s, most professionals in the sanitary fields did not believe in the miasma theory of disease; however, some still believed that gases coming off of filth were unhealthy.

The Lomb Prize Essay by Dr. George M. Sternberg was published in the proceedings of the APHA annual meeting held in 1899.[64] As one of the most important professional associations that supported the germ theory of disease, APHA conducted its business to further understand how bacteria caused disease. Sternberg's essay can be described as nothing less than schizophrenic. On the one hand, he clearly understood how bacteria are responsible for disease transmission.

> "... there is good reason for believing that the infectious diseases of man are also caused by pathogenic—disease-producing—organisms of the same class. Indeed, this has already been proved for some of these diseases, and the evidence as regards several others is so convincing as to leave very little room for doubt."[65]

On the other hand, Sternberg still had one foot firmly planted in the vestigial miasma camp when he brought up both ptomaines and bad air.

> "The offensive gases given off from decomposing organic material are no doubt injurious to health; and the same is true, even to a greater extent, of the more complex products known as ***ptomaines***, which are a product of the vital—physiological—processes attending the growth of bacteria of putrefaction and allied organisms.... ***persons exposed to the foul emanations from sewers, privy vaults, and other receptacles of filth, have their vital resisting power lowered*** by the continued respiration of an atmosphere contaminated with these poisonous gases, and are liable to become the victims of any infectious disease to which they may be exposed."[66] (emphasis added)

The idea of a person's "vital resistance" being weakened contributed to people's reluctance to fully embrace the germ theory of disease. Three of the expert witnesses who testified for the plaintiffs in the two Jersey City trials believed in the archaic concept of the weakening of "vital resistance." In the appendix to his book, William T. Sedgwick discussed why sewer gas did not cause diseases like typhoid, dysentery, diphtheria, or scarlet fever. However, he bows to the compromise theory.

> "If, now, we turn to stagnant sewage, such as might result from broken drains, or such as commonly exists in cesspools, we may reasonably expect to find more dangerous and more concentrated gases. . . . In such cases, however, the sickness may be expected to take either the form of sudden, sharper attacks, suggestive of poisoning [e.g., hydrogen sulfide], or else *the form of malaise and a general lowering of the vital resistance, lassitude, weakness, etc.*"[67] (emphasis added)

One of Sedgwick's former students, Charles-Edward A. Winslow, expressed the view in 1910 that water may contain properties that are not affected by disinfection and that could negatively affect a person's vital resistance.[68]

George C. Whipple, one of Sedgwick's students, was a civil engineer and bacteriologist. Nevertheless, Whipple believed filth and sewer gas could weaken people and make them vulnerable to disease. His astonishing testimony in 1909 at the second Jersey City trial indicated his beliefs.

> "To what extent . . . decomposing organic matter predisposes to other diseases cannot be stated very definitely. But we have an analogous case in that of sewer gas, where the breathing of sewer gas has been shown to be deleterious by way of predisposing to other diseases. . . . I think the future studies are going to show that there is some very important relation between filth and the public health, even though the filth is in water and may not be infectious."[69]

In 1902, John L. Leal, who testified for the defendants in the two Jersey City trials, had a strong opinion about the mistaken belief that miasma, exhalations, or emanations somehow weakened the human constitution and rendered it more prone to disease.

> "When we discovered the true causes of these diseases, and that filth, bad water and air, and telluric conditions could not directly cause them, as it was hard to break away abruptly from the ideas and mental attitude

induced by long belief and training, we conjured up certain vague, indefinite bodily conditions, which we attributed to the same as causes, and held that such conditions, through weakening the natural resisting powers of the body to disease, tended to make infection more probable in the event of exposure.

"This position was based upon the principle that whatever weakens or lowers the tone of the bodily organism also diminishes the power of resistance to disease, with which power such bodily organism is endowed by nature. There is no reason to believe today that our last attitude had any firmer foundation than the first [miasma]. Practical experience, as well as theory based upon our present knowledge, teaches us that *exposure to filth, filthy water and filthy air does not in itself necessarily lower the tone of such bodily organism.*"[70] (emphasis added)

Leal stated his dramatic condemnation of the vital resistance argument even more forcefully at the end of the second Jersey City trial.

Ceremonies of Incantation. Disinfection of households where victims of infectious diseases lived was one of the results of the lingering belief in miasma. In 1902, Leal discussed both the useful and the ridiculous aspects of disinfecting a household after removal of the infected person.

"Disinfection, then, is the process of destroying such infection by the destruction of the disease germs there existing. . . . Too often, however, it is intrusted [sic] to one whose training possibly has made [him] capable of distinguishing a pile of filth or an unpleasant odor, but who has no true conception of the cause of the disease, how it is possible to destroy it, and the means to be employed. In such hands it is more a *ceremony of incantation* than a scientific process."[71] (emphasis added)

Ceremonies of incantation persisted for decades. Charles V. Chapin, a health officer for Providence, Rhode Island, lamented in a paper published in 1923 that cities were loath to give up what he called "terminal disinfection," which referred to the disinfection of surfaces or the atmosphere in a dwelling where a person had suffered from a contagious disease. He emphasized that by the time of his writing, everyone was pretty sure contagious diseases were spread by people (and their emanations) and not things. Swabbing a house down with formaldehyde, burning sulfur, or heating pans of chloride of lime provided impressive special effects but were of little use to prevent transmission of epidemic diseases.[72]

Yet cities felt compelled to continue the tradition because the public expected it. The city of Providence stopped terminal disinfection for cases of diphtheria in 1905, but it was not until 1908 that Chapin was able to stop terminal disinfection for cases of scarlet fever. New York City eliminated virtually all terminal disinfection in 1913, and at that point many other cities followed suit.[73] Some of the resistance to eliminating terminal disinfection stemmed from the timidity of public officials to change something people were used to, but a huge part of the problem was the deeply ingrained belief that if someone was sick, they probably infected the air and the bad air had to be cleansed.

By 1918, Charles-Edward Winslow seemed a lot less sure that emanations from putrefaction were in any way related to disease. He published an article describing research originally intended to replicate the 1895 study by Giuseppe Alessi, who purported to show that putrefaction odors reduced the ability of animals to resist disease. Winslow noted that three other studies appeared to support Alessi's findings. In Winslow's research, conducted over three years, 261 guinea pigs were exposed to odors arising from the putrefaction of feces in specially constructed chambers. Unexposed control animals were used in all experiments.[74] The paper concluded by stating simply: ". . . our results entirely fail to substantiate Alessi's claims of a reduction in resistance against bacterial inoculation."[75]

If more evidence of the persistence of the miasma theory is needed, it is only necessary to inspect the literature describing the great influenza pandemic of 1918 to learn that as late as that date, miasma was considered a possible cause.[76]

The New Public Health

More than any other person, Charles V. Chapin was responsible for shaping the new view of public health after 1900.

> "At the end of the first decade of the twentieth century Chapin elaborated his reappraisal of principles and measures for the control of communicable disease in a book

which became the bible of the new era. *The Sources and Modes of Infection* provided the 'definitive synthesis,' which few other persons could have made, of the vast amount of laboratory and epidemiological findings collected during the previous forty years."[77]

Chapin's book was a landmark publication in the field of public health, and health officers across the country began to adopt his principles.

"Thus, in the years after 1900 the public health movement dramatically burst its old bounds of environmental sanitation and a limited concern with a few infectious diseases. A host of new voluntary organizations, national and local, rose ... to help public health officers shape a new pattern of public health work."[78]

While scientists and engineers were developing the new public health credo, parallel actions were being taken to enlist the public's cooperation. In her book *Gospel of Germs*, Nancy Tomes made the connection between public campaigns that educated people on personal hygiene—including anti-spitting laws that were effective in helping to stop the spread of tuberculosis—and the public's eventual acceptance of the germ theory of disease.

Although no magic wand was waved, and no gigantic headline was published in the *New York Times*, the ascendance of the germ theory of disease over miasma and superstition was largely complete in the United States by 1920. But an assessment of the situation by public health historian John Duffy was not accurate: "The medical profession in these years [1880–1900] was somewhat ambivalent about bacteriology and the development of public health agencies. If one can judge by newspapers and popular journals, the public was far quicker to accept the germ theory than physicians were."[79] The truth is that compared with the medical establishment, the public took a lot longer to embrace the germ theory of disease.

Unfortunately, other parts of the world still struggle with vestiges of the miasma theory in the twenty-first century. The front page of the *New York Times* on October 31, 2009, carried a photo of a worker who was clad in all-white protective clothing and was setting off "disinfection bombs" in a Turkish classroom in a school where swine flu caused by the H1N1 virus had broken out.[80] Presumably, the purpose of the multiple columns of toxic smoke depicted in the photo was to purify the "bad air" in the classroom or to "disinfect" the facility. There are many effective measures for controlling the spread of the H1N1 virus, including vaccinations and frequent hand-washing. Setting off airborne disinfection bombs is not one of them.

1   Rosen, *History of Public Health*, 264.

2   In the early 1800s, another term was used to describe epidemic, endemic, and contagious diseases: zymotic. The cause of zymotic diseases could be miasmatic fermentations or, later, germs; for a detailed discussion, see Sedgwick, *Principles of Sanitary Science*, 34–7. As the germ theory of disease began to take hold, the terminology changed to zymotoxic, which covered the theory of ferment-poisons produced by bacteria. Sedgwick, *Principles of Sanitary Science*, 56–8.

3   Vinten-Johansen, et al., *Cholera*; Budd, "Typhoid Fever."

4   The term "fomite" is currently defined as an inanimate object (clothing, bedding, books, or toys) that comes in contact with a person with infectious disease and, when used by another person, spreads the disease. Kate Winslett's character in the recent movie *Contagion* used the term to explain one way that the killer virus was spread.

5   Rosen, *History of Public Health*, 82.

6   Debre, *Louis Pasteur*; Geison, *Private Science of Louis Pasteur*; these are only two of many books on the subject.

7   Geison, *Private Science of Louis Pasteur*, 36–7.

8   Pasteur and Lister, *Germ Theory and its Application to Medicine*.

9   Ibid., 129–30.

10 Kuhn, *Structure of Scientific Revolutions*; any Google search of relevant words leads to thousands of hits of essays, publications and high school term papers on this topic.

11 Godlee, *Lord Lister*, 595.

12 Pasteur and Lister, *Germ Theory and its Application to Medicine*.

13 De Kruif, *Microbe Hunters*, 97.

14 Chick, "Laws of Disinfection." Every sanitary/environmental engineer who has taken a course in water treatment has heard of Chick's Law, but most do not know who this amazing woman was. After her groundbreaking work on disinfection, she made major contributions in the field of nutrition and ultimately was made a Dame of the British Empire. Her career was long and accomplished, and she died at the age of 102 in 1977, spanning a century that evolved from the miasma to the microchip.

15 Tomes, *Gospel of Germs*, 33.

16 To be fair, Tomes gives better chronologies of the involvement of the players in the development of the germ theory in other parts of her book.

17 Brock, *Robert Koch*, 32.

18 Debre, *Louis Pasteur*, 101, 504.

19 Tomes, *Gospel of Germs*, 28.

20 Ibid.

21 Godlee, *Lord Lister*, 11.

22 Brock, *Robert Koch*, 67–8.

23 Ibid., 62–3.

24 Brock, *Robert Koch*, 97.

25 Frankland, "New Aspects of Filtration," 698–9.

26 Bischof, "Notes on Dr. Koch's Water Test," 119.

27 Walther Hesse developed his idea from his wife's suggestion that agar was an effective thickening agent used in jams and jellies. Fannie Hesse was also Walther's laboratory technician and artist illustrator; Brock, *Robert Koch*, 102.

28 Brock, *Robert Koch*, 101–3.

29 Ibid., 103.

30 Crittenden, et al., *Water Treatment: Principles and Design*, 1037.

31 Sternberg et al., *Disinfection and Disinfectants*.

32 Wolf, "Coliform Count as a Measure of Water Quality," 333.

33 Keen, *Medical Research and Human Welfare*, 23.

34 Brock, *Robert Koch*, 290.

35 Payment et al., "History and Use of HPC," 20–48.

36 Wolman, "George Warren Fuller," 5.

37 Frankland, "New Aspects of Filtration."

38 Ibid., 708.

39 Nutrient gelatin, as noted in *Standard Methods,* differed from the method originally published by Koch in that the meat infusion was prepared by a cold infusion process with 500 grams of lean meat in one liter of water kept in a refrigerator for 24 hours. No salt was used in the preparation. Much higher amounts of peptone and gelatin were used—1% and 10 %, respectively. Incubation temperatures were 20 and 37 degrees Celsius, with incubation times noted in the methodology. Any deviations from the specified recipe or temperatures were to be noted by a researcher presenting results. *Report of Committee on Standard Methods*, 108.

40 *Report of Committee on Standard Methods*, 81–2. In the quote is referenced: Fuller and Johnson, "On the Question of Standard Methods for the Determination of the Numbers of Bacteria in Waters," 574.

41 *Standard Methods*, Second Edition, 77.

42 *Standard Methods*, Twentieth Edition, 9-34 to 9-38.

43 Means et al., "Evaluating Mediums and Plating Techniques," 585–90.

44 Escherich, *Enterobacteria of Infants*.

45 It is important to distinguish the group of organisms under the umbrella B. coli or total coliforms from the specific bacterial species, *E. coli*.

46 AWWA, *Water Quality and Treatment*, 1971, 5.

47 Rochelle and Clancy, "Evolution of Microbiology in the Drinking Water Industry."

48 Quote is from Tyndall, *New Fragments*, 198; quoted in Sedgwick, *Principles of Sanitary Science*, 60–1.

49 Kennedy, *Brief History of Disease*, 162.

50 Johnson, *Ghost Map*, 213.

51 Tomes, *Gospel of Germs*, 6.

52 Hayward, "The Causation of Typhoid Fever," 741.

53 Ibid., 742.

54 Ibid.

55 Hill, "Is Our Drinking Water Dangerous," 131–2.

56 Ibid., 132.

57 Brock, *Robert Koch,* 182.

58 Hill, "Is Our Drinking Water Dangerous," 134.

59 Fuller, "Sand Filtration of Water," 64–71.

60 Swarts, "Discussion," 83–4.

61 Leal, "Facts vs. Fallacies," 133.

62 LeFevre, "Ptomaines and Ptomaine Poisoning," 400.

63 Ibid., 403.

64 Sternberg, "Disinfection and Individual Prophylaxis," 624.

65 Ibid., 630.

66 Ibid., 625–6.

67 Sedgwick, *Principles of Sanitary Science,* 349.

68 Winslow, "Water-Pollution and Water-Purification," 15.

69 Between Jersey City and Water Company, May 21, 1909, 6509.

70 Leal, "Facts vs. Fallacies," 130.

71 Ibid., 138.

72 Chapin, "Disinfection in American Cities," 92–4.

73 Ibid.

74 Winslow and Greenberg, "Effect of Putrefactive Odors," 759–62.

75 Ibid., 767.

76 Barry, *Great Influenza,* 256.

77 Cassedy, *Charles V. Chapin,* 123.

78 Ibid., 141.

79 Duffy, *The Sanitarians,* 196.

80 *New York Times,* October 31, 2009.

4

Progress in Disinfection
and Filtration

*"Such poisonous materials should not be permitted to
be used on water intended for public supplies."*
— Maignen, "Discussion," 286

Early examples of drinking water disinfection influenced the decision to introduce a continuous application of chlorine into the Boonton Reservoir supply in 1908. Filtration as well as disinfection was linked to the conquest of waterborne disease. In the United States, the rise of filtration involved two competing versions—slow sand filters and mechanical filters.

The sanitary engineering literature from 1890 to 1910 is filled with articles describing the disinfection efficiencies and bench-scale test results for numerous potential disinfectants, including ozone, ultraviolet light, bromine, iodine, copper, and heat. Other chroniclers of the late nineteenth and early twentieth century have described the theoretical advantages and disadvantages of these disinfectants and even some small, short-term applications.[1] However, chlorine was the only disinfectant ready for practical use.

SWEDISH CHEMIST CARL W. SCHEELE has been credited with the discovery of chlorine in 1774. Along with Scheele, French chemists Antoine Lavoisier and Claude Louis Berthollet originally described chlorine as oxygenated muriatic acid. It was not until 1810 that British chemist Sir Humphrey Davy proved it was an element and named the element "chlorine" after the Greek word for green. The first use of chlorine as a disinfectant (not in drinking water) was credited to Guyton de Morveau in France and William Cruikshank in England in about 1800.[2]

American authors F. E. Turneaure and H. L. Russell of the University of Wisconsin reported three early investigations of water disinfection by calcium hypochlorite. European researchers Traube, Bassenge, and Lode noted that addition of the chemical

71

readily disinfected the water and that excess amounts of chlorine could be removed by sodium sulfite or calcium bisulfite. Success with the method was briefly described in three papers published in 1894 and 1895 in German. "Either this [chlorine] or the preceding method [ozone] is applicable on a large scale and may be of special service in times of epidemic where wholesale treatment is demanded."[3]

IN THE SUMMER OF 1897, Percy Adams, deputy registrar of diseases for Maidstone, England, noted the city's high incidence of deaths caused by diarrheal disease. On September 15 of that year, he issued a notice for the public to boil all drinking water and take other precautions.[4] The disease affected more than 1,900 people, according to the approximate timeline of new typhoid fever cases shown in Figure 4-1.[5] Another source listed

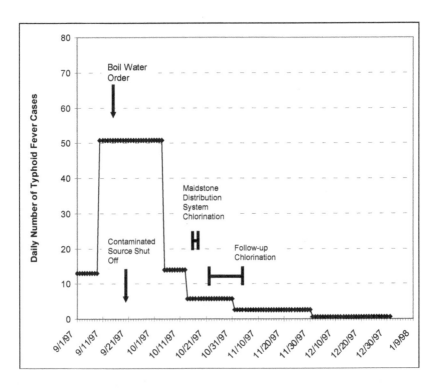

Figure 4-1   Typhoid Fever Incidence in Maidstone, England, September–December 1897

the total number of typhoid fever cases comprising the Maidstone epidemic as 1,808.[6] The infection rate equaled about *five percent of Maidstone's population,* an extraordinarily high proportion of the customers served by the water company. Ultimately, about 150 people died from the Maidstone epidemic.[7]

In 1897, Maidstone received its drinking water from three sources. One of the sources was a combination of three "springs" from the Farleigh area a few miles up the Medway River valley. One of the so-called springs was labeled the Tutsham source. Water from this area was collected in open-jointed pipes about two feet below the ground surface. In August and September 1897, 250 to 350 itinerant hop pickers were encamped in the water-gathering area. No sanitary facilities were provided for the workers, and contamination of the "spring" at Tutsham was easily predictable.[8] As one chronicler of the event stated, "What more could anybody want in the way of inviting disaster?"[9]

The Tutsham spring as a source of water for Maidstone was shut off on September 19. This decision, coupled with the boil-water order issued on September 15, slowed the spread of disease. Figure 4-1 shows that the high point of the epidemic was followed by a rapid decline in typhoid fever incidence over the next few weeks. The graph also illustrates that the prompt actions by Adams probably averted a much bigger disaster. Because the incubation period for typhoid fever is one to three weeks with an average of two weeks, secondary transmission by person-to-person contact could explain the low infection rate three weeks after the peak.[10]

Compounding the lack of protection for the Tutsham spring, Maidstone's local governing board had decided years before the epidemic that spending £50 per year for source water bacterial monitoring was too much and should be discontinued. After that, the frequency of monitoring was modest—only once per fortnight. As stated in a summary of the Maidstone epidemic, "They have learnt by this time that there is something a good deal more expensive than analyzing water, and that is not analyzing it."[11] A critical article that found no excuse for a typhoid fever epidemic noted: "At Maidstone they all went to sleep together and let the ship drift on the rocks."[12]

On October 11, 1897, the water company hired Dr. Sims Woodhead to survey the situation and determine what ". . . precautions should be taken at once to prevent, as far as possible, any further dissemination of the disease through the agency of the water supply. . . ."[13] Dr. Woodhead was a well-respected physician who, at the time of the Maidstone epidemic, was director of the laboratories of the Conjoint Board of the Royal Colleges of Physicians and Surgeons in London. Shortly after the Maidstone episode, he was appointed professor of pathology at Cambridge University. He published extensively in the medical literature of the time on bacteriology topics.[14] It is significant that the person responsible for the first use of chlorine to disinfect a distribution system was a physician who thoroughly understood the bacteriological cause of typhoid fever.

A contemporaneous description of the event summarizes what happened.

> "Under the personal supervision of Dr. Sims Woodhead, acting on behalf of the water company, the reservoir and mains of the Farleigh area of [the] water supply at Maidstone were on Saturday night [October 16, 1897] disinfected with a solution of chloride of lime. About ten tons of the [chloride of] lime were mixed in the reservoir with 200,000 gallons of water and afterwards the mains throughout the town were charged at full pressure with the solution. At a certain hour the liquid was allowed to flow through the whole of the house connexions [sic] in the districts concerned, and eventually the solution left in the mains was got rid of through the various fire hydrants in the streets."[15]

The dates on which chloride of lime was added are shown on Figure 4-1.[16] The chlorine dosage under the conditions described was about 4,200 parts per million (ppm). A subsequent article by Woodhead and W. J. Ware described the chlorination incident in detail. The original intent of the chloride of lime application was to achieve a concentration of chlorine of about 3,500 ppm. The reason for the high concentration of chlorine was to ensure that the typhoid bacillus was thoroughly destroyed and to use the strong odor of chlorine at the high concentration as a warning to residents to refrain from drinking the water during what engineers would now call "superchlorination" of the distribution system. Disinfection of the system was complete after about three days.[17] It is extraordinary that chlorination of the distribution system was carried out so quickly—five days after the decision was made. Maidstone serves as an excellent example of how an emergency led to the application of a known technology (chloride of lime disinfection) in an unusual circumstance (a water distribution system). Sometimes technological progress is incremental, and sometimes it leaps forward. The use of disinfection at Maidstone was a leap forward.

Woodhead and Ware recognized the importance of their actions when they published their paper describing the chlorination incident. "We believe that this is the only instance in which disinfection of water mains has been carried out on anything like so large a scale, and, although the difficulties to be surmounted were by no means small, we were thoroughly satisfied with the results."[18] It is unlikely, however, that they had any idea their actions would continue to be quoted more than 110 years after the event.

IN HISTORIES OF THE EARLY USE OF CHLORINE in drinking water, the application of a disinfectant at the resort town of Ostende, Belgium, is sometimes mistakenly listed. A mechanical filtration plant treated the water from a contaminated canal that served as Ostende's source of supply. About 1900, the water was treated for a relatively short time with something called peroxide of chlorine. In effect, the compound was chlorine dioxide. The chlorine dioxide gas was produced by the combination of potassium chlorate and oxalic acid. As we know today, mixing chlorate (or chlorite) ion and an acid as concentrated chemicals will produce chlorine dioxide gas. The process was difficult to control, expensive, and dangerous. As George C. Whipple put it, "The process is not one to be commended."[19]

THE DISINFECTION OF THE WATER at the resort town of Middelkerke, Belgium, is often cited as the first continuous use of chlorine in a water supply Although Middelkerke's permanent population was only 2,000, an influx of summer visitors brought the population up to 12,000.[20] Beginning in July 1902, Ferrochlore, a combination of chloride of iron (either ferric or ferrous chloride) and chloride of lime, was added to the water supply before it went through a small mechanical filtration plant on the Belgian coast. During the test period 40,000 gallons of water per day were treated; the full-scale facility treated about 100,000 gallons per day. The raw water supply for the plant was taken from a nearby canal whose water quality was marginal. The Ferrochlore process was named by its inventor, Maurice Duyk, a chemist with the Belgian Ministry of Public Works. Once the mixture of chemicals was added to water, the iron salt acted as a coagulant and the chloride of lime as a disinfectant. A report describing the process noted the presence of oxygenated compounds of chlorine (dominated by hypochlorous acid) after addition of the chemical mixture. The quality of the water produced was described as "bright and palatable," although no bacteriological results were included in the report.[21]

The raw water was also described colorfully as ". . . a bright brown, which [was] compared to the tint of burnt sienna, so that it had the appearance of a fairly pale sherry."[22]

George Whipple visited the Middelkerke plant and reported disinfection data from the initial trials, which began in July 1902. Four samples collected from August 22, 1902, to October 17, 1902, showed that total bacteria counts in the raw water were reduced to 10–20 per milliliter from 4,000–6,000 per milliliter. B. coli were found in all of the raw water samples and were absent in all tap water samples after the treatment process. The chlorine dosage at Middelkerke was very high—5 ppm. Whipple's results also illustrated two other advantages of using chlorine in a challenging water supply. After chlorination, the yellowish raw water (no doubt resulting from high concentrations of humic material) became clear and colorless, and a pronounced marshy odor became very slight.[23]

Whipple noted that the reaction of chlorine with potassium iodide and starch produced a blue color, allowing traces of chlorine to be qualitatively detected in the treated water.[24] Treatment of the Middelkerke water supply with Ferrochlore continued until 1921, when the town began using "pure" spring water as its supply.[25]

The principles of Ferrochlore treatment were presented at a 1904 international conference in St. Louis, and a discussant described how well the process worked. "The advantages which Ferrochlore possesses over all other substances is that it is a powerful sterilizer, and, at the same time, a coagulant." The discussant obviously knew the mechanism that was occurring and why it had advantages over other treatment methods: ". . . it is invaluable for the purification of water drawn from rivers or lakes contaminated with sewage or other refuse waters."[26]

Expanding on the use of chlorine at Middelkerke, Whipple, at the same 1904 conference, went on to describe the then-current state of acceptance of the use of chemical disinfectants in drinking water.

"The use of ozone, the hypochlorites, ferrochlore, etc., which have attracted considerable attention abroad have received but little practical attention in this country [U.S.] and cannot be considered as having yet emerged from the experimental state. Whether or not they ever will remains to be seen. . . . *there are certain objections to the use of chlorinated compounds, which will prevent them from ever taking an important part in the art of water purification.*"[27] (emphasis added)

Whipple should have known better than to make so definitive a statement during a time of rapid technological change. Within five years, he would be faced with just such an application of chlorine to a water supply when he was on the witness stand in the second Jersey City trial.

At that same international conference, renowned American engineer Allen Hazen called for more U.S. investigations of drinking water disinfection: "The treatment of water, even experimentally, by extremely powerful oxidizing agents, such as ozone, compounds of chlorine and oxygen, and the more recently described 'ferrochlore' . . . has thus far been confined almost entirely to Europe. . . . It is hoped that opportunities will be afforded to test these processes thoroughly in this country [the United States] by experiments comparable to those which, ten years ago, served to establish the fundamental facts regarding mechanical filtration."[28] Hazen's allusion to establishing the fundamentals of filtration was a reference to George Warren Fuller's work in Louisville and Cincinnati and his own work in Pittsburgh.

Hazen then sounded a cautionary note that reflected the inbred conservatism of sanitary engineers. "It is hoped that something substantial will be gained from these processes; but it must be remembered that, although new processes of water treatment have been proposed with regularity and frequency, the proportion of such new processes leaving a permanent impression upon the art of water purification is comparatively small."[29]

In the case of chlorine disinfection of drinking water, progress would not be incremental. The revolution would take place four years later in New Jersey.

IN THE EARLY TWENTIETH CENTURY, Lincoln, England, was a manufacturing town on the banks of the River Witham.[30] The city, an asylum, and three suburban areas were served water from the Lincoln Corporation Water Works. In 1904–05, the estimated population of these areas was about 54,000. As in other areas of England, meticulous records of the incidence of enteric (typhoid) fever were kept during the second half of the nineteenth century. In the case of Lincoln, the records were available from 1867 to 1905, the year of the epidemic. Slow sand filtration of the water supply began in the 1870s, with improvements added over time. As a result, the death rate from typhoid fever in Lincoln fell from an average of 51 per 100,000 during 1871–1880 to 6 per 100,000 in 1904.[31] Even so, the Lincoln epidemic provided an example of how filtration did not eliminate all waterborne disease problems in the late nineteenth and early twentieth centuries. Slow sand filters had to be properly designed and operated in order to protect against disease from contaminated sources.

Dr. Richard J. Reece, the main investigator of the Lincoln typhoid fever epidemic, chronicled it in great detail.[32] Figure 4-2 shows the weekly number of new typhoid fever cases in the area served by the water utility. A total of 125 people died from typhoid fever during the epidemic. If this number had been calculated as an annual death rate for the exposed population, it would have been an astonishing 231 per 100,000.[33]

The water supply for the city and associated suburban districts was taken from "land drains" that collected surface water from agricultural lands and from the upstream reaches of the

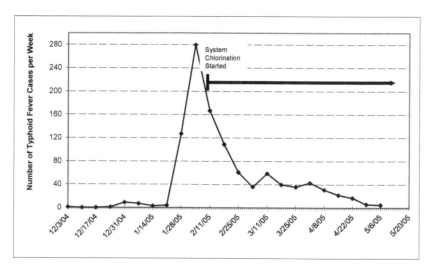

Figure 4-2   Weekly Cases of Typhoid Fever at Lincoln, England, 1904–05

River Witham before it entered the city limits. The water was filtered through two slow sand filters covering a total surface area of about 3,400 square yards and distributed to the city as well as to two large reservoirs located at higher elevations.[34]

Operation of the slow sand filters at Lincoln was substandard. For reasons known only to the "water engineer" at the Lincoln treatment plant (Mr. Teague, Jr.), the top layer of the filters was scraped off and piled up in a small area whenever filter throughput decreased. New or washed sand was added only to a small area on the top of the filter, creating a "step." Also, the engineer noted that the filtration process worked a lot faster if the top layer (the *Schmutzdecke*) was disturbed by raking it so that water could be forced through the filter beds.[35]

The decreased filtration efficiency caused by these peculiar operational procedures undoubtedly contributed to the passage of pathogens through the filters and into the distribution system. Another treatment defect was observed when the water used to clean the dirty sand was returned, in part, to the slow sand filter influent. Mr. Teague resigned his appointment as water engineer shortly after an investigation by Dr. Alexander Cruickshank Houston.[36]

An appendix to the report on the Lincoln epidemic referred to previous findings that the city's source of supply was seriously polluted.[37] Also, an article published in *The Lancet* noted that seven years before the epidemic, the Lincoln Medical Society had passed a resolution on November 29, 1897, calling attention to problems with sewage discharges into the city's source of supply. The society urged that additional land be purchased for larger water treatment works and that some of the recently developed tools for monitoring bacteriological quality be employed to ensure proper operation of the slow sand filters. The Medical Society noted that,

"... such water must be filtered, but there is the risk that the filters may get out of order. It is to check this risk that there should be constant analysis. Such examinations will indicate whether the water is up to its normal standard, and any departure from that standard will suggest the urgent necessity of immediately examining the conditions of the filters. By such means it may be possible to nip the mischief in the bud."[38]

All of the recommendations were rejected by the local governing board, which was criticized severely.

> "It seems absurd to spend large sums of public money in building reservoirs, filters, &c., and then to stint the small cost of analyses. This policy of spoiling the ship for the sake of a pennyworth of tar is the most foolish of all policies."[39]

The Lancet article was horribly prescient. Monitoring for B. coli and total bacteria in the bizarrely operated slow sand filters would have shown that operation of the filters was deficient. The typhoid fever epidemic could have been avoided.

More than a month into the epidemic, it was clear to Dr. Reece that the water supply was compromised and would have to be treated before it would be safe for the people of Lincoln to drink. Reece first suggested boiling the drinking water in bulk and distributing it with carts. But boiling in bulk was not feasible and was not accepted by the populace. Eventually, each household was advised to boil water to be used for drinking and cooking.

Reece then did what every smart consulting expert has done when faced with an impossible water quality problem—he called in other experts. Fortunately for continued progress in the implementation of drinking water disinfection, Reece called in the right experts: Dr. Alexander Cruickshank Houston was, at the time of the Lincoln investigation, a bacteriologist with the Royal Commission on Sewage Disposal. The lesser-known Dr. George McGowan was a chemist with the same organization.

Shortly after working on the Lincoln project, Dr. Houston became director of water examination for the London Metropolitan Water Board. He contributed to the knowledge of water quality in England for many years and ultimately was knighted for his service. When he died on October 29, 1933, an obituary stated ". . . the world of water purification has suffered a very severe loss. During the first three decades of the present century his name has dominated all questions relating to the scientific treatment and purification of water, and the worldwide reputation which he rapidly came to enjoy was abundantly earned and more than deserved."[40] In 1918, Joseph Race, the bacteriologist and chemist for the city of Ottawa, Canada, dedicated his classic book on water chlorination to Dr. Houston.[41]

Dr. Houston arrived in Lincoln on February 8, 1905, and was able, somehow, to begin treating the water with chlorine three days later—on February 11. Figure 4-2 shows the initiation of chlorine treatment in relation to the progression of the typhoid fever epidemic. Houston had been studying the addition of chlorine to sewage and surmised that under the emergency conditions in Lincoln, the use of chlorine in the water supply was worth a trial. With the example of Maidstone in his mind, Houston considered feeding chlorine to a treated water reservoir on a continuous basis, but that was not feasible given the layout of the Lincoln water system.[42] Ultimately, he chose to feed chlorine into the water being applied to the slow sand filters.

In his sewage "sterilization" studies, Houston had used a commercial preparation of alkaline sodium hypochlorite called "Chloros." The chemical had an active chlorine content of 10–15 percent. Houston believed that Lincoln's slow sand filters were probably "infested" with typhoid bacteria and that disinfection of the influent water would, over time, kill pathogens in the filter beds.[43]

The feed system for Chloros was neither sophisticated nor precise in the beginning. Carboys of Chloros were arranged so that the chemical could drip continuously into the water by means of a siphon action, but because the siphon was not arranged in a constant head tank, the chlorine dose is unlikely to have been consistent. In addition, the planned dosage of Chloros applied to the slow sand filter influent changed over time. It started out as several parts per million but the dosage settled on was a sustained concentration of 1 ppm as active chlorine.[44]

The advisory for residents to boil their drinking water, even with the application of chlorine to the slow sand filter influent, extended for several months past the initiation of treatment. Wisely, Houston recommended that boiling be continued while he worked out the details of Chloros addition.[45]

Because of his experiments with sewage disinfection, Houston knew that Chloros was a powerful oxidant. He also knew about the ability of chlorine to oxidize potassium iodide to iodine, which would react with starch in solution to produce a blue color. Putting this knowledge to good use, he was able to report that the water applied to the top of the slow sand filters produced a dark blue color as a result of the potassium iodide–starch test and that water after filtration had only a trace of the blue color.[46]

The taste of the water after chlorine was added was objectionable to many of the people who drank it. One of the descriptors used to characterize the taste was "mawkish,"[47] which is not a word normally used in the United States to describe the flavor of drinking water. The word mawkish is derived from the Middle English word, *mawke*, which meant maggot—not a good start to describing the quality of a drinking water supply. Mawkish is defined variously in dictionaries as sickening, unpleasant, or insipid in taste.

The other flavor characteristic of the water after chlorine treatment was more descriptive. "The water at Lincoln after treatment with 'chloros,' though quite clear and free from turbidity, had a distinctly musty smell and taste. . . . The scent given off from the water somewhat resembled that of nuts in an earthenware vessel which had been brought up from storage in a damp cellar."[48]

Contemporary experience with off-flavors in water over the past 30 years tells us that this quote from 108 years ago was describing the off-flavor caused by one or both of two organic compounds—geosmin or 2-methyisoborneol. Adding Chloros to the water before it entered the slow sand filter beds likely caused the chlorine to kill and lyse cells of blue-green algae in the *Schmutzdecke* on top of the filters. Lysis of the algae that produced these earthy–musty odor compounds would have released high concentrations of the offending materials into the treated water. Earthy–musty odors are particularly offensive to drinking water consumers,[49] and, apparently, the people of Lincoln were not used to being exposed to them. In addition to the fear of typhoid, the earthy–musty off-flavors made acceptance of the water even more difficult.

Addition of Chloros to the Lincoln water supply continued until 1911, when use of a new, protected water supply was instituted. However, even six years after the typhoid epidemic, the public still did not trust Lincoln's water supply.[50]

Dr. Reece's report on the typhoid epidemic at Lincoln and the use of chlorine to disinfect the water was discussed at length in the second Jersey City trial by Dr. John L. Leal under direct and cross-examinations.[51]

IN 1903, LIEUTENANT VINCENT B. NESFIELD of the British Indian Medical Services published a remarkable paper in a British public health journal.[52] In the paper, he described his search for a chemical disinfectant that could purify drinking water and that would be suitable for use in the field as part of a military campaign. He came up with the idea of producing chlorine gas by electrolytic cells and then compressing the gas with six atmospheres of pressure until it liquefied, which facilitated its storage in lead-lined steel tanks that held about 20 pounds of liquid chlorine. He treated 50-gallon batches of water by submerging the gas valve of the chlorine cylinder and opening it slightly to bubble the chlorine gas into the water.

In a later paper, Nesfield stated that about 5.4 ppm of chlorine (2 grams per 100 gallons of water) killed all typhoid and cholera bacteria in the water. After a 5-minute contact time, he added sodium sulphite to the treated water to remove the excess chlorine and prevent taste problems.[53] To say that he was ahead of his time is a vast understatement. It would be nine years before liquid chlorine in pressurized cylinders was widely available in the United States for water utilities to use as an alternative to chloride of lime.

References to Nesfield's unique treatment method can be found in some early twentieth century publications. In a discussion of two 1911 papers on chlorination of water and sewage, chemistry professor Leonard P. Kinnicutt mentioned Nesfield's method of liquid chlorine addition and described an iodine tablet, also developed by Nesfield, that was more portable (and undoubtedly caused more taste problems).[54] Therefore, there was at least some early knowledge in the United States of the use of liquid chlorine to disinfect drinking water. There was one mention of Nesfield's system of purification in a 1920 encyclopedia section on water supply,[55] and in 1907 a note in a journal devoted to tropical medicine described how successful chlorination was for a unit of the British colonial army marching toward Agra, of Taj Mahal fame.[56]

Three authors mentioned Nesfield and his groundbreaking work on chlorine disinfection in histories of drinking water disinfection. In Joseph Race's remarkable 1918 book on chlorination of water, he gave Nesfield credit for the first use of liquefied chlorine to disinfect water.[57] In his classic book, *The Quest for Pure Water*, Moses N. Baker devoted a few sentences to Nesfield's contributions.[58] In a later summary of the progress of drinking water disinfection, in 1950, Race again gave credit for Nesfield's unique application of chlorine technology.[59]

Early Disinfection in the United States

Several miscellaneous publications mention the temporary use of disinfection to purify drinking water in the United States prior to 1908, but none of these operations was continuous. Two examples show how disinfectants were used intermittently to solve specific problems but not as an overarching treatment for contaminated water supplies.

Chemistry professor William P. Mason reported on the temporary disinfection of a reservoir lining at Buffalo, New York, in 1894. The disinfectant was a bromine solution. Water in the reservoir had become contaminated with typhoid germs, so it was drained in preparation for applying a disinfectant to the contaminated lining. Physicians in the city disagreed as to which disinfectant would be best—chlorine or bromine. As noted in a contemporaneous report in 1894, "Bromine Boy" won out over the "Chlorine Cyclone." The bromine solution was sprayed on the lining, and the reservoir was refilled, supposedly removing the threat of typhoid germs from the water supply.[60]

Several sources cite the early use of chlorine for drinking water disinfection in the small town of Adrian, Michigan (population about 9,000 in 1900). In 1899, either chlorine gas or chloride of lime was used, depending on the season of the year and the availability of chemicals. Adrian is located in the southeastern part of the state about 10 miles north of the Michigan–Ohio border. The city's surface water supply was treated with early versions of mechanical filters, but the treated water quality could not meet the bacteriological standards of Michigan's state health department. At some point, chlorination was discontinued and not employed continuously until about 1915.[61]

Chemophobia—Adding a Poison to Drinking Water

One of George Warren Fuller's earliest papers (1894) reflected the prevailing view that water treatment professionals could not even consider adding chemicals to water for disinfection.

> "Bacteriology teaches us that water may be sterilized in three ways, by means of chemicals, by means of heat, and by means of filtration. While chemicals have been of much aid in surgery by bringing about antisepsis and asepsis, it is very improbable that people would allow their drinking water to be drugged with chemicals, even with the view of removing dangerous bacteria—indeed, such a method might prove very dangerous in many cases."[62]

Fuller was undoubtedly reiterating the views of his professors, including Massachusetts Institute of Technology (MIT) professor Thomas M. Drown, who declared that "... the idea itself of chemical disinfection is repellent."[63]

No matter that germs were known to cause disease, no matter that germs could be killed by disinfectants, and no matter that water supplies were known to contain pathogenic organisms; public fear of adding chemicals to water was strong enough to prevent disinfection from being considered in the United States in the 1890s and early twentieth century.

THE FEAR WAS FOSTERED BY THE FOODS people ate that were being laced with any chemical that was cheap and would bring profit to the many unscrupulous food manufacturers at the time. Medicines were made from chemical stews that rarely cured anything. In 1904, Upton Sinclair described the problem of food and medicine adulteration in his novel *The Jungle*.

> "How could they know that the pale-blue milk that they bought around the corner was watered, and doctored with formaldehyde besides? When the children were not well at home, Teta Elzbieta would gather herbs and cure them; now she was obliged to go to the drugstore and buy extracts—and how was she to know that they were all adulterated? How could they find out that their tea and coffee, their sugar and flour, had been doctored; that their canned peas had been colored with copper salts, and their fruit jams with aniline dyes?"[64]

> "There was never the least attention paid to what was cut up for sausage; there would come all the way back from Europe old sausage that had been rejected, and that was mouldy and white—it would be dosed with borax and glycerine, and dumped into the hoppers, and made over again for home consumption. There would be meat that had tumbled out on the floor, in the dirt and sawdust, where the workers had tramped and spit uncounted billions of consumption germs."[65]

When the novel appeared in January 1906, it created a sensation.

> "President Roosevelt reportedly threw his breakfast sausages out his window. ... In spite of the desperate

denials by the meat industry, within six months the Pure Food and Drug Act and the Beef Inspection Act were passed [by the U.S. Congress]."[66]

CHEMOPHOBIA IN RELATION TO DRINKING WATER was not a new fear on the part of the public. Whipple described the problem in a discussion of papers at that 1904 international conference in St. Louis. "The popular prejudice against the use of chemicals seems to be gradually passing away, yet in certain places it is still strong. Thus in St. Louis the popular prejudice against the use of alum in clarifying the water is said to be so intense that a local engineer has said 'it is very doubtful if alum could be used, no matter how excellent the results which might be obtained.' One reason for this prejudice was illustrated by the following expression that appeared in an editorial in a local newspaper: 'We don't want to drink puckered water.'"[67]

Chlorine as a Poison

At the 1906 annual conference of AWWA in Boston, George Whipple started a review of the current state of knowledge of drinking water disinfection on a negative note:

> "There is a growing feeling on the part of some sanitary engineers and chemists that disinfection of water and sewage by the use of chemicals is destined to be a matter of some importance in the future. The idea of adding *poisonous* chemicals to water for the purpose of improving its quality for drinking purposes has generally been considered as illogical and unsafe, and unfortunately most of the substances which have the power of disinfecting are *poisonous* to a greater or lesser extent."[68] (emphasis added)

Because Whipple believed disinfection should be considered, he summarized a number of full-scale and experimental uses of chlorine, chlorine dioxide, and ozone. He related details of the application of disinfection at Ostende and Middelkerke in Belgium. Interestingly, Whipple recognized that chlorine dioxide gas (and not chlorine) was being produced by the Ostende process and cautioned that the gas was explosive.[69]

Whipple also described tests with chloride of lime on Seine River water near Paris and the disinfection of sewage with chloride of lime in London. He briefly mentioned the chlorination of the Maidstone system and went into an exhaustive discussion of ozone. He listed 58 references in a comprehensive review of ozone technology, which he called ". . . extremely disappointing in many respects." Production of ozone for water treatment was not yet available on a reliable, commercial scale.[70]

Whipple was cautious to make no recommendation on the use of chemical disinfectants in water. Indeed, he stated the belief that there was ". . . need of a more thorough study of the subject."[71]

Nonetheless, the discussion after the presentation amounted to a blistering attack on Whipple for considering the use of chemical disinfectants. William P. Mason stated his opposition to chemical disinfection clearly. His words carried considerable weight with water practitioners because of his two books on water analysis and treatment and his involvement in the top ranks of AWWA. He would become president of the organization in two years. At the time, he was also a respected professor at Rensselaer Polytechnic Institute.

> "The idea of disinfection of water supplies seems to me to belong more to the future than to the present. I very much question if the public at large would be willing to disinfect water to-day. We are scarcely driven that far yet."[72]

Mason's final comments reflected the conventional blindness to reality that was prevalent during this period. He urged that good raw water supplies were still available and that the public was not ready yet for disinfection.[73]

Three years later, Mason's change of opinion, delineated in a discussion of papers by Leal, Fuller, and George A. Johnson and in his testimony at the second Jersey City trial, had a significant impact on the court proceedings.

Professor Kinnicutt's discussion of Whipple's 1906 AWWA presentation showed more openness to the idea of disinfecting drinking water with chemicals, and he referred, obliquely, to the work of Lieutenant Nesfield.

Another discussant of Whipple's presentation expressed strong feelings on the subject of chemical disinfectants.

> "Among the so-called 'disinfectants' tried may be cited copper, chlorine and oxalic acid. All these substances are *poisonous*. They certainly can kill the bacteria, but they also kill the fish and cannot fail to be hurtful to

man. Such *poisonous* materials should not be permitted to be used on water intended for public supplies."[74]

Summing up the discussion, Whipple did some serious backpedaling.

"... I do not want to be misunderstood. With regard to the use of these new methods I am in some respects *more conservative* than any of the gentlemen that have spoken; but I think that it does us good sometimes to think outside of the ordinary lines, and these various methods have been talked about a good deal of late."[75] (emphasis added)

In the published version of his remarks, however, he went on to plead for scientific study of the subject. However, in a contemporaneous magazine article containing the unedited stenographic transcript of the discussion of his paper, Whipple called, instead, for a commission to "... take up the subject."[76]

Twelve years before Whipple's presentation and the discussion that resulted, a brave hydraulic engineer stood up at a meeting of the New England Water Works Association and said aloud what many people were thinking but did not have the gumption to state publicly because of their reluctance to go against their colleagues.

The declaration was prompted by a paper presented by renowned professor Thomas M. Drown of MIT, which at the time was the center of excellence for sanitary engineering work in the United States. As noted earlier, Professor Drown had strong feelings against chemical disinfection. In the discussion following his paper, the second discussant agreed with Drown that chemical disinfection or any treatment of contaminated water was a bad idea: "... I fully agree with him that we don't want to take dirty water for this purpose, and clean it before using."[77]

It was obvious from the next exchange of views that discussant Clemens Herschel could not contain himself, and he jumped into the fray. Herschel was a well-respected hydraulic engineer, who later applied another technology, the Venturi meter, on a then-unimagined scale to measure water flow for the East Jersey Water Company at Little Falls, New Jersey. Herschel's candid remarks in support of chemical disinfection, or any technology that would protect the public from waterborne disease, were highly unusual in 1894.

"I am a little disappointed by the professor [Thomas M. Drown] casting cold water upon the idea of, or discouraging the purification of drinking water by any means that at all would promise success. . . . Now, what are you going to do about it? . . . I want to hear a word of encouragement for anybody and everybody who will attempt to purify water in some practical way."[78]

Professor Drown did not answer Clemens Herschel.

Filtration

Filtration of drinking water was developed during the same period when the fundamentals of disinfection were beginning to be understood. Ultimately, the combination of filtration and disinfection was the key to protecting drinking water through application of the "multiple barrier" principle of drinking water treatment.

The cost of filtration dissuaded many cities from raising the funds and approving the construction of this important barrier to waterborne disease. William T. Sedgwick stated, "Most American cities have hitherto shrunk from the heavy pecuniary outlay involved in filtration, and have preferred to rely for protection simply upon the general good character of the water-shed from which their supply is taken."[79] The result of depending on "the general good character of the water-shed" was endemic typhoid fever, high rates of illness and death caused by diarrheal diseases, and occasional widespread typhoid fever epidemics.

In writing *The Quest for Pure Water*, Moses N. Baker recorded the most complete history of filtration in Europe and North America.[80] Only a few examples of filtration that are relevant to the New Jersey trials are elucidated here.

The terminology used to describe filtration processes makes reading early filtration histories somewhat confusing. Here, the term "slow sand" filtration is used to describe the European filtration method that is many times referred to as "filtration" in nineteenth-century publications. "Mechanical" filtration, developed primarily in the United States, is used to describe filters that treated water at a higher filtration rate and that were usually preceded by sedimentation or treatment with coagulants such as alum. Mechanical filters were also called American filters. The more modern term "rapid sand filtration" refers to the same process as mechanical filtration, even though the filter media is composed of more than just sand.

SLOW SAND FILTERS were first developed in Europe, especially in England, and were tried on U.S. water supplies with varying success. The first attempt to build and operate a slow sand filter in this country was in Richmond, Virginia, in 1832. It was not successful, and other filtration installations were not attempted for several decades.[81] The early uses of slow sand filters were to remove turbidity and algae. Their utility as barriers against waterborne disease was not recognized until after the germ theory of disease was being accepted.

Filtration rates for slow sand filters were on the order of 2 million gallons per acre of filter surface area per day. Large areas of land were needed for slow sand filters for large cities, making this technology difficult to implement in many instances because of the high value of land near the water supply.

Slow sand filters performed differently in the United States compared with England and Europe for three reasons—weather, geography, and geology. Average annual rainfall in London around 1895 was about 28 inches. Along the crowded, polluted Eastern and Midwestern rivers of the United States, average annual rainfall was typically 40 to 50 inches per year, with rainfall intensities far greater than in England and continental Europe. U.S. river watersheds were also much bigger than those in England, and rivers had a tendency to pick up more fine particles during their travels.

Intense storms, spawned routinely by the interaction of warm air from the Gulf of Mexico with cold Canadian air, occurred far more frequently in the United States than in Europe. The intense North American rainstorms also churned up clay deposits that were more common in U.S. geology than on the other side of the Atlantic. The huge, brown flows of water in the Ohio, Mississippi, Missouri, and other rivers were emblematic of U.S. geology and hydrology. A high proportion of clay in the influent to slow sand filters made them operate poorly and for only short periods of time before manual scraping and cleaning were required. Fuller's investigation of filtration at Cincinnati proved that with major American rivers, pretreatment of slow sand filter influent with both presedimentation and coagulation followed by sedimentation was needed to remove the fine clay from the raw water supplies.[82]

In 1865, the St. Louis Board of Water Commissioners gave civil engineer James P. Kirkwood the task of traveling to Europe to learn all he could about filtration because the process was being considered for the city's water supply—the Mississippi River.

Kirkwood's report, published in 1869, is a landmark in U.S. understanding of slow sand filtration.[83] His detailed drawings and engineering information were used by U.S. cities in later years as part of the process of designing slow sand filters. But St. Louis was not ready for filtration and opted instead for simple settling basins to remove Mississippi mud from the water. This decision meant that St. Louis would suffer through severe epidemics of typhoid fever caused by its contaminated water supply until filtration and chlorination were implemented in 1915.

DR. ROBERT KOCH BECAME AWARE of the cholera epidemic in Hamburg shortly after it started in August 1892. Within a few weeks, more than 8,000 deaths had been recorded, and the Senate of Hamburg asked Koch to provide assistance.[84]

A major port on the Baltic Sea, Hamburg is located on the Elbe River in northern Germany and at the time of the epidemic had a population of about 560,000. Its water supply was unfiltered and taken from the river upstream of the city but not before significant contamination had been introduced by upstream sewage discharges. Altona, a smaller city with a population of about 143,000, was located adjacent to Hamburg. Altona also took its drinking water from the Elbe River, but its withdrawal point was located downstream of both cities and was influenced by the sewage discharges from both. Because the Altona water supply was obviously contaminated, slow sand filters were used to treat the water before it was distributed.

The cholera death statistics told the story. In 1892, Hamburg and the suburbs to which it provided water experienced almost 20,000 cases of cholera and 7,582 deaths for a death rate of 1,350 per 100,000. Altona had 572 cases of cholera with 328 deaths for a death rate of 230 per 100,000, better than Hamburg's but still shockingly high. Filtration of a more polluted supply provided only some protection for the citizens of Altona.[85]

Koch supplied the answer to the question of how the filter was "protecting" the citizens of Altona. Using his recently developed total bacteria count method, which involved using gelatin on glass plates, Koch demonstrated that the bacteria counts coming out of the slow sand filters were lower than the bacteria counts in the filter influent. A few years later, Koch helped create regulations requiring bacteriological examination of water supplies and specifying the proper operation of slow sand filters.[86]

Hamburg installed slow sand filters shortly after the devastating epidemic subsided. In 1910, to further protect its citizens against the contaminated Elbe River, Hamburg installed a chlorine feed system.[87] A footnote to the story of filtration in Altona demonstrated that slow sand filtration of a contaminated water supply did not always protect users from waterborne disease. An appendix to Allen Hazen's famous book on filtration noted that manual cleaning of Altona's slow sand filters in the winter caused frost to disturb the filtration process, allowing the passage of pathogens. Altona experienced typhoid epidemics in 1886, 1887, 1888, 1891, and 1892. Its citizens were fortunate that the cholera epidemic of 1892 occurred in the summer, when the slow sand filters were presumably operating properly.[88]

Hazen's appendix also noted that Altona suffered an increase in infant deaths caused by diarrhea in both winter and summer. Coupling the observation of frost damage in the slow sand filters in winter with this increase in infant mortality shows that at least some of the infant deaths caused by diarrhea were specifically linked to contaminated drinking water.[89]

SOME OF THE EARLIEST MECHANICAL FILTERS were installed in industrial settings to remove particulate matter. The advantage of higher filtration rates requiring smaller tracts of land attracted the interest of U.S. filtration investigators in the late 1800s. Mechanical filters could do the same job as slow sand filters (or better with the addition of chemicals) with filtration rates of 125 million gallons per acre of filter per day (2 gallons per minute per square foot [2 gpm/sq ft]) compared with only 2 million gallons per acre per day achieved by slow sand filters. In other words, the footprint of the filters could be reduced by a factor of more than 60 to 1. However, mechanical filtration could not be successful in drinking water applications until engineers found a way to agglomerate the finely divided clay particles and the bacteria in the raw or settled water before running the water through a mechanical filter.

In 1885, Professors Peter T. Austen and Francis A. Wilbur conducted early research on coagulation at Rutgers College in New Brunswick, New Jersey, not far from the center of operations of John L. Leal and George Warren Fuller. Reports of Austen and Wilbur's work stated that 1.5 grains per gallon (26 ppm) of alum were sufficient to clarify turbid water if left quiescent over two days. The clear supernatant could then be filtered to remove any remaining turbidity. The researchers speculated that this process could remove not only particles but also albuminoid (organic) nitrogen and other organic material.[90]

The work on mechanical filtration in Providence, Rhode Island, over the period 1892–94, seldom gets the credit it deserves for marking advances in the science of drinking water filtration.[91] After an epidemic of typhoid fever in Providence in 1888, Charles V. Chapin began to seriously investigate filtration for use on the city's water supply. "This Providence experimentation provided the first careful tests anywhere of the mechanical type of water filtration."[92] Chapin published a paper claiming that bacteria removals from mechanical filtration were typically 98.7 percent, and he recommended that mechanical filtration be installed on Providence's source of supply.[93] The city council, however, was not ready for such a new technology. A slow sand filter was installed instead.[94]

ADVANCES IN THE SCIENCE AND ART of filtration associated with the work of George Warren Fuller were evident as early as 1902, when he designed a water treatment plant for the East Jersey Water Supply Company at Little Falls, New Jersey. The plant's design, construction, and operation, using alum coagulation, sedimentation, and mechanical filtration, was a major milestone in public health protection.

Figure 4-3 is a schematic of the treatment processes at Little Falls in 1903.[95] Part of the genius of Fuller's design was his arrangement of the treatment processes on the very limited land area available for construction of a treatment plant on the banks of the Passaic River. A lot of land space was saved by building the filters above the clearwells. Also, the roof of the sedimentation basin formed the floor of the main building.

Water was extracted from the Passaic River above Beattie's Dam and entered a head-race canal. The water was then pumped up to a standpipe where chemicals were added.[96] From the standpipe, the water flowed by gravity to the "coagulating and subsiding basin," or sedimentation basin as it is called today. At the plant's nominal capacity, 30 million gallons per day (mgd),[97] the 1.75-million-gallon basin provided only about 1.4 hours of detention time.[98] The sedimentation basin could be bypassed during the time needed to clean out the accumulated sludge.[99] Sludge and used washwater from the filters were directed to a catch basin or discharged directly into the Passaic River.

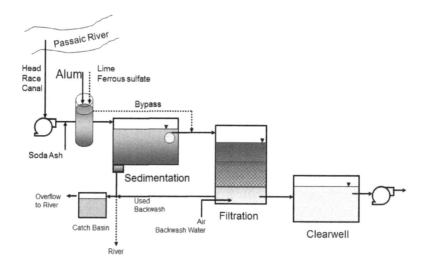

Figure 4-3   Treatment schematic of the Little Falls Water Treatment Plant

Alum was added to the standpipe at dosages ranging from 5 to 34 ppm during the period Fuller reported on in 1903. Alum was maintained at an approximate concentration of 2 percent (20,000 ppm) in mixing tanks before being metered into the water from orifice tanks. The tank and pump system for feeding alum to the water was critically important to the successful operation of the entire treatment plant. Details of the alum system will be discussed in Chapter 10, where it is compared with the chloride of lime feed system Fuller designed for Boonton Reservoir. Turbidity removal problems experienced during Fuller's testing of the plant almost always resulted from plugging of the pump that moved dilute alum from the mixing tanks to the orifice tanks. Usually, wood splinters from the barrels used as alum shipping containers were the culprits.

A chloride of lime feed system was subsequently added to the Little Falls plant, and chlorine was being fed to the water prior to the sedimentation basin as of February 4, 1909.

The plant had 32 filters with a capacity of 1 mgd per filter.[100] The filtration rate was typically 2 gpm/sq ft, which was later adopted by many state public health boards as the maximum filtration rate allowable.[101] In a later article, Fuller expressed regret that this filtration rate had acquired such an exalted level of importance; he had never intended such a result. He urged that higher filtration rates be investigated and, if warranted by the data, approved by state health boards in specific situations.

The filters consisted of 30 inches of sand on top of 7 inches of crushed quartz. "Strainers" were embedded in the crushed quartz at the bottom of the filters to collect the filtered water and to provide a way to introduce backwash water to the filters during cleaning. The sand was carefully graded to meet exacting specifications, which were overseen by Allen Hazen, the world's most knowledgeable expert on sand gradation.[102]

A major innovation exemplified by the filters at Little Falls was that the filter boxes were rectangular and constructed out of reinforced concrete. Mechanical filters up until this time typically had been round and constructed of wood and metal. Rectangular filter boxes allowed for significant capital cost savings because the walls between the filters were common. Engineers began to refer to such economic construction as "common-wall construction," which became a design standard that continues to this day.

Another innovation at the Little Falls plant was a combined air and water backwash used to clean the sand after turbidity removal caused head loss in the filter to reach a preset level. Backwashing the filters was a delicate operation because there was only 12 inches from the top of the filter sand to the edge of the backwash gullet,[103] and this meant the sand could only expand about 10 percent (a 3-inch rise compared with 30 inches of sand). The limited bed expansion meant that the sand was not cleaned properly, and the time periods between filter backwashes were short, ranging from 6 to 18 hours and averaging 9.7 hours.[104]

The filters were operated at a constant rate by controlling the flow downstream of the filter with a butterfly valve and an orifice plate structure.

Turbidity removal was acceptable, but judging the plant's efficiency is difficult because of the unfamiliar turbidity results reported in Fuller's paper—turbidity was measured in ppm, not in nephelometric turbidity units or even Jackson candle units. Color removal appeared to be excellent. The plant was able to produce water that almost always had color values of less than 10 ppm and were often about 5 ppm. Again, these are not the units we use to express color concentrations today. Bacteria removal was also excellent. Typically, total bacteria counts in the plant effluent were less than 100 per milliliter. Influent bacteria levels varied greatly but were usually about 6,000 per milliliter.

WHILE FULLER WAS DOING MOST of the developmental work for mechanical filtration, other talented engineers were deeply engaged in making the technology work. Hazen was a colleague of Fuller's and part of the team from the Lawrence Experiment Station. In 1896, a commission established by the city of Pittsburgh asked Hazen to investigate treatment options for the city's source of supply—the Allegheny and Monongahela rivers. After extensive investigations of both mechanical and slow sand filtration, Hazen recommended in 1899 that slow sand filtration be used. The commission accepted this recommendation.[105] Unfortunately, the slow sand filter would not be operational until 1909, and the delay resulted in an unconscionably high rate of typhoid fever deaths (well over 100 per 100,000) for another 10 years. Chlorination of the water supply began in 1910—just one year after filtration was initiated. The Pittsburgh filter plant was one of the last, large slow sand filter facilities built in the United States.

Disinfection of Sewage

Joseph Race credits the chlorination of sewage in Brewster, New York, in 1893 with being the first use of chlorine for disinfection—meaning the destruction of bacteria. However, other known examples preceded this date. In any event, the Brewster application of chlorine appears to be the first instance of sewage disinfection, even though the waste stream was generated from only a few homes.[106]

After the devastating cholera outbreak in Hamburg in 1892, hypochlorites were studied for their germicidal properties. In 1905, English chemist Samuel Rideal demonstrated that small quantities of chlorine could be effective for disinfecting sewage—the target for disinfection being the elimination of pathogenic bacteria and not complete sterilization, which is virtually impossible in sewage.[107] Because of its ease of use, chloride of lime was initially the form of chlorine most frequently applied in these investigations. Dr. Alexander C. Houston experimented with disinfection of sewage in London, England, in the early 1900s.

Researchers at the Lawrence Experiment Station studied sewage disinfection from about 1906 to 1909. Earle B. Phelps, a professor of chemistry at MIT, is usually credited with expanding the early knowledge base for this new treatment scheme through his publications and his practical experience at Red Bank, New Jersey, and Baltimore, Maryland.[108] According to Phelps, 3 to 5 ppm of available chlorine dosed into "average-to-strong sewage" was expected to give satisfactory disinfection results—about a 95 percent reduction in bacteria concentrations.[109]

It is somewhat surprising that applying chlorine to sewage effluent did not become popular in the United States until after it had first been used in a large-scale drinking water application at Boonton Reservoir. By 1910, disinfection began to be installed in a number of sewage plants, and by 1913 chlorination of sewage was being used extensively.[110] By 1926, more than 400 sewage plants disinfected their effluent with chlorine. By that time, liquid chlorine had largely replaced chloride of lime.[111]

1   Whipple, "Disinfection as a Means of Water Purification," 276-80; Baker, *Quest*, 321–56.

2   Race, *Chlorination of Water*, 1–3.

3   Turneaure and Russell, *Public Water Supplies*, 1st ed., 493.

4   "Local Government Board Inquiry into the Maidstone Epidemic," 391.

5   "Local Government Board Report on the Epidemic of Typhoid Fever at Maidstone," 50.

6   "Maidstone Epidemic," 238.

7   Soper, "Role of Public Water Supplies in the Spread of Typhoid Fever," 85.

8   "Local Government Board Inquiry into the Maidstone Epidemic," 391.

9   Shadwell, "Suicide by Typhoid Fever," 721.

10 "Local Government Board Inquiry into the Maidstone Epidemic," 391.

11 Shadwell, "Suicide by Typhoid Fever," 721-2.

12 Ibid., 723.

13 Stanwell-Smith, "The Maidstone Typhoid Outbreak of 1897."

14 Plarr, "Woodhead, German Sims," 1189.

15 "Typhoid Epidemic at Maidstone," 388.

16 "Local Government Board Report on the Epidemic of Typhoid Fever at Maidstone," 50.

17 Woodhead and Ware, "Disinfection of the Maidstone Water Service Mains," 52.

18 Ibid., 58.

19 Whipple, "Disinfection as a Means of Water Purification," 269.

20 Ibid.

21 Kemna, "Purification of Water for Domestic Use," 167.

22 Boby, "Discussion on Water Filtration," 682.

23 Whipple, "Disinfection as a Means of Water Purification," 270

24 Ibid.

25 Baker, *Quest*, 336.

26 Howatson, A., Discussion of "Purification of Water for Domestic Use," 192.

27 Whipple, G.C., Discussion of "Purification of Water for Domestic Use," 199.

28 Hazen, Discussion of "Purification of Water for Domestic Use," 248.

29 Ibid.

30 Reece, "Epidemic of Enteric Fever in Lincoln, 81–2.

31 Ibid., 84–5.

32 Ibid.

33 Ibid., 89–90.

34 Ibid., 103.

35 Houston, "Studies in Water Supply," 62.

36 Reece, "Epidemic of Enteric Fever in Lincoln," 105–6.

37 Ibid., 113–4.

38 "Lincoln Water-Supply," 172.

39 Ibid.

40 "Sir Alexander Houston," *Nature,* 810–1.

41 Race, *Chlorination of Water,* iii.

42 Later in February, Houston applied Chloros to the Westgate and Cross Cliff Hill reservoirs to ensure that the contents and the facilities of the reservoirs were disinfected.

43 Reece, "Epidemic of Enteric Fever in Lincoln," 116.

44 Ibid.

45 Ibid., 117.

46 Ibid.

47 Ibid., 141.

48 Ibid., 117.

49 McGuire et al., "Early Warning System for Detecting Earthy-Musty Odors."

50 Houston, "B. Welchii, Gastro-Enteritis and Water Supply," 484.

51 Between Jersey City and Water Company, February 5, 1909, 5059–61.

52 Nesfield, "Chemical Method of Sterilizing Water,"601–2.

53 Nesfield, "Simple Chemical Process of Sterilizing Water," 624–5.

54 Kinnicutt, Discussion of "Sterilization of Public Water Supplies," 44–5.

55 Hill, "Water Supply," 51.

56 "Pure Water," *Journal of Tropical Medicine and Hygiene,* 30.

57 Race, *Chlorination of Water,* 89.

58 Baker, *Quest,* 341.

59 Race, "Forty Years of Chlorination," 479.

60 Mason, *Water Supply,* 317; "Buffalo," *Pharmaceutical Era,* 431.

61 Baker, *Quest,* 335; Lanier, "Historical Development of Municipal Water Systems," 180.

62 Fuller, "Sand Filtration of Water," 64.

63 Drown, "Electrical Purification of Water," 185.

64 Sinclair, *The Jungle,* 79.

65 Ibid., 136.

66 Ibid., 344.

67 Whipple, Discussion of "Purification of Water for Domestic Use," 199.

68 Whipple, "Disinfection as a Means of Water Purification," 266.

69 Ibid., 269.

70 Ibid., 276–80; Kinnicutt agreed with Whipple in the related discussion of the paper.

71 Ibid., 267.

72 Mason, Discussion of "Disinfection as a Means of Water Purification," 282–3.

73 Ibid., 283.

74 Maignen, Discussion of "Disinfection as a Means of Water Purification," 286.

75 Whipple, Discussion of "Disinfection as a Means of Water Purification," 287.

76 "American Water Works Association Convention," *Fire and Water Engineering*, 412. It was common during this period for discussants to edit their remarks before they were published. In the 1906 discussion, there were substantial differences between the AWWA Proceedings and the stenographic transcription published in *Fire and Water Engineering*. Most of the differences were the result of the discussants wanting to sound a bit more erudite compared with their oral remarks.

77 Smith, Discussion of "Electrical Purification of Water," 186.

78 Herschel, Discussion of "Electrical Purification of Water," 186–7.

79 Sedgwick, "Data of Filtration," 69.

80 Baker, *Quest*, 29-272.

81 Ibid., 125.

82 Fuller, *Purification of the Ohio River for Cincinnati*.

83 Kirkwood, *Report on the Filtration of River Waters*.

84 Brock, *Robert Koch*, 231.

85 Pollitzer, *Cholera*, 39.

86 Brock, *Robert Koch*, 231.

87 Ibid.

88 Hazen, *Filtration of Public Water-Supplies*, 147.

89 Ibid., 226–7.

90 Notes, *American Monthly Microscopical Journal*, 138; Notes, Knowledge, 147.

91 Swarts, "Discussion," 82.

92 Cassedy, *Charles V. Chapin*, 58.

93 Chapin, "Filtration of Water," 13–4.

94 Cassedy, *Charles V. Chapin*, 58.

95 All of the information for this section came from Fuller's 1903 paper describing the Little Falls treatment plant. This is the paper for which he received the only award in his lifetime, the Thomas Fitch Rowland Prize from the American Society of Civil Engineers; Fuller, "Filtration Works at Little Falls."

96 Although there was turbulent water in the standpipe, high energy was not added to the water to facilitate good mixing of the coagulant with the water to be treated. Nor was any slow mixing added to facilitate the formation of aluminum hydroxide floc. Both of these innovations would come much later.

97 Fuller noted that the plant could be operated at up to 48 mgd for short periods of time, which were not defined.

98 Typical detention times for classical sedimentation today are 1.5–4 hours. The only sedimentation basin at Little Falls was 130 feet long, 42 feet wide, and 43 feet deep. To treat an average daily flow of 30 mgd (48 mgd maximum flow) today, a possible design would be 5 rectangular plain sedimentation basins, with each one 12 meters (39.4 feet) wide and 57.3 meters (188 feet) long; this would give each basin a length-to-width ratio

of 4.8, which is within the typical recommended range for this factor (4:1–5:1). Depth is typically fixed at 4 meters (13.1 feet) for today's sedimentation basins because removal of particulates is a function of surface area, and sufficient depth is needed only to contain the inlet/outlet structures and sludge removal equipment and to allow some storage of sludge. The site area for the Little Falls plant was severely limited, which explains the large depth of Fuller's sedimentation basin.

99   Sludge was stored in the bottom of the basin at depths of 6–8 feet. The basin was cleaned about every two months, which led to the development of anaerobic conditions in the deep sludge layer.

100   Each filter box was 24 feet by 15 feet and 8 feet deep. There were four rows of eight filters each.

101   It was fairly well established at this early point in the development of mechanical filtration that the higher the filtration rate, the more likely turbidity and bacteria were to pass through and end up in the filter effluent. Many conservative design practices were incorporated into the Ten State Standards, which became design requirements in the ten states that adopted them and unofficial guidelines in many other states. The Little Falls filtration rate of 2 gpm/sq ft was one of the most important design factors in the Ten State Standards.

102   The effective size of the sand ranged from 0.35 to 0.42 millimeters (mm) with a uniformity coefficient of 1.50. Less than 1 percent by weight of the sand could be finer than 0.25 mm, and less than 0.2 percent could be finer than 0.2 mm. After much effort to grade the sand outside of the filter, the best material was placed into the filter and backwashed until the fines were removed. Hazen and Fuller were very experienced in the problem of sand fines on filter surfaces as a result of their work at the Lawrence Experiment Station and their filtration studies at Louisville, Cincinnati, and Pittsburgh. Fuller even had the fines that had accumulated on the surface of the filters scrapped off after a few months of operation. Such attention to detail would greatly improve filter operations today.

103   The backwash water velocity was only 1 foot per minute, which is the equivalent of 7.5 gpm/sq ft. Typical backwash rates today are 20–25 gpm/sq ft, but that is only possible because the backwash gullets are much farther from the sand surface and bed expansions of 37–50 percent are typical.

104   Today, filters that have to be backwashed more often than once per day reduce the plant output because of the large amount of washwater they use. At Fuller's plant, even with the frequent washings, the washwater used for each filter ranged from about 3.5 to 4.5 percent of each filter's daily production of water.

105   *Report of the Filtration Commission of Pittsburgh*, 3–4.

106   Race, *Chlorination of Water*, 6.

107   Rideal, *Sewage and the Bacterial Purification of Sewage.*

108   Winslow, "Field for Water Disinfection," 2.

109   Phelps, "Disinfection of Water and Sewage," 1–17; Phelps, "Disinfection of Sewage and Sewage Effluents," 1–8.

110   Fuller and McClintock, *Solving Sewage Problems*, 15.

111   Ibid., 348–50.

5

Paterson and the Passaic River

"The Passaic is our most valuable stream from every
point of view."
— Vermeule, *Report on Water-Supply*, 150

The city of Paterson, New Jersey, was developed on the Passaic River and by the Passaic River; the histories of the city and the river are inexorably linked. Similarly, Dr. John L. Leal's career began in Paterson, New Jersey, and was deeply connected to the Passaic River. The river and the city molded him, and he greatly influenced the city and the river.

Every history of Paterson begins with a discussion of the birth of the American Industrial Revolution there and the role played by Alexander Hamilton, George Washington's first Secretary of the Treasury.

In 1791, the nascent U.S. Congress gave Alexander Hamilton the task of assessing likely sites to focus on industrial development. Hamilton concentrated on the land around the Great Falls on the Passaic River, partly because of the plentiful water power provided by the 77-foot drop in elevation.

In 1810, Paterson was described as a hamlet with a few hundred people.[2] One hundred and ten years later, the population had grown to 136,000.[3] Rapid growth in the 1800s put a significant strain on all municipal services, especially the water supply and the sewers.

Large numbers of immigrants were needed to run Paterson's industries, which were expanding rapidly in the mid- to late 1800s. The human element of this influx of immigrants was described in a contemporaneous report.

"During the whole summer of 1880 there was an unprecedented increment in the foreign population of Paterson, and it was an interesting sight to see the arrivals every night by the 'emigrant train' of quaintly attired newcomers from the Old World—from England, Scotland,

Ireland, Holland, France, Germany, Italy, and other countries—with their quainter luggage, who had come straight to Paterson, attracted hither by the enthusiastic reports sent 'home' by friends who had come earlier to try their fortunes in the 'Lyons of America.'"[4]

The growth of manufacturing in Paterson was the story of the original availability of water power, conversion of water power to steam, and ultimately the introduction of electric power. Early industries that made use of the abundant water power included ". . . paper, calico, cotton duck, thread, rope, tools, and machinery."[5] In Paterson and the rest of the country, water power was the dominant form of power until after the Civil War.[6]

As efficient steam power became available in the mid-1800s, industries sometimes added it to the buildings next to the raceways (artificial channels built to carry swiftly moving water to the factories, also known as mill races) so the factories could use either technology. Of greater importance for the development of the city, steam power made it possible for factories in Paterson to locate away from the raceways.

Although electric power was not widely available for industries in Paterson until the early twentieth century, in the 1890s a limited supply of electric power was provided to industries by two companies: The Edison Electric Illuminating Company and the Paterson Electric Light Company.[7] Electric motors made the industrial might of Paterson truly portable because factories could be located anywhere in the city. Electric power was supplied to the motors from a distribution grid or generated on site.

In 1890, 73 different industries were reported in 597 Paterson establishments, and these industries employed 24,135 persons. Manufacture of silk and silk goods was one of the largest industries, employing more than 11,000 people. Because of this concentration of industry, Paterson was often called the "Silk City" or the "Lyons of America" after Lyons, France, which was known for its own concentration of silk manufacturing.[8]

THE PASSAIC RIVER was an important resource for the state and an object of much desire for water developers and speculators alike. As stated in a survey of all New Jersey water supply resources in 1894:

"The Passaic is our most valuable stream from every point of view. By a fortunate coincidence, its headwaters afford our very best gathering grounds for public water-supply, and at the same time are the most accessible to the points of greatest demand."[9]

Figure 5-1 shows the Passaic River watershed and its relation to the boundaries of New Jersey. The basin is located in the northern part of the state. Northern New Jersey is bounded on the west by the Delaware River, which constitutes the state line shared with Pennsylvania, on the northeast by the New York–New Jersey state line, and on the east by the Hudson River, which forms part of the border with New York.

In general, the topography of the northern part of the state ranges from well over 1,000 feet in the Appalachian region in the extreme northwest to sea level in the southeastern part of the state, where the land meets the Hudson River and the Atlantic Ocean.

The mouth of the Passaic River (located on the southeastern side of the watershed) empties into Newark Bay, which leads to the lower Hudson and ultimately to the Atlantic. The watershed area above the mouth of the Passaic encompasses 949.1 square miles.

Table 5-1 lists the Passaic's major tributaries and their drainage areas, plus drainage areas above selected locations on the main stem of the river. Sixteen percent or 148.6 square miles of the watershed lie in the state of New York. One of the major schemes for capturing and exporting water from this basin to New York City made use of this geographic circumstance.

Figure 5-1 also shows the Passaic River's major tributaries and the important cities that used the watershed for a source of water supply. The Saddle River, in the eastern part of the drainage basin, is not labeled because it was not a significant source of water supply during this period. The three northern tributaries (Ramapo, Wanaque, and Pequannock) merge into the Pompton River just below Pompton, New Jersey. The Pompton River flows only for about six miles before joining the Passaic. Little Falls is on the main stem of the Passaic River about three miles below the confluence of the Passaic and the Pompton.[10]

The tributaries of the Passaic River were developed into major water supplies for New Jersey cities. The Passaic above and just below Great Falls was the location of Paterson's early water supply (1857–1899). In 1899, its source of supply was moved to the Passaic River at Little Falls. Little Falls also temporarily supplied Jersey City with water from 1896 to 1904. In 1904, the Rockaway River above Boonton (with the addition of Boonton

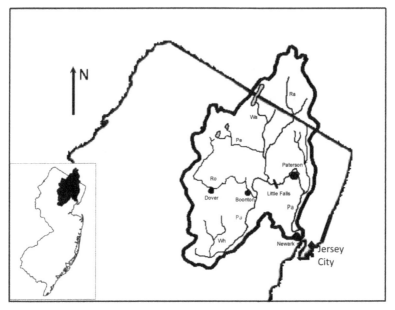

Legend: Ra-Ramapo, Wa-Wanaque, Pe-Pequannock, Ro-Rockaway, Wh-Whippany, Pa-Passaic

Figure 5-1   Passaic River watershed, major tributaries, and cities

Reservoir) became the water supply for Jersey City. Reservoirs on the Pequannock and Wanaque rivers supplied Newark during 1890–1910 (and beyond).

The main stem of the Passaic River travels from southwest to northeast as it approaches the city of Paterson. After the Great Falls, the river continues on for about three miles before taking a hard right turn and heading south-southeast. This bend in the river created a tongue of land on which Paterson was established and grew.

Long-term rainfall records for the Passaic River watershed were limited up to the late 1800s. On the basis of a report prepared by engineer Cornelius Clarkson Vermeule, Sr., average annual rainfall for the Passaic River at Dundee Dam can be calculated as 46.6 inches for the period 1877–1893.[11]

The equation Vermeule developed to predict flow in the Passaic River determined that the yield without storage would be 84.7, 36.8, and 14.5 million gallons per day for the Passaic at Little Falls, the Pompton River at Two Bridges, and the Ramapo River at Pompton, respectively.[12] With storage, the yield from these rivers could be increased by more than a factor of 10. Thus, the potential

Table 5-1   Passaic River drainage areas and major tributaries

Passaic River Locations	Drainage Area square miles
Passaic River Locations	
Passaic River—total watershed	949.1
Passaic River above Dundee Dam	822.4
Passaic River above Great Falls at Paterson	796.9
Passaic River above Little Falls	772.9
Passaic River in New York	148.6
Northern Tributaries	
Pompton River—total watershed	379.9
Ramapo Rive—total watershed	160.7
Wanaque River—total watersehd	109.6
Pequannock River—total watershed	84.8
Southern Tributaries	
Rockaway River—total watershed	138.4
Rockaway River above Boonton	118.2
Whippany River—total watershed	71.1

supply of water from the Passaic River Basin totaled hundreds of millions of gallons per day—far more than was being used by northern New Jersey cities at that time (1894).

The Passaic's huge flow has been coveted by industrial barons, developers, and politicians from the early 1800s to the present. An 1828 report by a civil engineer who was consulting on the question of the Morris Canal made a plea to provide a water supply to New York City from the Passaic River at Great Falls.[13]

THE MORRIS CANAL was a key transportation route in New Jersey in the mid-1800s. The Morris Canal and Banking Company was incorporated in 1824 for the purpose of building a canal between the Passaic and Delaware rivers. Ultimately, the canal stretched 102 miles from Jersey City on the Hudson River to Phillipsburg on the Delaware River. A concise description of the canal was published in an 1885 U.S. Census Bureau report.

"The canal, which was completed in August, 1831, and connects the city of Newark with the Delaware river at Phillipsburg, follows the Passaic and Rockaway rivers, rising, by means of 12 inclined planes (with a total rise of 758 feet) and 16 lift-locks (with a total lift of 156 feet), to its summit, near lake Hopatcong, 914 feet above mean tide. West of the summit it follows the Musconetcong valley, descending by 11 planes (with a total fall of 691 feet) and 7 lift-locks (with a total fall of 69 feet) to the Delaware, where it is 154 feet above tide. The canal is fed entirely from lake Hopatcong and from Greenwood lake."[14]

In addition to its value as a transportation corridor, the Morris Canal had water rights in the Passaic River Basin, and these rights were important components of the first and second Jersey City trials. Also, contaminated water (including some sewage wastes) from part of the city of Boonton drained into the canal. Water rights originally held by the Morris Canal and Banking Company had to be sorted out in the second trial.

Paterson's Water Supply

The early settlement at Paterson used water taken directly from the nearby river and the raceways and from hand-dug wells. It would be the closure of these shallow urban wells that Dr. John L. Leal fought so hard to accomplish as a public health protection measure in the 1890s.

The Passaic Water Company was the first organized enterprise to serve Paterson. The initial meeting of the company directors was held on February 11, 1854. Shortly after its incorporation, the company secured franchises for the sale and distribution of water to the town of Paterson and surrounding areas. In 1857, when the company's water source was the Passaic River below Great Falls, water was pumped from the pool below the falls directly into pipes laid by the company to serve the town.[15]

Because of contamination problems with the lower intake and the required pumping from the pool, the withdrawal point was moved above the Great Falls in 1861. This was the source of supply for Paterson while Leal served as the city's health officer.

The company underwent a transformation in 1877, with an almost complete change-out of the board of directors. In 1885, Garret A. Hobart, future Vice President of the United States under William McKinley, joined the board and two years later was elected president of the company. One source described Hobart as representing a syndicate of New York capitalists.[16]

In December 1899, Clemens Herschel, who was by this time chief engineer for the East Jersey Water Company, completed the construction of pipes and pumps so the intake for Paterson's water supply could be moved four to five miles upstream to Little Falls above Beattie's Dam.[17]

In addition to the intake structure at Little Falls, a water-power pumping plant was built and put into operation during 1899. Larger than necessary for Paterson, Passaic, and Clifton, the facility at Little Falls was designed to provide water temporarily to Jersey City and on a more permanent basis to Newark. At this point, municipal water supplies in New Jersey were untreated in any fashion. The Passaic Water Company was still providing water under contract to the city of Paterson, but the water was obtained from the East Jersey Water Company at the Little Falls source.

Many large U.S. cities during this period had opted for municipal ownership of their water utilities. However, Paterson voted in 1852, 1871, 1877, and 1906 to retain private ownership of its water works.[18] Private ownership of water utilities had a strong tradition in New Jersey at that time and has continued to the present. In many ways, it was private ownership of the Jersey City Water Supply Company that made possible a full-scale trial of chlorine for drinking water disinfection. The private company had to answer only to itself and its shareholders. There was limited interference by city staff or politicians in decisions about capital investment or operations.

The East Jersey Water Company was formed on August 1, 1889, for the stated purpose of supplying Newark, New Jersey, with a safe water supply. All of the men who were shareholders of the new company were identified with the Lehigh Valley Railroad Company,[19] and the company's vision extended far beyond a water supply for Newark.

The company began as a confidential syndicate composed of businessmen interested in executing grand plans for water supply in northern New Jersey and New York City.[20] The early planners included stock broker and naturalist Delos E. Culver, who had dreams of supplying not only Jersey City but also using the rich water resources of the Passaic River to supply the lower part of Manhattan and all of Brooklyn. He teamed up with John R. Bartlett, a financier, entrepreneur, and schemer, who has been described as "aggressive and wealthy."[21] Bartlett immediately pursued Culver's vision by obtaining water rights to the Passaic River.

Bartlett also secured the rights to a tunnel partially excavated under the Hudson River and designed to connect Hoboken, New Jersey, with Manhattan. He began excavating the tunnel further so it could be used to pipe New Jersey water into New York City. All of this activity was explained in a slick report that Bartlett and his associates prepared. Bartlett also used a series of public meetings and speeches to promote his plan to supply New York City with water from the Passaic River. Of course, Bartlett stated in his talks that there was plenty of water to serve all of the New Jersey cities as well as New York City. [22]

Despite Bartlett's enormous efforts, one major barrier could not be surmounted. Many New Jersey leaders of the day believed it would be illegal to export New Jersey water to the state of New York for the profit of a private company. Bartlett lost interest in the water export scheme when it became clear that he could not overcome this impediment. In 1906, the New Jersey Court of Errors and Appeals held that the East Jersey Water Company could not export water from the Passaic River to serve Staten Island in the state of New York.[23]

The law firm of Collins & Corbin represented the East Jersey Water Company in this lawsuit as it did for all the water company's legal troubles. (This is the same law firm that later defended the Jersey City Water Supply Company in the two Jersey City trials.) The U.S. Supreme Court affirmed the New Jersey court's decision, establishing a rock-solid precedent.[24] A news article of the day summarized the importance of the decision.

> "The law prohibiting the diversion of the potable waters of New Jersey to other States has been upheld by the U.S. Supreme Court. The decision was rendered in a suit involving the right of a subsidiary organization of the East Jersey Water Co. to sell water for use in the Borough of Richmond (Staten Island), New York City. It is thought that the decision, combined with that of the

state courts, lessens the claims of the large water corporations of New Jersey that they have a monopoly right in the waters of the state." [25]

On October 30, 1923, both the Passaic Water Company and the East Jersey Water Company ceased to exist. They and two other companies (Acquackanonk Water Company and Montclair Water Company) were combined into the Passaic Consolidated Water Company, with the approval of the New Jersey Public Utility Commission.[26] In 1927, the state legislature passed a law creating the Passaic Valley Water Commission. The function of the commission was to acquire the Passaic Consolidated Water Company and to provide reliable sources of supply to the towns of Paterson, Passaic, and Clifton. Several major upgrades of George Warren Fuller's mechanical filtration plant at Little Falls would make the modernized treatment plant the cornerstone of the commission's supplies for local cities up to and including the present day.

1   Shriner, *Paterson*, 120.

2   Clayton and Nelson, *History Bergen and Passaic*, 407.

3   Gibson, "Population of the 100 Largest Cities;" Shriner, *Paterson*, 120.

4   Clayton and Nelson, *History Bergen and Passaic*, 407.

5   Development Proposal, Paterson, October 8, 1974.

6   Malone, Waterpower in Lowell, 1.

7   Shriner, *Paterson*, 106–7.

8   *Appleton's Cyclopaedia*, 170–1.

9   Vermeule, *Report on Water-Supply*, 150.

10   Ibid., 151.

11   Vermeule, *Report on Water-Supply*, 57–62.

12   Ibid., 159, 170.

13   Sullivan, *Paterson Manufactories*, 56.

14   Trowbridge, *Reports on Water Power*, 134.

15   Brown, "Paterson's Water Supply," 74.

16   Nelson and Shriner, *History of Paterson*, 410.

17   *Transactions of the Medical Society of New Jersey—1900*, 243–4.

18   Nelson and Shriner, *History of Paterson*, 410.

19   *New York Times*, August 2, 1889.

20   It was a ". . . syndicate which has been working systematically and persistently to control as large a portion as possible of the available water-supply sources of northeastern New Jersey." Colby and Peck, *International Year Book—1899*, 858.

21   Nelson and Shriner, *History of Paterson*, 411.

22   Youmans, *Popular Science Monthly*, 282.

23   Rich and Farnham, *Lawyers Reports Annotated*, 197–209.

24   *Hudson v. New Jersey*. 629.

25   "U.S. Supreme Court Prohibits NJ Diversions" *Engineering News*, 454.

26   *New York Times*, October 31, 1923.

❦

6

Leal—Hero of Public Health

"It is the misfortune of those having in charge the
public health to have always with them a certain small
proportion of the population who either from ignorance
and prejudice, or on account of selfish interest, are very
far from appreciating the work done in their behalf."
— Leal, "Annual Report of Board of Health 1892," 239

John L. Leal is the public health hero whom no one knows about. He was a quiet man with an ordinary appearance, but he was not quiet or ordinary when he challenged outmoded and entrenched views that caused conditions that killed people. His specialized knowledge in the fields of medicine, bacteriology, and public health catapulted him to a position of leadership in the critical field of drinking water treatment. Leal was not bound by the conservatism of engineers, who during this period were reluctant, even unwilling, to put chemical disinfectants in drinking water. Leal's heroism was a product of his daring and doggedness, his father's struggles with gastrointestinal disease, his education and training, and the lessons he learned as a city health officer. His leadership in the medical field and the courage of his convictions enabled him to take on the toughest public health problem of the day and achieve what no other person had done in the United States to stem the catastrophes of waterborne disease.

John L. Leal's father, John Rose Leal, was born on October 20, 1823 in Meredith, Delaware County, in southern New York.[1] John Rose Leal received his medical training under Dr. Almiran Fitch of Delhi, New York, and completed his medical degree at Berkshire Medical College in Pittsfield, Massachusetts, in 1848. Shortly thereafter he opened a medical practice in Andes, New York, a small farming community in Delaware County.[2] Dr. Leal continued his education with a postgraduate course at the Columbia College of Physicians and Surgeons in New York City[3]—an institution that would figure prominently in his son's education.

Historical records contain limited information about John L. Leal's mother, Mary Elizabeth Laing. Born in 1837 in Andes, after her family moved there from Argyle, New York, she was the fourth child of eight children. Her father, Rev. James Laing, was pastor of the Presbyterian Church of Andes.[4]

John Rose Leal and Mary Elizabeth Laing were married in Andes on August 29, 1855.[5] Mary Laing was only 18 when she married the successful country doctor. The couple had three sons; John L. Leal was the eldest. Census records from 1860 show that another child, William G. Leal, was born to the couple about 1859 in Andes. A third son, Charles E. Leal, was born much later in Paterson, New Jersey, about 1870. There are no records showing that William G. Leal survived into adulthood. Charles E. Leal lived to the age of 24 and died in Paterson in 1894, two years after the death of his father.

THE LEALS' SIMPLE RURAL LIFE in Andes was changed forever by the Civil War when in 1862 the 144th Regiment, New York Volunteers, was formed in Delaware County. John Rose Leal was appointed regimental surgeon, and over the next three years he was promoted to surgeon at the brigade, division, and corps levels. Toward the end of the war, he held the title of Medical Director in the Department of the South. According to an obituary, Dr. Leal was wounded twice[6] and was with his regiment at the battle of Johns Island, South Carolina, in July 1864.[7]

The 144th Regiment was stationed on Folly Island in 1863 as part of the siege of Charleston, South Carolina. According to the history of the regiment, "very nearly every man in the Regiment got sick . . . with bad and unhealthy water to drink."[8] The only treatment at the time for the debilitating dysentery that overwhelmed the regiment was the administration of "opium pills" by Dr. Leal. The pills did not cure anything, but they made the recipients feel somewhat better.[9] Dr. Leal became so ill with dysentery that he received medical leave for a time, but the records show that he never fully recovered.

The casualty list for the 144th Regiment demonstrates that disease was far more devastating to these soldiers than Confederate bullets. One officer and 18 enlisted men died of wounds received in combat. Four officers and 174 enlisted men died of disease (and other causes) during the regiment's service in the war.[10]

Dr. Leal was mustered out of the 144th Regiment on June 25, 1865,[11] after which he returned to his simpler life in Andes.

However, he brought a dreadful souvenir of the war home with him and suffered with it for the next 17 years.

One obituary stated: ". . . his death, which resulted from an attack of peritonitis of an asthenic character, sequel to an attack of dysentery, which at the outset did not indicate an unusual degree of severity, but was undoubtedly aggravated by the chronic diarrhea from which he had been a sufferer more or less constantly since his retirement from the army."[12]

Another obituary was equally clear as to the cause of his death: "He never recovered from the effects of disease contracted on Folly Island, and this induced other complications, resulting in his death."[13]

AFTER THE WAR the Leal family left Andes and moved to a small town (Purdy[14] or North Salem[15]) in the northern part of Westchester County, New York, where Dr. Leal resumed his rural medical practice. However, he had to give up the rural practice because his chronic gastrointestinal illness prevented him from riding his horse for extended periods to visit his patients.[16]

In 1867, the Leal family moved to Paterson, New Jersey, where Dr. Leal maintained a city medical practice requiring minimal travel.[17] Located in the Passaic River Valley, a center of the manufacturing engine that was creating wealth in the United States at the time, Paterson was one of the larger cities in the new country. In 1880, its population was about 60,000. Moving from a rural environment to this crowded, bustling city must have been a shock to the Leal family.

There is little mention of John Rose Leal in public records. He was a member of the Medical Society of Passaic County[18] as well as a number of New Jersey medical societies, serving as an officer of several.[19] He was also a member of the First Presbyterian Church of Paterson.[20] It was said that he established a lucrative medical practice in Paterson.[21]

JOHN LAING LEAL was born on May 5, 1858, in Andes, New York. Although no details are available, his birth was presumably in good hands because of his father's profession. His parents named him John after his father and his grandfather. He was given the middle name Laing after his mother's maiden surname.

In an interesting coincidence of history, John Laing Leal was born 42 days before the death of Dr. John Snow, the famous English doctor, anesthesia researcher, father of epidemiology, and cholera expert. In many ways, Dr. John L. Leal carried on the tradition of public health protection championed by Dr. John Snow.

John L. Leal attended Paterson Seminary in Paterson, New Jersey, as part of his early education. He was a student at Princeton (then known as Princeton College) from 1876 to 1880. A contemporary newspaper story listed John Laing Leal as one of the 74 individuals who received a Bachelor of Arts degree on June 23, 1880.[22] As a reminder of the scourges of disease that some people lived through during that time, the newspaper account of Princeton's 1880 graduation ceremony mentioned the recent deaths caused by an outbreak of malaria.[23]

Princeton's records include the only known physical description of Leal.[24] His height was listed as 5 feet, 9.25 inches and his weight at 165 pounds. Later photos showed that he had dark hair (parted near the middle) and round cheeks and that he sported a full mustache. These photos show a man who was not physically imposing or intimidating, but his face manifests a kindness in keeping with his role as a physician and protector of public health. He was a Presbyterian and identified himself as a member of the Republican political party.[25]

John L. Leal entered Columbia College of Physicians and Surgeons in 1880. There are no records of his academic performance at Princeton, but he must have done pretty well to be admitted to one of the most prestigious medical schools in the country at that time. Columbia College had a system of preceptors, or mentors. Dr. John Rose Leal was his son's preceptor during the first two years of his medical studies. The function of preceptor was important in medical education at this time, and John L. Leal must have admired his father's medical expertise enough to have him fill this role.

DR. JOHN ROSE LEAL died on August 28, 1882,[26] before John L. Leal had finished his medical education. To the young man about to embark on his career, his father's death must have been a terrible blow. Perhaps the kind words in the obituaries for Dr. John Rose Leal were of some comfort to the family:

> "He was a man of quiet, even temperament, dignified in his manners; but with a genial warmth of interest in all those with whom he came in contact that drew hearts to him and made strong and enduring friendships."[27]

"The doctor was of a genial and cheerful disposition, always ready to respond to the call of suffering humanity, though suffering himself."[28]

Although it is not possible to determine a definitive diagnosis merely on the basis of written records, it seems likely that Dr. John Rose Leal contracted amoebic dysentery while serving with the 144th Regiment at Folly Island, South Carolina. At the time, there were no antibiotics to treat this debilitating disease, which is otherwise incurable. It is hard to imagine the suffering Dr. Leal experienced in dealing with chronic dysentery for more than 17 years, but it seems clear from the 144th Regimental History that contaminated water was the source of the dysentery and, ultimately, the cause of his death.

The agonizing death of his father from a waterborne disease must have had a profound impact on John L. Leal. His later actions as Paterson's health officer and his career as a sanitary adviser to private water companies suggest that this seminal event dominated the trajectory of his career.

In 1883, John L. Leal was granted a Master of Arts degree from Princeton College.[29] Not content with his three academic degrees, Leal continued his education in 1884 by taking a course at the Post Graduate Medical School, where he obtained a certificate for his work.[30] Much later (in 1898), he took a special course in bacteriology and physical chemistry from Dr. Cyrus Fogg Brackett at Princeton College.[31] Leal maintained an avid interest in the exploding field of bacteriology. Of particular importance is an exchange during his testimony at the second Jersey City trial, when Leal described his interest in and experience with bacteriology.

"Q: Did you continue your studies in bacteriology and chemistry after you began practice?

A: I did, yes, sir.

Q: Did you yourself make experiments in bacteriology?

A: I did, for some years.

Q: For how long a period did you yourself practice as a bacteriologist?

A: Well, the work which I did personally was done mostly between the years 1897 and 1901."[32]

Leal's practical experience prior to his trial testimony on February 4 and 5, 1909, supported his role in determining the quality of the Boonton Reservoir water supply.

"Q: Have you kept up your study and kept abreast of the learnings on the subject of sanitation of waters, and of bacteriology, at the present time?

A: That has been a very important essential of my business, to know what was going on, and to keep in touch with it, in order to advise the large interests with which I am associated; and in order to do that I have been in the habit of attending meetings of various associations devoted to this purpose; I am a member of them, and have been president of the New Jersey State Sanitary Association, vice president of the American Public Health Association, and a member and officer of various other organizations of this character."[33]

AMY LUBECK ARROWSMITH and John Laing Leal were married in 1888. Little is known about their marriage except that they had one child, Graham Arrowsmith Leal, on December 4, 1888. Amy Leal spent the latter part of her confinement away from Paterson, and Graham Leal was born in Yonkers, New York, at the home of his maternal grandparents. According to the obituary of Amy Leal's mother, Elizabeth M. Arrowsmith, she lived with her daughter's family in Paterson at the time of her death in 1900 at the age of 79.[34] The 1900 census was taken at the Leal house at 555 Broadway in Paterson on June 9, only one month after Amy's mother's death. At the time of the census, John, Amy, and Graham were living at the Broadway house, and so was Amy's father, Augustus Toplady Arrowsmith.

Records show that John L. Leal's mother also died in Paterson—on November 21, 1906, at the age of 69. It is likely that Leal's mother also lived with the family part of the time. It must have been a busy household.

Amy Arrowsmith Leal died on June 1, 1903, at 41 years of age.[35] Graham was only 15 years old when his mother died. Census records show that in 1910, Amy's aunt (Graham's great-aunt), Anna L. Laing, was living with Graham and John at 517 28th Street in Paterson (only four blocks away from their long-time home at 555 Broadway). Anna L. Laing was born in 1847 and never married. With John Leal's ever-increasing workload during 1900–1910 and the demands of the two trials during 1906–1909, it is likely that most of the responsibility for raising Graham fell to his great-aunt.

PROFESSIONAL ORGANIZATIONS occupied much of John L. Leal's time. He evidently viewed his involvement with these organizations and their committees as an important part of his career. He was described as an "indefatigable worker [who] went most deeply into every subject which interested him."[36]

In 1884, Leal was elected a member of the Medical Society of New Jersey. He was also an active member of the Passaic County Medical Society,[37] serving as its vice president in 1900. Among the members of this society was Andrew F. McBride, a prominent physician in Paterson. Later, as mayor of the city, McBride would get Leal involved for a second stint in the work of the Paterson Board of Health.[38] Despite Leal's concentration on water treatment after 1899, he continued to be involved in the Passaic County Medical Society, serving on its Legislative Committee in 1900, and in 1901 he was the society's president.[39] He was still active in the state Medical Society in 1905, serving as a permanent delegate from Passaic County along with McBride, and he was a councilor from the second district in 1906.[40] Leal was also a member of the American Medical Association.[41]

In 1903, Leal became president of the New Jersey Sanitary Association. This association gave him many opportunities to interact with the leading public health authorities of the day, and by all reports, he was active in the organization for many years. At the Sanitary Association's meeting December 4–5, 1903, Leal gave the president's address on the "Present Attitude of Sanitary Science." Moses N. Baker was chair of the association's Garbage Disposal Committee, and at a session on the topic of Sewage Disposal in New Jersey, the speakers included Rudolph Hering, Allen Hazen, George M. [sic] Fuller, and the New Jersey water supply expert Charles A. Vermeule.[42]

The American Public Health Association (APHA) was the premier professional public health organization in the United States during Leal's lifetime. During this period, members had to be elected to APHA, and Leal was elected early in his career, in 1891. Bound reports, produced each year, listed the association's officers and the composition of its committees and contained the proceedings of its annual meetings. At the 1897 annual meeting,

Leal presented a paper titled, "House Sanitation with Reference to Drainage, Plumbing, and Ventilation."[43]

The 1899 annual meeting was held in Minneapolis, Minnesota, October 31–November 3. The annual report that year listed John L. Leal as a member of APHA's Executive Committee and as New Jersey's designated member on the organization's Advisory Council. Leal was also a member of the Pollution of Public Water Supplies Committee, which was chaired by George Warren Fuller. At this meeting, Leal presented a paper on a typhoid fever outbreak in Paterson. Allen Hazen was also a member of that committee. Rudolph Hering chaired the Disposal of Refuse Materials Committee. The Committee on Standard Methods of Water Analysis had George C. Whipple as a member and George Warren Fuller as the chair.[44] All of these public health leaders knew each other, worked with each other, and socialized with each other. They would all be involved in the Jersey City trials, though some testified on behalf of the plaintiffs and others testified for the defendants. At the 1902 APHA annual meeting in New Orleans, Leal was elected second vice president of the organization.[45]

In addition to his duties at Paterson and his many obligations to professional organizations, Leal was a member of the board of examiners for Rutgers College in New Brunswick, New Jersey, in the late 1890s. Members of the board of examiners were expected ". . . to conduct examinations and grant certificates in municipal hygiene. . . . The examinations are designed to test the fitness of persons who may be called upon to engage in the execution of the health laws. . . ."[46] Leal must have believed it was worth his time to help develop the next generation of professionals who would protect public health in the state.

Like most professional men of his day, he also belonged to a number of social organizations, including the Hamilton and Elks clubs of Paterson, the Princeton and Republican clubs of New York City, and the Union League Club of Jersey City. On the many nights he spent in Jersey City during the two trials, he obtained lodging and meals at the Union League Club.

WHEN LEAL STARTED TO PRACTICE MEDICINE in Paterson in 1883, there were two hospitals in the city. St. Joseph's Hospital was founded in 1867 by the Sisters of Charity of Saint Elizabeth. In 1871, the Ladies Hospital Association of Paterson was formed by concerned citizens to increase the capacity of hospital care in the growing city. Leal's father, Dr. John Rose Leal, was listed as one of the founding members of the new hospital's medical staff.[47] In 1887, the hospital's name was changed to Paterson General Hospital Association, and five years later a new building was opened for patient care.

The transcript of the second Jersey City trial notes that Leal was an intern at "Elizabeth General Hospital" in Paterson right after medical school, which would have been approximately 1884 to 1885.[48] In a later recitation of his background and experience as a physician, Leal stated that he practiced medicine at both Elizabeth General Hospital and Paterson General Hospital.[49] However, Elizabeth General Hospital did not exist in Paterson at that time or at any time. Leal was undoubtedly referring to St. Joseph's run by the Sisters of Charity of Saint Elizabeth. The good reputation of his father in the medical community of Paterson must have made it somewhat easier for John L. Leal to start a private practice and become associated with local hospitals.

Leal was an active member of the Paterson chapter of the New Jersey Medical Society and was noted for his participation. He ". . . showed a specimen of ulcerated rectum [1884]," "read of a case of multiple abscess of the liver extending into the omentum [fatty structure that is part of the intestines] [1885]," ". . . presented a paper on 'Typhus Fever,' [1892]," and ". . . reported cases of [ptomaine] poisoning by . . . [cream] cake [1894]. . . ."[50]

In 1887, Leal and several of his medical colleagues founded the Outpatient Clinic at the newly named Paterson General Hospital. He served as a physician in that department from 1887 to 1891. On an 1890 roster of the hospital staff, Leal was listed as a member of the "Out-Patient Medical Staff" along with five other physicians.[51] In 1891, he was added to the medical staff of the hospital and made a member of the medical board. During this period, several of the hospital's annual reports listed him as a member of the Surgical Department of the hospital along with his colleague Dr. Bryan C. Magennis.[52] In the early 1890s, he was also an assistant surgeon at the Paterson Eye and Ear Infirmary, founded to care for the indigent in Paterson and environs. In one of many connections among Paterson's leading citizens, a founding member of the Eye and Ear Infirmary's Board of Governors was Garret A. Hobart, mentor to Leal and future Vice President of the United States.[53]

According to his trial testimony, Leal maintained his private medical practice at least up to the end of the second trial in late 1909.[54] With all of his obligations as health officer for Paterson, sanitary adviser for the East Jersey Water Company and the Jersey City Water Supply Company, and water supply consultant to other cities, it is extraordinary that he continued his work as a physician.

John Leal obviously felt a calling to public service that extended beyond his career as a physician. From about 1885 to 1899 and later from 1911 to 1914, Leal played an important role in Paterson's city government, specifically with regard to public health protection. Fortunately, city officers and departments prepared annual reports that were printed and bound during the time of Leal's service.[55] His testimony at the second Jersey City trial indicates that his first position in Paterson's city government was as police surgeon in 1885 or 1886. It is likely that he held this position only for one year, possibly less.[56]

As a public servant, Leal was said to be known for ". . . his attention to duty and for his insistence on the enforcement of the law . . . vigorous administration . . . his persistency in following up investigations, his requirement that inspections be complete and thorough, and his general detailed supervision of the sanitary work of the city."[57]

Paterson's 1887 annual report contained an account that Leal wrote as city physician for the fiscal year ending March 20, 1887. The city physician was responsible for the care of the indigent at the city offices, at the almshouse, and through home visits. In that report, Leal noted that he was sworn in as city physician on May 5, 1886 (his 28th birthday!). His 1887 report to Paterson's mayor and Board of Aldermen, the city's legislative body, was thorough, professional, and long,[58] and his reports during the rest of his service to the city would be notable for their adherence to these three characteristics.

In 1890–1891, Leal's position changed from city physician to Board of Health inspector. Leal signed the Paterson Board of Health's report for the fiscal year ending March 20, 1891, as secretary of the Board of Health. His reputation for writing thorough reports undoubtedly got him appointed to that post, especially because the board had not published a report for the previous six years. The report contained a number of interesting bits of information: John L. Leal's salary as health inspector was $1,200 per year. Deaths from all causes were reported in detail, including 69 cases of typhoid fever resulting in 18 deaths. Disinfectants, purchased for $42.91, were used to "disinfect" or fumigate 445 premises where contagious diseases had occurred. (Fumigation was a carryover from beliefs in miasma as a cause of disease.) The death rate for children under one year of age was astonishing; the primary cause was diarrheal diseases, including dysentery. During the year reported, 128 children under the age of one year died of diarrheal diseases. Of all the other diseases in the compilation of vital statistics, only tuberculosis killed more people—204 people from all age brackets.[59]

Although these annual reports were usually dry, boring recitations of statistics and monies expended, the 1891 annual report indicated that trouble was brewing between the recently reconstituted Board of Health and some of the city's powerful residents. The mayor's summary remarks addressed the issue: "Although in some quarters there has been criticism of some of the actions of the Board of Health, it must be remembered that no body of men can govern to the satisfaction of all, especially if that body is entrusted with duties, the discharge of which may at times have the appearance of interfering with individual rights. The safety of the public health is of such supreme importance that the board having charge thereof should meet with the cordial co-operation of every department of the city government."[60]

In the Board of Health section of the 1891 annual report, John L. Leal, M.D., was identified as the author, and a further allusion to the board's problems was carefully worded. "We are fully alive to the responsibilities resting upon us as guardians of the lives and health of the people of a city of nearly 80,000 inhabitants. To the citizens in general we wish to express our indebtedness for the *almost universal support* which they have given us in the performance of our duties."[61] (emphasis added)

The 1893 annual report of the Paterson Board of Health described quick action by Leal and the New Jersey Board of Health when cholera threatened the state. The fifth worldwide cholera pandemic lasted from 1881 to 1896. One of the largest cholera outbreaks in Europe (8,000 deaths) occurred in Hamburg, Germany, in August 1892. This outbreak was the subject of Robert Koch's famous study of the incidence of cholera in populations using filtered versus unfiltered water supplies.[62] Four passenger ships bringing European immigrants to New York City were detained at Quarantine Island in New York Harbor in late August and early September of 1892. Ultimately, thousands of passengers were quarantined, and more than 100 died from cholera.[63] But the lessons of quarantine had been learned from previous epidemics, and the health authorities used all of their powers to prevent infected passengers from entering the country.

LEAL FIGHTS RESISTANCE TO CLOSING CONTAMINATED WELLS

During the fiscal year ending March 20, 1892, Leal and the Paterson Board of Health confronted an increasing death rate from all types of disease. However, Leal's main challenge during this year was to gain control over shallow public and private wells that were contaminated with sewage.[1]

Contaminated wells were not a new problem in Paterson—shallow wells provided some of the city's first water supplies—but past efforts to close the wells had been ineffective and people continued to use them. Leal figured out that the weight of his office and his authority under local laws were not enough to turn around this serious public health problem. He believed some hard data might convince the skeptics and critics. He took samples from four wells that he knew from past evidence were contaminated and sent the samples for chemical analysis to Shippen Wallace, Ph.D., chemist to New Jersey's state Board of Health and dairy commissioner.[2] Neither routine nor special microbiological testing was available in these years of developing awareness of the role of drinking water quality in public health protection. Instead, sewage contamination was detected on the basis of elevated chloride concentrations, the presence of ammonia, and high concentrations of nitrite and nitrate.

The analytical results convinced Wallace that the four wells should be "condemned," and the handles for the four pumps were subsequently removed. (This was a typical public health measure, echoing the era of Dr. John Snow and the Broad Street pump). Unfortunately, the specter of politics intruded on what should have been a public health judgment. Individual aldermen pressured the Inspector of Lamps, Wells and Pumps to restore the four pump handles. Leal wasted no time taking on the Board of Aldermen, invoking the power (and somewhat limited prestige) of the Board of Health.

"The water from the wells . . . was condemned and declared to be in a condition dangerous to the health of anyone drinking it. This Board therefore

requests you to either close said wells or to put them
in such a condition as to furnish pure water."[3]

The aldermen were in no mood to knuckle under to the
Board of Health, whose members nominally reported to
them. "In answer to your request as to action . . . the committee . . . decided on motion to deposit the communication [from the Board of Health] into the waste basket."[4]

Leal raised the ante by again removing the handles from the
four pumps in a direct exercise of the authority of the Board of
Health. He also took a second round of samples from the four
wells and asked Shippen Wallace to provide another opinion
of the quality of the water from the wells. Wallace minced no
words. ". . . it would be difficult to find water of a worse character. . . . it is a source of wonder, they have not been the means of
producing an epidemic. . . . the water from these wells is contaminated and totally unfit for potable purposes."[5] Fully exercis-
ing the power of the Board of Health, Leal stated: "Those wells
are therefore closed; and closed they will remain as long as the
power to keep them so remain[s] with this Board of Health."[6]

Because the annual report of Paterson's Board of Health was
an official document in which Leal laid out the controversy
in full public view, he was casting a lot of sunshine on a situation that had been controlled by political forces. The courage Leal showed in taking on the political forces in his city
was an indicator of the courage he would later demonstrate
in implementing chlorination at Boonton Reservoir in the
face of possible condemnation by the court and his peers.

Leal did not restrict his annual report to a presentation of the
facts of the controversy. As he often did in his career, he took the
opportunity to state a philosophical basis for his actions with an
eye toward posterity and the future of public health protection.

"It is the misfortune of those having in charge the
public health to have always with them a certain
small proportion of the population who either from
ignorance and prejudice, or on account of selfish
interest, are very far from appreciating the work
done in their behalf. However we live in the hope

that time and advancing intelligence and education
may bring even to this class more of a realization
of the importance of sanitary science and a more
through [sic] understanding of its principles. Surely
it is not visionary to expect a day when the sanitary officials of a city shall guard the public health
and fight disease with as universal a public support
as is given to those whose duty it is to guard such
city against fire and disorder. Until that happy day
does come we must go on accomplishing as much
as possible with the means and the support at our
disposal, turning neither to the right nor to the left,
but fearlessly and firmly performing the duties which
alone have called us into existence as a Board."[7]

An interesting parallel occurred with regard to Leal's closure
of wells in Paterson and the fate of the handle on the Broad
Street pump during the 1854 cholera epidemic in London.
Dr. John Snow did not march out to Broad Street and physically remove the pump handle himself, as has been described
by others. Matters such as these were controlled by the St.
James Board of Commissioners of Paving, so Dr. Snow presented his findings to the commissioners, plainly demonstrating that the Broad Street well was responsible for the local
cholera epidemic. On September 7, 1854, the commissioners ordered that the handle be removed. Structural defects
in the Broad Street well sump and the cross-connection to
the nearby house sewer were not corrected until 1855.[8]

The residents of Broad Street petitioned the commissioners to
reopen the well that had caused hundreds of deaths in their
neighborhood. This was partly a consequence of the official
linkage of the severe, isolated epidemic in the Broad Street area
to miasma (foul air). Incredibly, the commissioners voted 10 to
2 to reopen the well on September 26, 1855. According to contemporary reports, there was much rejoicing in the street that
the Broad Street well had been reopened. The polluted well was
not permanently closed until the cholera epidemic of 1866.[9]

Doubtless, many other health officers during this time experienced the same type of interference by public officials and

citizen wrath when they tried to protect public health by doing something unpopular. In the 1890s, applying the germ theory of disease to real-world public health emergencies was not easy, nor was it easy for the protectors of public health to fend off political interference. However, Leal's courage and persistence in defending the authority of the Board of Health paid off. None of Paterson's annual reports following these episodes mentioned any challenges to the authority of the Board of Health in general or to the defender of public health, Dr. John Leal, in particular.

---

1   Leal, "Annual Report of Board of Health 1892," 240.

2   Ibid., 231.

3   Ibid., 232.

4   Ibid.

5   Ibid., 233.

6   Ibid., 240.

7   Ibid., 239.

8   Vinten-Johansen, et al., *Cholera*, 294.

9   Ibid., 310, 316–17.

Cholera was also reported in Mexico, and inevitably a few cases appeared in the United States. Dr. Leal landed in the middle of the search for cholera victims in September 1892, when it was reported that a William Wiegmann, who had died of cholera in New York City, had visited friends only three miles from Paterson. Leal was dispatched to the home of the Windhurst family, where he found no disease but quarantined the family immediately.

Leal had all of the authority he needed because New Jersey's Board of Health authorized him to exercise jurisdiction anywhere within the state over the reported case of cholera and its potential spread.[64]

In the Paterson Board of Health's 1893 annual report, Leal's description of a smallpox outbreak was professional but lacking in the human dimension. He stated that he was pleased that isolation and quarantine had prevented any secondary infections.[65] Another source of information, however, presents a more human narrative of Leal's role in the smallpox outbreak.

"Dr. Leal was in office when the smallpox epidemic broke out in [1893], and did work which will long be remembered. . . . Several people suffering from the dread disease were taken [to the old pest house], and cared for by a nurse who volunteered for the dangerous work. Dr. Leal aided the late Dr. Smith, then attending physician at the isolation hospital, in attending the patients. Supplies were left near the house by grocers, druggists and others. When the first patient during the epidemic died there, Dr. Leal got a man to dig a grave nearby, and himself, with the assistance of the nurse, carried the man out of doors, where an undertaker and James Fitzpatrick, sanitary inspector, conducted the burial on the almshouse grounds."[66]

The 1894 annual report of Paterson's Board of Health included statistics on the number of people who were interred at the more than 12 cemeteries in the Paterson area. In 1894, 346 people were buried in Cedar Lawn Cemetery.[67] The yearly statistic had special meaning for John Leal. His brother, Charles E. Leal, age 24, died during that year and was buried in the family plot at Cedar Lawn Cemetery.[68]

Also in 1894, there were 146 cases of typhoid fever in Paterson and 28 deaths from the disease. "Of typhoid fever we have had about the usual number of cases, some forty cases of which were traced directly to the water of a badly polluted 'artesian' well opposite 87 Jersey Street. The death rate has been somewhat higher than has been usual with us in this disease."[69] The routine nature of this report indicates that no great import was attached to typhoid fever deaths. Death from typhoid fever was pretty much business as usual in large cities with unfiltered surface water supplies. Calculated on the basis of Paterson's population of about 90,000, the death rate for typhoid fever in 1894 was 31 per 100,000 people. As the two New Jersey trials would demonstrate, a death rate of 25 per 100,000 was considered "acceptable" by public health experts during this period.

In the 1895 annual report, Leal urged the construction of an isolation hospital. He was instrumental in getting the state legislature's authorization for the city to sell bonds to finance the hospital's construction.[70]

In this report, Leal raised serious concerns about the future safety of the current water supply, which was being extracted from the Passaic River at the Great Falls. The watershed was becoming more populated above the intake, and it was inevitable that contamination of the city's water supply would increase. Leal knew that Passaic Water Company's contract to supply Paterson with water was coming up for renewal. "I would therefore recommend that in the event of a new contract being made with said company, provision be made, that whenever the water supplied can be shown to be polluted and dangerous to the public health, said company shall purify such water by sand filtration or some other approved means, or shall secure their supply at some point above [the] source of pollution."[71]

Leal was reflecting the dominant view of public health officials and sanitary engineers at the time—either find a purer source or filter the water. In the case of Paterson, the water company eventually did both (along with adding a disinfectant for good measure). The water supply intake was moved four to five miles upriver to Little Falls, and the world's first modern mechanical filtration plant was built and put into operation seven years after this annual report. As Leal's career evolved—he changed from being a city employee responsible for all aspects of public health to becoming a sanitary adviser focusing on improving drinking water supplies and safety—he would continue to play a major role in the protection of Paterson's water supply.

During 1897 and for the next two years, Leal must have wondered if he was in the right business. His 1897 report encompassed 21 pages, by far the longest report of his tenure in Paterson. As part of the positive news, he described in detail the recently completed isolation hospital. It was an impressive facility capable of housing about two dozen patients who might be afflicted with up to three different contagious diseases.[72] The hospital was cited as a model for facilities of its type.

> "This isolation hospital has been selected by the American Committee of Hygiene at the Paris Exposition of 1900, as the model isolation hospital for plans and illustrations, to represent this country's progress in that department of public hygiene through health boards."[73]

The number of typhoid fever cases in Paterson in 1897 was 278, resulting in an estimated 39 deaths. For Paterson's population of about 105,000, the death rate from typhoid fever was 37 per 100,000—an uncomfortably high figure.[74] The typhoid fever cases were scattered around the city, and an analysis of the drinking water used by the afflicted people led Leal to conclude that ". . .

the Passaic River above the falls was the common source of infection." In other words, the water supply was killing a larger number of people than usual. Leal's temporary solution: "The only way in which the public can assist in stamping out the disease is by boiling all drinking water, and that I would most earnestly advise." He advised boiling drinking water for 30 minutes,[75] much longer than necessary, but he had no way of knowing how long was sufficient.[76]

Because of the deteriorating sanitary quality of Paterson's water supply, Leal wrote and had introduced into the New Jersey legislature a bill for the protection of the Passaic River above Paterson. The bill was passed unanimously by both chambers of the legislature and signed into law by the governor, John W. Griggs. The act provided for daily fines for any violations and 30 days of jail time if the fines were not paid. Most interesting, the law stated that a suit seeking an injunction against any continued pollution of the river could be brought in the New Jersey Court of Chancery.[77] Leal would take full advantage of the law in conducting his duties as protector of the Passaic River watershed while working for Paterson and as protector of the river above Boonton Reservoir while working for Jersey City Water Supply Company. From 1897 until he left employment with Paterson, Leal enforced the law, removing overhanging privies and other obvious sources of pollution above the water intake for Paterson.[78] To enforce cleanup of Passaic River contamination, Leal brought two lawsuits in the Chancery Court and about a dozen in a lower court. He emphasized that "moral suasion" was used most often to achieve compliance with the law.[79] The law written by Leal is referenced in a review of New Jersey water supply protection statutes as part of an assessment of the quality of the Passaic River downstream of Paterson.[80]

During this period, Leal had many other obligations. He had a wife and son and one, or more, older relatives living in their house at 555 Broadway. He had his own private medical practice. He was on the board of the Paterson General Hospital. He was a member of many associations and committees and attended their meetings throughout this period. He was also beginning his affiliation with the private water companies serving New Jersey.

Paterson's 1898 Board of Health report did not transmit much good news to the mayor and the Board of Aldermen. Leal described in detail the progression of two typhoid fever outbreaks that occurred in Paterson from November 1897 to January 1898.

Distribution of the disease among the population caused him to suspect that the milk supply was compromised. In particular, several cases of typhoid fever appeared to be associated with the creamery of F. W. Fulboam of Branchville, Sussex County, New Jersey, about 35 miles northwest of Paterson and not far from the New Jersey–Pennsylvania state line.[81] The creamery was shut down until the sanitary problems were fixed.

Significant developments regarding Paterson's water supply also occurred in 1897–98.

> "During the year measures have been taken to change the source of our water supply. As is known, the intake has been, and is as yet, at the Great Falls. . . . It has therefore been decided to remove the intake to Little Falls, which is above the sources of pollution which now threaten us."[82]

One source listed the date that the intake was moved as December 7, 1899.[83]

Paterson's 1899 annual report, the last one published while Leal was the city's health officer, did not include a narrative by Leal. The statistics of disease and death in the 1899 report showed a continuing serious problem with typhoid fever.[84] Despite the Paterson report's lack of explanation for the typhoid problem that year, APHA's annual report published a paper in which Leal discussed Paterson's typhoid epidemic in some detail.[85]

From September 1, 1898, to April 1, 1899, 314 cases of typhoid fever were reported in Paterson. This incidence of disease was approximately six times higher than in previous comparable periods. After determining that the cases were spread throughout the city and that there was no apparent source of contaminated milk or ice, Leal concluded that the public water supply was the source of the disease-causing organisms.[86]

During a search of the nearby watershed, Leal found one person who had typhoid fever prior to the onset of the epidemic. This person's wastes were discharged from a privy to a tributary of the Passaic River about 8,000 feet upstream of Paterson's intake. Leal believed this one infected person had shed millions of typhoid bacilli, which, after transport down a tributary to the main Passaic River, could have infected consumers of the untreated drinking water. He opined that the infected fecal wastes were sufficient to contaminate the entire water supply even on a river with an average flow of 500 million gallons per day. High volumes of rainfall and runoff helped to flush the contaminated wastes from the tributary into the Passaic River.[87]

In late 1899, Leal resigned from service to the city and was replaced as health officer and as a member of the Board of Health by his friend and colleague, Dr. Bryan C. Magennis.[88] A short article in *The New York Times* specified the date of his resignation as December 18, 1899.[89]

In a report to the Medical Society of New Jersey, Dr. Leal's service was remembered with an emphasis on how he solidified the authority of the Paterson Board of Health in all matters dealing with public health.

> "The Paterson Board of Health has lost its well appreciated, active and progressive worker in matters of public health, Dr. John L. Leal, who had spent many years of service as health officer and who was the prime mover in most of the recent aggressive movements and acquisitions of authority and facilities vested in this body."[90]

Leal became a member of the Paterson Board of Health again near the end of his life. He was appointed by Mayor Andrew F. McBride during McBride's first term (which extended from 1908 to 1912, making Leal's appointment start about 1911). He served as president of the board from 1912 until his death in 1914.[91]

LEAL BECAME INTENSELY INTERESTED in the purity and protection of water supplies in the mid- to late 1890s. "I continued it and went still deeper into it as a personal adviser to Garret A. Hobart, who was interested in various water enterprises, about 1897; and then about 1899, I became identified with the East Jersey Water Company as a sanitary adviser."[92]

According to the transcript of the first Jersey City trial, Leal began employment with the Jersey City Water Supply Company in March 1902, at the time the New Jersey General Security Company took charge of that entity.[93]

> "Q: Are you in the employ of the Jersey City Water Supply Company?
>
> A: Yes, sir; I am.
>
> Q: In what capacity?
>
> A: Sanitary adviser.

Q: How long have you been employed by them?

A: Why, for the Jersey City Water Supply Co. since the spring of 1902.

Q: What in general are your duties in that employment?

A: My duties consist in removing, as far as possible, any unlawful source of pollution in the water shed and on the river and its tributaries, and also to watch the quality of the water and to see if it answers the requirements of a safe potable supply."[94]

Leal's responsibilities for protecting the Passaic River watershed were to remove contamination sources upstream of Little Falls and the new Boonton Reservoir. Records from the Little Falls treatment plant indicate that Leal had a significant presence at the plant in 1903 and 1904.[95] Specifically, his watershed work involved:

- Moving overhanging privies and cesspools back from the river bank,
- Preventing the building of new privies directly discharging into the river,
- Cutting off pipes carrying sewage directly into the river and its tributaries,
- Examining bacteriological data produced by consulting bacteriologist Dr. McLaughlin to ensure a safe potable water supply,
- Working with local Boards of Health to eliminate pollution sources,
- Coordinating with inspectors of the Morris Canal to report polluters, and
- Educating people in the watershed to refrain from polluting the water supply.[96]

Leal's involvement with these water companies and with Garret A. Hobart significantly influenced the development of private water supply companies in New Jersey at this time.

A major part of his duties during the preliminary phases and throughout both Jersey City trials was to advise William H. Corbin, lead attorney for the defendants. During the first trial, which extended over 40 days of testimony and depositions during the period February 20, 1906–May 29, 1907, Leal sat at Corbin's elbow every day of the trial. He also testified numerous times during the first trial.

When Leal left employment with the East Jersey Water Supply Company is not known. During the last years of his life, he apparently operated as a consultant to other cities regarding their water supply development.[97]

Leal acted as a consultant to many cities with regard to the sanitary quality of their water supplies. An assignment that has been thoroughly documented involved his assessment of the sanitary quality of the watershed for the Bristol and Warren Water Works Company in Rhode Island. In 1902, the Rhode Island Board of Health issued a letter to the water company stating that the sanitary quality of the water supply was "a danger to the health of consumers." Leal and his fellow consultants (two Brown University professors and a physician) investigated the watershed and issued a report, which was quoted in a December 1, 1902, letter from the water company to the Board of Health. According to that letter, Leal's opinion was that ". . . the conditions of the water and the water sheds 'do not in any way justify the action of the State Board of Health.'"[98]

On March 28, 1910, Leal delivered a presentation on drinking water disinfection to the Hartford, Connecticut, water department.[99] A detailed description of the Hartford experiments with chloride of lime was published in *Engineering News*. Leal designed the chemical feed system and supervised its construction. The tanks and pipes were arranged in a manner similar to the treatment facilities at Boonton Reservoir. The water to be treated came from the Connecticut River, which had been abandoned as too contaminated for use as a water supply, but the addition of about 0.4 parts per million of available chlorine was sufficient to produce bacteriologically safe water.[100]

Leal was also directly involved in the design of a mechanical filtration plant for Newport, Rhode Island. The plant went into operation in 1910 and contained facilities for sedimentation, coagulation, mechanical filtration, and disinfection with hypochlorite solution. Leal was identified as an "adviser to the water works and supervised the design of the plant."[101]

In addition, Leal was responsible for installation of the chloride of lime feed plant in Omaha, Nebraska. The design of this operation was based on the principle of feed tanks and orifice plates used at Boonton Reservoir.[102]

Just before his death, Leal consulted for a water company in South Carolina.[103] He was also known to consult for "water supply companies in several of the largest cities in the country, New York being among them."[104]

PROTECTING PUBLIC HEALTH and improving drinking water quality were Leal's obsessions. He devoted an extraordinary amount of time to these endeavors.[105] One can only speculate about how the influences in his early life drove him to become so committed as a physician, public health champion, and water expert. His predominant personal attribute was a quiet courage. In the face of certain disapproval from his peers and possible condemnation by the public, he did what he thought was right—he added a chemical disinfectant to drinking water. It is extraordinary that he found the fortitude to follow this path when all of the experts railed against it.

Nothing in Leal's appearance hinted at his steel spine or his determination to follow through with his plans to protect public health. One definition of the word hero reads: "a man of distinguished courage or ability, admired for his brave deeds and noble qualities." These days, many people feel the word "hero" is overused. Implementing a revolutionary water treatment process that has saved millions of lives qualifies John L. Leal to be known as a hero of public health.

❦

1   Atkinson, *Physicians*, 383 and 389; or possibly 1825, Watson, "Necrology," 406; or possibly 1827, McKee, *Back in War Times*, 272. John Rose Leal's parents, John Leal and Martha McLaury, were descended from early settlers of the county. Records indicate that John Rose Leal's great-grandfather, Alexander Leal, was born in Scotland in 1740 and immigrated to the British colonies in North America, landing in New York City on April 13, 1774, Watson, "Necrology," 406. Little information is available on John Rose Leal's early years. According to one source, he received his preliminary education at the Literary Institute in Franklyn, Delaware County, New York, and at the Delaware Academy in Delhi, New York, Watson, "Necrology," 406.

2   "John L. Leal," 427; Alumni file, John L. Leal; "John Laing Leal," *Who's Who in America*, 1233.

3   Watson, "Necrology," 406.

4   Atkinson, *Physicians*, 389; Watson, "Necrology," 406; Reverend James Laing: His Legacy.

5   Ogborn and Miller, *Index to Deaths & Marriages*, 35-6; Delaware County, Marriages from *Bloomville Mirror*.

6   *New York Times*, August 29, 1882.

7   McKee, *Back in War Times*, 273.

8   Ibid., 132.

9   Ibid.

10 Phisterer, *New York in the War of the Rebellion*, 3667–8.

11 McKee, *Back in War Times*, 304.

12 Watson, "Necrology," 406.

13 McKee, *Back in War Times*, 273.

14 Watson, "Necrology," 406.

15 Atkinson, *Physicians*, 383.

16 McKee, *Back in War Times*, 273.

17 Ibid.

18 Butler, *Medical Register and Directory*, 460.

19 Atkinson, *Physicians*, 383.

20 Watson, "Necrology," 407.

21 *New York Times*, August 29, 1882.

22 *New York Times,* June 24, 1880.

23 Ibid.

24 The average height for a man in the 1860s to 1890s was 5 feet, 7.7 inches, and Leal was about 20 pounds heavier than an average man of his day, Sargent, "Physique of Scholars," 248–56.

25 Alumni file, John L. Leal.

26 *New York Times*, August 29, 1882.

27 McKee, *Back in War Times*, 273–4.

28 Watson, "Necrology," 406–7.

29 *General Catalogue of Princeton University*, 260.

30 "Deaths Leal," 208–9.

31 Between Jersey City and Water Company, February 4, 1909, 4923.

32 Ibid.

33 Ibid., 4925.

34 *New York Times*. May 8, 1900.

35 *New York Times,* June 3, 1903.

36 "Deaths Leal," 208–9.

37 Nelson and Shriner, *History of Paterson*, 432.

38 *Transactions of the Medical Society of New Jersey—1900*, 22–3.

39 Ibid., 45; Nelson and Shriner, *History of Paterson*, 432.

40 *Transactions of the Medical Society of New Jersey—1905*, 3.

41 Between Jersey City and Water Company, February 5, 1909, 5012.

42 Mitchell, *Twenty-Seventh Annual Report*, 203.

43 Leal, "House Sanitation," 401–8.

44 *American Public Health Association, 1900,* ix, xi, xv.

45 *Washington Post*, December 13, 1902.

46 *Evening Times*, February 16, 1898.

47 Nelson and Shriner, *History of Paterson*, 434.

48 Between Jersey City and Water Company, February 4, 1909, 4923.

49 Between Jersey City and Water Company, October 14, 1909, 6731.

50 *Transactions of the Medical Society of New Jersey—1894*, 200, 204, 205.

51 *Twenty-Second Report Paterson General Hospital*, 1893.

52 *Twentieth Annual Report Paterson General Hospital*, 1891; *Twenty-Third Report Paterson General Hospital*, 1894; *Twenty-Fourth Report Paterson General Hospital*, 1895.

53 *Transactions of the Medical Society of New Jersey—1894*, 212.

54 Between Jersey City and Water Company, October 14, 1909, 6731.

55 The Paterson Public Library maintains a collection of these annual reports in its rare book room.

56 Between Jersey City and Water Company, February 4, 1909, 4923.

57 "Deaths Leal," 208.

58 Leal, "Report of City Physician," 1887, 94–6.

59 Leal, "Annual Report of Board of Health 1891," 209–13.

60 Beveridge, "Mayor's Message," 11–12.

61 Leal, "Annual Report of Board of Health 1891," 214.

62 Brock, *Robert Koch*, 229–32.

63 Blanck, "Cholera in New York Bay."

64 *New York Times*, September 16, 1892.

65 Leal, "Annual Report of Board of Health 1893," 254.

66 "Deaths Leal," 208–9.

67 Leal, "Annual Report of Board of Health 1894," 238.

68 Cedar Lawn Cemetery, Lot and Grave Index, Leal.

69 Leal, "Annual Report of Board of Health 1894," 242.

70 Ibid., 243–4.

71 Ibid., 245.

72 Leal, "Isolation Hospitals."

73 *Transactions of the Medical Society of New Jersey—1900*, 243.

74 Ibid., 250–1.

75 Ibid., 251–3.

76 We now know from the study of bacterial, virus, and protozoan pathogens that a one-minute rolling boil will kill all pathogens of concern.

77 Leal, "Annual Report of Board of Health 1897," 256–7.

78 Between Jersey City and Water Company, April 18, 1907, 3029.

79 Ibid., 3031.

80 Curts and Hazen, *Joint Committee on Sewage Disposal*, 10.

81 Leal, "Annual Report of Board of Health 1898," 242–8.

82 Ibid., 249.

83 Department of Public Health, City of Newark, N.J., 106.

84 Leal, "Annual Report of Board of Health 1898," 212.

85 Leal, "Epidemic of Typhoid Fever," 166–71.

86   Ibid., 168.

87   Leal, "Epidemic of Typhoid Fever," 169–70.

88   Board of Health, *Statement of Mortality*, November 1899; Board of Health, *Statement of Mortality*, December 1899.

89   *New York Times,* December 19, 1899.

90   *Transactions of the Medical Society of New Jersey—1900*, 243.

91   "Deaths Leal," 208.

92   Between Jersey City and Water Company, February 4, 1909, 4923.

93   Ibid., 4924; in the spring of 1902 "when Mr. Gardiner became President," Between Jersey City and Water Company, December 27, 1906, 2257.

94   Between Jersey City and Water Company, April 18, 1907, 3028.

95   A sign-in book for the Little Falls treatment plant that was found in the Passaic Valley Water Company museum on the treatment plant site showed Leal signing in twice per week from March 5, 1903 to September 30, 1904.

96   Between Jersey City and Water Company, April 18, 1907, 3028, 3032.

97   "Deaths Leal," 208–9.

98   *Twentieth Annual Report Board of Health Rhode Island*, 1891.

99   *Hartford Courant,* March 29, 1910.

100 Peck, "Experiments with Hypochlorite of Lime," 395.

101 Milligan, "Mechanical Filtration Plant," 61.

102 Hooker, *Chloride of Lime*, 28.

103 *Paterson Morning Call*, March 14, 1914, 1.

104 *New York Times*, March 14, 1914.

105 I was able to locate Scott Leal Morehouse, grandson of Graham A. Leal and great-grandson of John L. Leal. We met in Washington, D.C., on October 25, 2009. Scott was exactly the right Leal descendent to speak with. He has kept track of members of his extended family, and he filled me in on who was living where. He personally knew "Baba," Graham A. Leal, who lived until 1970. Until I explained to him Dr. Leal's impact on drinking water disinfection and public health, Scott had no idea what his great-grandfather had accomplished. Graham Leal never spoke of his father except to say that he was a physician. One has to wonder if John Leal's obsession with his work affected his familial relationships.

IN 1998

"In 1998," an editoral cartoon that appeared in the *Jersey City Evening Journal* on April 30, 1909.

Undated photos of John L. Leal during a holiday at the Atlantic shore. *Photos reprinted with permission of Leal-Morehouse family.*

Amy L. Arrowsmith Leal and Graham Arrowsmith Leal, circa 1891. *Photo reprinted with permission of Leal-Morehouse family.*

John L. Leal with a doctor's bag, in front of an unidentified building. Date unknown. *Photo reprinted with permission of Leal-Morehouse family.*

John L. Leal at work in an office setting. Date unknown. *Photo reprinted with permission of Leal-Morehouse family.*

George Warren Fuller when he was President of the American Public Health Association in 1928.

George Warren Fuller (center, with left foot on the step), as chief chemist and bacteriologist for Louisville Water Company and the staff of chemists, bacteriologists, and engineers to hired to assist him, 1895–1896. Upper row, left to right: Robert Spurr Weston, assistant chemist; William, the janitor; George A. Johnson, bacteriologist; Harold C. Stevens, assistant engineer. Bottom row: Reuben E. Bakenhus, assistant engineer; J.W. Ellms, assistant chemist; Charles L. Parmelee, assistant engineer; Hibert Hill and Benton (first name unknown), assistant bacteriologists.
Photo courtesy of Louisville Water Company.

An outdoor privy, circa 1893, overhanging Stony Brook, near Lowell, Mass. Source: Sedgewick, William T. 1893.

The lower gatehouse for the Boonton Dam, built in the early 1900s, is still functioning today and, along with the dam, is an American Water Works Association Water Landmark. Photo courtesy of Jersey City Water Works.

A 1919 editorial cartoon drawn specially for *The American City* magazine by Zim (Eugene Zimmerman, 1862–1935).

George Warren Fuller's filtration experiment at Louisville Water Company, circa 1896. Photo courtesy of Louisville Water Company.

Members of Fuller's filtration team (bacteriologists George A. Johnson at the microscope and Hibert Hill, right) in the laboratory at Louisville Water Company, circa 1896. Photo courtesy of Louisville Water Company.

7

Fuller—The Greatest Sanitary Engineer

*"One of the most important points in water purifica-
tion is the removal of disease-producing germs, since it
has become clearly established that high death-rates from
diseases, caused by germs which can live in water, result
largely from drinking polluted water."*
— Fuller, "Progress in Sanitary Science," 73-4

George Warren Fuller was, quite simply, the greatest sanitary
engineer of his time, and his time as a practicing engineer
was long—lasting from 1890 to 1934. We have not seen his like
since. How did he reach the pinnacle of his field? What early influences led him to his career path?

Fuller's ancestors settled in the rural areas of Massachusetts.
His great-great-great-great-grandfather, Ensign Thomas Fuller,
was a member of the Massachusetts Bay Colony. On January 20,
1619, Ensign Fuller was baptized in Redenhall, England, and he
immigrated to America sometime around 1640-2.[1]

Ensign Fuller was active in his community and held several
official positions in the colony. He was chosen as one of the survey-
ors of the town around 1660. It was reported that he helped lay out
the best road between Cambridge and Dedham.[2]

George Warren Fuller was born in Franklin, Massachusetts,
on December 21, 1868—10 years after the death of Dr. John Snow
and the birth of Dr. John L. Leal. He was the son of George Newell Fuller and Harriet Martha Craig.[3] George Newell Fuller was
born in Franklin, Massachusetts, on November 22, 1819. Harriet
Martha Craig was born on February 2, 1841, and grew up near
Leicester, Massachusetts.[4] She married George Newell Fuller on
November 15, 1866, when he was 46 and she was only 25. They settled down in the Franklin–Medway area of rural Massachusetts
for a quiet life of farming on the ancestral Fuller family property.
They had two children, George Warren and Mabel B., who was
born in 1876.[5]

Place names in Massachusetts have changed over the past several hundred years as town boundaries expanded or contracted and new towns were carved off from old ones. Towns that figured prominently in Fuller's history—Dedham, Franklin, and West Medway—were all located in the same general area about 10 to 25 miles southwest of Boston.

One report describes Fuller's early education:

> "George Warren Fuller was at the head of his class when he attended the Dedham schools. His scholarship was, of course, a source of great satisfaction to his mother. At sixteen he passed the examination for entrance at MIT [Massachusetts Institute of Technology] but, his father having died a few weeks before, it was thought best for him to have a fourth year in high school. . . ."[6]

After his father's death on May 3, 1885, Fuller's mother moved 2,500 miles away to Claremont, California, where she lived until she died in 1915.[7] Fuller's first wife, Lucy Hunter, was born in October 1869 and died far too young on March 19, 1895.[8] She was 18 years old and Fuller was 20 when they were married. Fuller was only in his second year at university (1886–1890).

The couple had one son, Myron E. Fuller, born in Boston on June 4, 1889.[9] Little is known about the marriage. Fuller was issued a passport on May 2, 1890, for his upcoming trip to Germany to continue his studies.[10] There is no record that Lucy applied for a passport for herself or Myron or that they accompanied Fuller to Germany. Massachusetts death records listed her cause of death as "enteritis," a general term used for diseases caused by the ingestion of pathogens from food or water. The death records listed her as "married," which meant that her marriage to Fuller had not been dissolved before her death in 1895,[11] but there is no evidence that Fuller lived with her and their son after 1889.

A 1910 census report shows that Myron was living with his father in Summit, New Jersey. One documented connection between Myron and his father was mentioned in the preface of Fuller's 1912 book, *Sewage Disposal*.[12] Fuller acknowledged Myron (who was 22 at the time) for creating the book's index. One source showed that the firm of Fuller and McClintock employed him from 1911 to 1916 and again from 1919 until at least 1922.[13] In 1918, Myron registered for the draft and listed his occupation as civil engineer.[14] The same reference showed Myron working for the city of Philadelphia in the Bureau of Surveys—the same occupation as his great-great-great-great-grandfather, Ensign Thomas Fuller. He was living in Philadelphia with his wife and one child.[15]

While Fuller was in Louisville, Kentucky, conducting his celebrated filtration investigations, he met Caroline L. Goodloe, who came from a respected Louisville family. Fuller married her in Louisville in November 1899;[16] they were both 31 years old. In May of 1900, husband and wife went on a trip to Europe—a somewhat delayed honeymoon.[17]

Their son, Kemp Goodloe Fuller, was born on March 10, 1901.[18] A second son, Asa W. Fuller, was born on November 11, 1903, while they were living in New York City.[19]

Tragically, Caroline Goodloe Fuller died on June 21, 1907, while her husband was heavily engaged in numerous water and sewage disposal projects all over the United States. At the time of her death, Fuller was living at 309 West 84th Street in New York City with her and their sons. She was 38 years old.

The 1910 census form showed that Fuller was living at 160 Boulevard in Summit, New Jersey, with Alice C. Goodlow [sic], identified as his sister-in-law; Mary L. Goodlow [sic], identified as his mother-in-law; and his three sons, Myron, Kemp G., and Asa. Fuller's in-laws had come up from Louisville to help him raise the boys. Also listed at the same residence was a guest, Grace F. Thomson, age 43, born in China of English ancestry and claiming a trade of metal-working. Three servants (two Irish and one Greek), rounded out the household. The census form showed Fuller as widowed, so by 1910 he had not remarried.[20]

Several accounts suggest that George Warren Fuller was, in many ways, a big man. Physically, he was tall. A colleague described him as over six feet tall,[21] but passport application forms that Fuller filled out showed his height as 5 feet, 10 inches.[22] Photographs of him from 1903 until at least 1928 showed that he was, to use a descriptor from the time, stout.[23] One description had him at 285 pounds with a size-eighteen collar.[24]

His hair was dark brown and, in the style of the day, slicked down and parted in the middle. As time marched on, he began to gray at the temples, and then the gray seemed to take over his thinning head of hair. He was clean-shaven, except for his days in Louisville during the filtration studies when he sported a bushy mustache. He had blue eyes that could bore into someone who did not please him and twinkle when he was trying to charm a lady. The round spectacles he always wore did not detract from the intensity of his eyes.

FULLER WAS TRAINED AT MIT, graduating at the head of his class in 1890 with a Bachelor of Science degree in chemistry—not engineering.[25] His dominance in the field of sanitary engineering was the result of his intelligence and his immersion in and redefinition of the emerging discipline. He attended MIT with two men who would figure prominently in his career and who played an important role in the development of water treatment processes: Allen Hazen attended a special one-year course in chemistry at MIT, finishing in 1888,[26] and George C. Whipple graduated from MIT in 1889. As luck would have it, Fuller happened to attend the first school of public health, although it was not called that at the time. Through William T. Sedgwick, head of MIT's biology department, the school had close ties with the Massachusetts Board of Health. Fuller became closely associated with Sedgwick, who had a tremendous impact on his early career and his life.[27]

All successful people with an iota of humility acknowledge times when luck played a huge role in putting them on the path to success. Fuller's luck was being chosen as one of Sedgwick's students. "He [Sedgwick] picked his young assistants with care; he inspired them and criticized them. He gave them the opportunity and the stimulus to work."[28]

Fuller did not leave any archives or journals, and only a few of his letters have survived. His words do not reveal what he experienced in the rarefied atmosphere Sedgwick created at MIT. However, there are detailed and personal descriptions of what Sedgwick was like as a teacher and how he influenced his students.

> "Sedgwick was an inspired and inspiring teacher. As a member of a group, one noted only a genial personality with the forehead of a scholar, but on the platform he was a notable figure. He was an exponent of what is today almost a lost art—that of lecturing. His addresses to large classes were apparently extemporaneous but actually prepared with meticulous care to produce a desired effect. In these lectures, as in his writings, he showed an unusual gift for the apt and perfect phrase. 'The so-called 'germ' theory of disease is the child of fermentation and the grand-child of the microscope.' Or again, 'a [slow] sand filter is more than a mechanical strainer, and more than a chemical furnace, it is a breathing mechanism.'[29]

"Sedgwick taught us to think clearly and to think deeply. As I look back over the years there were two qualities of Sedgwick's mind and personality which made him a teacher of truly gigantic stature. The first of these qualities was intellectual. It involved an almost passionate interest in exactitude as to words. . . . The second—and less obvious but more important—quality of Sedgwick as a man and as a teacher was a moral and spiritual one."[30]

"Three of Sedgwick's students [Edwin O. Jordan, George C. Whipple, and Charles-Edward A. Winslow] have said that 'The really great teacher must give to his pupil three different things: a vision of the subject in hand in its relations to the evolving universe, a vigorously honest method of thinking and working so that the truth may be adhered to and if possible advanced, and an enthusiasm for service which will prove better even than the desire for fame as the compelling motive to make men 'scorn delights and live laborious days.' These three gifts, and in rich measure, Sedgwick bestowed upon his pupils.'"[31]

SHORTLY AFTER GRADUATING FROM MIT, Fuller traveled to Europe. The United States was still learning about bacteriology and water treatment from Europe. It is not known exactly what event or influences prompted Fuller to pack up and move to a foreign country, but it is likely that Sedgwick encouraged him to go to Germany to get hands-on experience with the engineering and operational details of slow sand filters.

In Berlin, Fuller also had a golden opportunity to learn about the exploding field of bacteriology at the center of the universe for those advances.[32] He was exposed to the revolutionary developments being led by Dr. Robert Koch at the Hygiene Institute of the University of Berlin.

"After my graduation there I spent six months in Europe, making studies along those lines [chemistry, mathematics, and engineering], including a course in bacteriology at the Hygienic Institute of the University of Berlin; also spent some time in the office and laboratory of the engineer of the Berlin Water Works, particularly with reference to the operation and tests of the old [Stralau] sand filters."[33]

In his many descriptions of his time in Berlin, Fuller never mentioned meeting Dr. Koch. In the fall of 1890, Koch was preoccupied with developing what he thought might be a cure for tuberculosis—"tuberculin." The treatment was hugely controversial, and tremendous pressure bore down on Koch from his colleagues, the press, and the public. In October 1890, Koch requested to be relieved of his official duties at the Hygiene Institute so he could focus his research on tuberculin. In the winter of 1891, wanting to escape the pressure cooker of Berlin, Koch went on an extended vacation in Egypt.[34]

Numerous sources have mentioned Fuller's stay in Berlin at the offices of Carl Piefke, who was the equivalent of chief engineer of the Berlin Water Works.[35] Piefke's early work on slow sand filtration was widely reported, including in journals that published translations and abstracts of his important papers.[36]

Fuller and Piefke evidently developed a personal relationship while Fuller was in Berlin. In 1892, just two years after Fuller left Germany, Piefke sent him an extraordinary letter commenting on the political situation in Germany: "Here in Germany, conditions have changed to be very uncomfortable. Militarism and bureaucracy are spreading over our very existence, and despite all effort, no success rewards him who does not occupy a high rank within this hierarchy." Twenty-seven years later, Fuller read the letter at a 1919 meeting of the New Jersey Sanitary Association—shortly after Germany demonstrated its militarism and bureaucracy to the world. After reading the letter, Fuller added that the later years of Piefke's career were unhappy. Piefke lost his position at the Berlin Water Works and ended up as a field technician collecting groundwater data.[37]

There is some disagreement about how much time Fuller spent in Berlin. Fuller said he was there for six months, but other accounts said he was there for one year.[38] The passport he applied for on May 1, 1890, was issued the next day, and he probably left for Germany shortly thereafter. He stated on the passport application that he intended to return to the United States within one year.[39] In one of his obituaries, he is identified in a detail of a group photo as "a Student in Berlin in 1890."[40]

The name George Fuller appeared on a passenger manifest of the steamship *State of Nebraska*, which sailed from Glasgow, Scotland, and arrived in New York City on November 5, 1890.[41] Fuller may have traveled from Berlin to Glasgow to see the filtration works in that city. The first slow sand filters supplying piped water to a city were built in Glasgow in 1807.[42]

In biographical notes on George Warren Fuller written by George C. Whipple and dated August 13, 1915, Whipple stated that Fuller joined the staff at the Lawrence Experiment Station in November 1900.[43] Thus, it seems likely that Fuller was in Germany with Piefke for about six months.

IMMEDIATELY AFTER RETURNING to the United States, Fuller went to work at the Lawrence Experiment Station, which became known as the leading research facility for investigating sewage treatment and drinking water filtration.

> "Upon my return to this country I spent five years in the employ of the Massachusetts State Board of Health. Most of that time was at Lawrence, Massachusetts, where a well known experiment station was operated. A large portion of that time I was in charge of that station which had for its purpose the determination of the best and most economic means of purifying water and sewage, under various local conditions in that state."[44]

Fuller was again fortunate that circumstances had created a bold new research opportunity for him. By 1880, the population of the Commonwealth of Massachusetts had grown to the extent that pollution of streams and rivers was a serious problem. In 1884, the commonwealth appointed the Massachusetts Drainage Commission to investigate methods to reduce pollution and protect water supplies. The commission's report in 1886 urged the legislature to appoint a commission with the power to control discharges of sewage and industrial waste and to advise communities on selecting safe water supplies. The Massachusetts legislature adopted the commission's recommendations and gave the state Board of Health all necessary powers, as recommended in the report.[45]

The legislature's action was extraordinary and farsighted. In effect, these lawmakers created a mechanism to develop expert sanitary advisers who would investigate treatment methods and advise cities in the state on the proper disposal of wastes and use of water supplies. They also provided sufficient money to accomplish these tasks.[46]

The state Board of Health was reorganized to assume these responsibilities, and a respected hydraulic engineer, Hiram F. Mills, became a member. Mills had access to a building on the left bank of the Merrimack River in Lawrence, and he converted it to a testing facility. One person who knew Mills noted: "And since, with him, to think was to do, this soon became the experiment station of the Board."[47] One of the first projects at the experiment station was to test whether intermittent filtration of sewage through granular filters would effectively stabilize contaminated water supplies and render them harmless.[48]

"This station, the Lawrence Experiment Station, [was] the first of the kind in America, if not in the world. . . ."[49] Creating such a facility was truly groundbreaking. It had access to the talents of Sedgwick, Hazen, and Fuller, and they were soon conducting investigations of sewage treatment and drinking water filtration that rivaled any water treatment research being done in the world.

Fortunately for historians, one of the requirements of the enabling legislation was for the staff to prepare a report each year on the progress of their investigations. These reports can be found in the annual reports of the Massachusetts Board of Health to the state legislature, Public Document 34.

A summary of the annual report for 1890, which has been published as an abstract, did not mention Fuller. The summary covered progress in the investigations of sewage filtration and precipitation.[50] Two reports written by Fuller were published in the twenty-third annual report, which covered a two-year period—November 1, 1889, to December 31, 1891.[51] Allen Hazen's report summarized the sewage purification and water filtration work conducted during these two years.

In the first of Fuller's two reports, he was identified as the "biologist in charge." His biological work was related to a variety of topics including:

- Standardizing the alkalinity of nutrient gelatin media so that total bacteria counts could be made with greater precision,
- Determining the survival of the typhoid bacillus in the Merrimack River,
- Devising a means to identify the different species (or class of species) of bacteria in raw and treated water,
- Determining that many of the bacteria measured downstream of the test filters were the product of regrowth in the pipes, and
- Determining the concentrations of *Bacillus colon communis* (the indicator of sewage contamination at the time) in Lawrence's tap water over time and distance.

The small amount of data in the report proved that the tap water in Lawrence was severely contaminated. Concentrations of Bacillus colon communis (now known as total coliforms) were in the neighborhood of 100,000 per hundred milliliters, astronomically high by today's standards.[52]

Developing laboratory methods to differentiate the typhoid bacillus from other bacteria was the focus of Fuller's second report.[53]

In the next annual report, covering the work done in 1892, Fuller was identified as being in charge of the biology department. Allen Hazen reported on the progress with sewage filtration studies, and Fuller reported on investigations of slow sand filtration for purifying drinking water.[54]

Fuller's first summary of his filtration work described his test filters in detail. The tanks were constructed of galvanized iron measuring 6 feet deep by 20 inches in diameter. The water was introduced into the top of the tank, flowed thought the filter material, and exited the bottom through a faucet. Sand of different sizes was typically used as the filtering material. The sand was supported by several inches of gravel of varying sizes. In the case of Filter No. 8, the filter used for many of the experiments, 5 feet of filtering material was used. Loam or fine soil was placed on top of the sand in depths of a few inches. Typically, 6 inches of water covered the top of the loam.[55]

In a summary of the filtration research conducted in 1892 and published in 1893, Fuller provided a broad perspective on the efficacy of slow sand filtration.

> "One of the most important points in water purification is the removal of disease-producing germs, since it has become clearly established that high death-rates from diseases, caused by germs which can live in water, result largely from drinking polluted water. The results of the Lawrence experiments show that it is possible to construct filters which will purify at least 2,000,000 gallons of water per acre daily and remove more than 99 per cent of the bacteria in the unfiltered water."[56]

Fuller's filtration rate was somewhat greater than but not a huge improvement over the filtration rate achieved by Piefke in Berlin—1,730,000 gallons of water per acre per day. However, Piefke's slow sand filters contained only 24 inches of sand, and he never reported the percentage of bacteria removal. After a typhoid fever outbreak in Berlin from January to April 1889, pilot filters of the same type used by Piefke were constructed and tested at the Lawrence Experiment Station. These slow sand filters were not found to be "germ tight."[57]

In the annual report covering work done in 1893, Fuller noted that Allen Hazen had left the Experiment Station on March 1, 1893, and that he (Fuller) was in charge of the station.[58] His filtration report focused on studies that examined coarser filter materials and higher rates of filtration (that remained in the realm of slow sand filtration). Filtration rates up to 9,000,000 gallons per acre per day resulted in bacteria removal efficiencies greater than 99.5 percent.[59]

As a result of research at the Experiment Station, a slow sand filter was installed to treat the water supply for the city of Lawrence in 1893. The filter design was based on intermittent operation instead of continuous operation, and this operational mode was not adopted by any other U.S. city. Even with the nontraditional design, the filter dramatically reduced the typhoid fever death rate in the city.[60]

Fuller left the Lawrence Experiment Station in 1895 at the age of 26, but his work there had prepared him for the challenges of his next two assignments. He now had the skills to plan experiments, build apparatus, collect samples, analyze samples for chemical and bacteriological parameters, summarize the data in tables and figures, integrate the data to figure out what it meant, and write reports describing his findings. His ability to manage a team of investigators would be honed in two subsequent filtration studies that forever changed water treatment in the United States.

BETWEEN 1895 AND 1899, Fuller was in charge of the two most important water filtration studies at that time—at Louisville, Kentucky, and Cincinnati, Ohio. The reports from these two investigations are milestones in our understanding of water filtration, and they launched Fuller's fame as a preeminent sanitary engineer.

Today, the filters studied at Lawrence Experiment Station would be classified as pilot-scale filters. Long experience has proved that the best way to investigate and implement a new technology is to first study it in the laboratory under controlled conditions (laboratory- or bench-scale testing). After the basic principles of the new technology are understood, larger treatment systems can be built to treat the actual water supply under consideration (pilot-scale testing). Typically, pilot-scale facilities can be constructed in a large room.

Before blundering forward and building a treatment plant, a careful engineer will conduct demonstration-scale studies using full-scale treatment systems in a test mode with the target raw water as the source of supply. The Louisville and Cincinnati filtration studies were definitely demonstration-scale studies. Finally, armed with all of the knowledge gleaned from these investigations, the prudent engineer will design a full-scale treatment plant. For the radical new system of mechanical filtration, Fuller's success with designing a full-scale plant would come much later with the treatment facility at Little Falls, New Jersey (see Chapter 4).

Fuller's investigation of mechanical filtration to treat Louisville's water supply was preceded by years of experimentation by others who had tested slow sand filtration on Ohio River water. Louisville was one of the last places where slow sand filtration would work on Ohio River water because of the heavy sediment load the river accumulated farther upstream. Muddy was not an unkind description of the river water available to the city. Also, more than four million people lived in the watershed of the Ohio River upstream of Louisville, and sewage from their activities was continuously discharged into the river. Louisville suffered from all of the problems of the Sewer Pipe–Water Pipe Death Spiral.

Mechanical filtration was characterized by many false starts, with competing companies and patented filters vying for a share of the market. The Louisville study was a chance for the leading filters of the day to be tested side by side.

The water works for the city were operated by a private enterprise, the Louisville Water Company. To conduct the needed studies, Fuller was hired by the company as chief chemist and bacteriologist. The company also hired a significant staff of chemists, bacteriologists, and engineers to assist Fuller. A famous photograph of the staff for the Louisville study showed Fuller in the middle with nine men surrounding him. Two of these men, George A. Johnson and Joseph W. Ellms, would later play important roles in the implementation of chlorine disinfection.[61]

Fuller began his duties on October 1, 1895. Three filter companies provided a total of four filters for testing, but the Louisville studies did not produce the information needed to design the optimal treatment for the turbid Ohio River. No engineer at this point recognized the critical importance of a presedimentation step—to remove the mud—followed by the addition of a coagulant and more sedimentation before the water was run through a filter.[62] But as with all early research in which an answer of "no" is sometimes as helpful as a successful study, the Louisville studies taught Fuller valuable lessons, which he would apply to the test platform in Cincinnati.

One conclusion that is never mentioned by authors who summarize the findings presented in the Louisville report is Fuller's statement that the best way to add sulphate of alumina (alum) to water for purposes of coagulation was to dissolve a known weight of alum into a known volume of water and then feed that solution by gravity into the water to be treated.[63] This protocol sounds simple, but simplicity and reliability were part of the success of the alum feed system at the Little Falls mechanical filtration plant that Fuller designed years later. This chemical feed system also laid the foundation for the simple application of chlorine to water at the Boonton Reservoir: dissolve a known weight of chloride of lime in a known volume of water and use gravity to feed the solution into the water to be treated.

Almost every history of drinking water disinfection mentions that a device for producing chlorine gas was tested at Louisville, but Fuller's report devotes only a few lines to the Jewell chlorine-generation device.[64] Moses N. Baker gave a more detailed description of the tests in *The Quest for Pure Water* and published a sketch of the Jewell chlorine gas generator provided by the inventor.[65] The test merely demonstrated that chlorine gas could be generated and dissolved in water. None of the test descriptions described the chlorine dosage. The device was operated for only a few days, and no bacteriological testing was done to determine the efficiency of disinfection. Nevertheless, Baker thought Jewell's test was a milestone. "It seems safe to say that this was the first use of chlorine gas to reduce the bacterial content of the effluent from a working-scale filter."[66]

William M. Jewell, a chemist with the O.H. Jewell Filter Company, noted in his communications with Baker that he had to discontinue his testing of the chlorine gas apparatus because he was told that using chlorine in the filter plant was jeopardizing the credibility of the entire project. Thus, the fear of chemicals and the controversy about adding chemicals to drinking water were so great that they even affected the course of scientific and engineering studies.[67]

The report of the Louisville research has been called a classic and has, rightly, received much praise. Right after the book was published, the most entertaining characterization of the work appeared in *The Lancet,* where British exuberance was exercised to its fullest splendor. "In short, it is a stupendous piece of work, the report containing thousands of practical and experimental results relating to filtration, sedimentation, coagulation, and the degree of efficiency of the purifying processes investigated." Further on in the review, *The Lancet* article called the work "admirable and colossal."[68]

The author of another friendly critique of the report observed that "One of the chief difficulties in employing this coagulant [aluminum sulfate] is the regulation of the quantity of the added chemical to known volumes of the water which is constantly exhibiting variations in composition at different seasons."[69] This was true in 1899, and it is true today. Hundreds of researchers have spent untold hours seeking a simple way to automatically pace the feed of aluminum sulfate. Today, dosage determinations for coagulants cannot be reliably automated strictly on the basis of raw water conditions. Feeding the chemical can easily be automated, but determining the dosage usually depends on the history of the coagulant's use and manual bench-scale studies.

Tests in Louisville concluded on August 1, 1897. The 500-page book that served as the project report was published in 1898.[70]

FULLER LEARNED A GREAT LESSON from the filtration studies in Louisville. For the Cincinnati work, he preceded the filtration steps with sedimentation. He started work on the project on December 3, 1897—four months after finishing the testing at Louisville and only weeks after completing the Louisville report. Originally, only a slow sand filter was going to be tested in Cincinnati. However, after the use of presedimentation and coagulation proved successful, a small mechanical filtration apparatus was brought to the test site and operated for several months.

Once again, George A. Johnson and Joseph W. Ellms, along with four others, assisted Fuller in testing and analyzing water samples. The demonstration-scale slow sand filter plant had a flow of 100,000 gallons per day.

The report on the Cincinnati studies concluded that slow sand filters without presedimentation were not feasible for treating water from rivers like the Ohio. Significant rainfall events caused high-volume flows in the watershed to transport large amounts of clay and silt to the river. These solid materials had to be removed, but slow sand filters alone could not do the job. Even with a three-day nominal detention time and presedimentation removing about 75 percent of the influent solids, the slow sand filters did not perform satisfactorily. A high fraction of clay content in the solids clogged the slow sand filters, requiring them to be manually cleaned too frequently. Also, three days of sedimentation time required a large basin occupying a lot of land. Most utilities did not have such large tracts of land near their water supply intakes. Fuller thought a detention time of one to two days might be satisfactory.[71]

To effectively remove the clay particles, Fuller found it essential to add a coagulant (alum being best) to the water before or after the presedimentation basin in order to minimize the solids load on the slow sand filter.

However, given that adding a coagulant was necessary, Fuller found that mechanical filters would be more economical if they were preceded by an intermediate sedimentation basin with a detention time of 0.5 to 6 hours. Interestingly, the bacteriological testing showed that neither the slow sand nor the mechanical filters was "germ proof." In other words, water treated by these filtration methods did not remove all of the bacteria measurable with the analytical methods of the day. It would take the addition of another barrier (chlorine) to eliminate bacteria from the effluent of a mechanical filtration plant.

The Cincinnati filtration report was submitted to the city on January 31, 1899.

IN 1899, AT THE TENDER AGE of 31, Fuller began a consulting practice at 220 Broadway in New York City, describing himself as a—". . . consulting expert on water purification and sewage disposal."[72] In 1901, he joined another giant in the field of water engineering, Rudolph Hering, to found the firm of Hering and Fuller. They moved into offices down the street at 170 Broadway, where Fuller maintained his consulting practice until his death in 1934. At the second Jersey City trial in 1909, Fuller summarized his consulting experience up to that point.

> "For the past ten years I have been in private practice in New York City, having devoted special attention to the matters which I have mentioned, that is, the question of the quality of water supply, methods of water purification, sewerage and sewage disposal. For the last eight years I have been in partnership with Mr. Rudolph Hering, and I suppose, during that time, we have looked into some sixty or eighty projects of magnitude in this

country, including those, for a large majority, of the more important cities of this country. We are designers and builders, that is, supervisors of construction, and operators of purification plants."[73]

"Q: Mention a few of the cities, the problem[s] of which you have helped to solve?

A: Well, New York city, with reference to the Croton water supply, and with reference to the design for the purification treatment works of the new Catskill supply. I was a member of the Engineering Commission appointed by the United States Senate, with reference to purification of the Washington water supply and I have been connected with the water purification plants at Columbus, Ohio, Indianapolis, Indiana, Cincinnati, Ohio, and New Orleans, Louisiana, Bangor, Maine, Burlington, Vermont, New Haven, Connecticut, Little Falls plant of the Passaic river, and the New Milford plant [of] the Hackensack water company [on] the Hackensack river; water plant at Columbia, Pennsylvania; water plant at York, Pennsylvania; South Pittsburgh, Pennsylvania, New Castle, Pennsylvania, Youngstown, Ohio, Muncie, Indiana, Burlington, Iowa, Norfolk, Virginia, and quite a number of others; those are the large ones for water."[74]

Joining the forces of the "young Turk," who had new ideas and modern skills, with the dean of sanitary engineers, Rudolph Hering, was pure genius. In 1901, Fuller was 33 years old, and Hering was 54. Hering had done it all in his career. In the late 1800s and early twentieth century, he provided fundamental water supply plans to New York, Chicago, Washington, New Orleans, Columbus, Montreal, and Philadelphia. Hering also devoted enormous amounts of time to addressing the problems of refuse disposal—a neglected area of sanitary engineering. Of course, he was involved in numerous professional societies and served on several committees of APHA, in which both John L. Leal and George Warren Fuller were active.[75] Hering would become president of APHA in 1913.

Fuller and Hering ended their partnership in 1911, 10 years after forming it, and Fuller opened a new consulting firm under his name alone. In 1916, Fuller and James R. McClintock founded the firm Fuller and McClintock, and Fuller ran that firm until 1934. "Mr. Fuller advised more than 150 cities, commissions, and corporations with respect to their water supply and sewerage problems. . . ."[76] One assignment was particularly long-term and important. From 1906 until his death, Fuller was a consultant to the New York Board of Water Supply for a variety of projects related to the Catskill reservoir and water supply system.[77]

James C. Harding, an employee of Fuller and McClintock, worked for Fuller for ten years and was Fuller's colleague in engineering projects for an additional seven years. His article on Fuller provided valuable insights into Fuller's personality and described events that were memorable.

> "He was calm and unruffled. It took quite a bit to get him really steamed up. When this did happen, it was best to get as far away from him as possible. I think the only occasion that I ever saw him really embarrassed and upset was one time when we made an inspection of a new sewage treatment plant that we had built for the city of New Rochelle [New York]. Present at this inspection was the mayor, a very dapper, well dressed individual. . . His honor was standing at the point where the screenings dropped into the hopper, with Fuller off to one side explaining just how things worked. Unfortunately, the operator chose that moment to throw the air into the wrong ejector. There was a great roar, and those who knew what was going to happen ducked. Of course, the mayor did not know, and he was plastered from the waist up with a layer of filth 2 in. thick. After his face was sufficiently wiped off . . . he called the operator a number of foul names finishing up with 'Democrat,' told him he was fired, and then turned to the highly embarrassed Fuller saying these words, 'Great engineer, huh—Fuller and McClintock—Fuller [s#!+]!'"

BOTH HERING AND FULLER TESTIFIED at a public hearing held on January 4, 1901, to help the U.S. Congress decide on the best treatment method for the Potomac River supply to Washington, D.C. By this time, the choice involved only two alternatives—slow sand filters or mechanical filters—and was driven by the quality of the surface supply to be treated. After the hearing, Hering and Fuller teamed up with Allen Hazen to write a short report to the chairman of the U.S. Senate Committee on the District of Columbia, James McMillan. The report was sent on February 18, 1901. The team of incredibly talented engineers made the "safe" recommendation. They recommended slow sand filters for the Washington, D.C., water supply.[78]

Although Fuller had just completed the Louisville and Cincinnati filtration studies, they had been conducted on an extremely turbid water supply. Hazen's just-finished studies for Pittsburgh had ended in a recommendation for slow sand filtration. The water of the Potomac did not contain particularly high concentrations of suspended solids, but the team recommended the treatment method with the fewest moving parts and the greatest simplicity of operation.[79]

Fuller had not yet designed and built the Little Falls mechanical filtration plant, which turned out to be so successful. Had the trio received the Washington, D.C., assignment after 1910, there is little doubt they would have chosen mechanical filtration.

Details of the design and processes of the Little Falls plant were covered in Chapter 4. In May 1900, the East Jersey Water Company asked Fuller to assess the feasibility of building a filtration plant at the Little Falls site.

After Fuller obtained some plans and cost estimates for the filtration equipment from the Jewell Filter Company, the water company hired him to oversee the design of the plant in September 1900. At this time he was operating as a consulting engineer under his own name, but before construction of the plant was finished, Fuller and Hering would be partners. Delivery of treated water from the plant began in September 1902.[80]

The plant was known for many firsts.[81]

- It was the first mechanical filtration plant that incorporated coagulation and sedimentation prior to filtration for the purposes of solids removal and bacteriological protection.
- Filters were constructed out of reinforced concrete and laid out in a rectangular pattern, initiating the economical design of common-wall construction used ever since.
- Filters were cleaned by means of a combined air-and-water backwash applied through the same underdrain structure.
- Filter valves were operated hydraulically for operator convenience and ease.

The feed system that Fuller designed for alum, the critical coagulant in his mechanical filtration process, was based on his experience at Louisville. The chemical feed design would be a crucial element in the success of chlorine application at Boonton Reservoir.

Commenting on other cities' adoption of Fuller's Little Falls design, one author pointed out that "The Little Falls plant did not, however, lead to the sudden emergence of a generic technology."[82] Slow sand filter plants continued to be recommended and built. The last, large slow sand filter plant was built by Philadelphia to treat water from the Delaware River.

In his 2006 book *Water Treatment Unit Processes,* civil engineering professor David W. Hendricks described the development of New York City's water supply and quoted Fuller expressing some reservations about recommending mechanical filtration in 1907 because only the Little Falls and Hackensack plants had been built by then. By 1912, Fuller was comfortable recommending mechanical filtration for the Croton supply because the state of the art of mechanical filtration plant design was more advanced.[83]

Fuller's reluctance to recommend a new filtration technology five years after a plant using this technology had been successfully built and operated says a lot about the conservative nature of engineers. Fuller wanted to see others adopt and successfully implement his design before recommending that New York City invest millions of dollars constructing such a plant. Besides saving money, the most important reason for conservatism by engineers is that their designs directly affect public health.

Leal and Fuller team up. On July 19, 1908, Dr. John L. Leal hired Fuller to build a chloride of lime feed system at Boonton Reservoir to treat the water supply for Jersey City, New Jersey, as recounted in Chapter 1. Fuller designed and built the first continuous chlorination system for drinking water (capable of treating 40 million gallons per day) in 99 days. The need for haste stemmed from an impossible deadline set by the New Jersey Chancery Court. However, as Fuller's testimony at the second New Jersey trial shows, the firm of Hering and Fuller was extremely busy. Fuller undoubtedly pushed other projects aside to take on this important task.

In 1916 and throughout World War I, Fuller changed his location but not his career path.

"During the World War, he was a member of a sanitary committee at Washington regulating the engineering planning and sanitation of the various Army camps in this country. As consulting engineer to the U.S. Public Health Service and to the Construction Division of the Army, he was responsible for a considerable part of the practices which resulted in the unprecedented low typhoid fever death rate in the Army camps."[84]

FULLER MARRIED HIS THIRD WIFE, Charlotte Bell Todd, on July 28, 1913, in Spring Lake, New Jersey. No narrative explains who Charlotte was or where she came from. The marriage ended in divorce on July 18, 1918. According to a story in the *New York Times*, she was to receive $250 per month in alimony, and Fuller paid the premium on a $10,000 life insurance policy for her benefit.[85]

Three months after the divorce was final, Fuller married Eleanor Todd Burt in October 1918 in Philadelphia.

"The story is told that at his fourth wedding he asked his bride, 'Eleanor, what time shall we set for the wedding?' Her reply, 'I don't care; what time do you usually set for your weddings, George?'"[86]

In addition to having three sons of his own, he adopted Eleanor's three sons, Kenneth B. Fuller, Gordon B. Fuller, and George B. Fuller.

Two summaries of Fuller's career and personal life published during 1917–19 reveal a familial relationship between Charlotte Bell Todd and Eleanor Todd Burt. They had the same parents. Presumably, they were sisters.[87]

Hints that Fuller enjoyed life to the fullest and had complex relationships with women are apparent throughout the short biographies of and reminiscences about Fuller. There was his mother's abandonment of him and her retreat to California after his father died; he married early and had a total of four wives; and there was the presence of his house guest in the Summit, New Jersey home.[88]

Abel Wolman, another pioneer of modern sanitary engineering, knew Fuller well and commented on his relationships with the opposite sex. "Fuller was married four times and had three sons and three stepsons. As in all other phases of his life, his conjugal experiences were multiple, complex, and invariably energetic."[89]

In a footnote, Wolman added, "The more precise family records give some clue to the nature of Fuller's heritage and provide some grist, perhaps, for latter-day psychoanalysts interested in this highly complex man."[90] James C. Harding, Fuller's employee and colleague, was more explicit, "I do not know whether he was a 'great lover' in the normally accepted sense of the phrase, but he was certainly very fond of the opposite sex."[91]

Fuller was a man who was larger than life in many ways. His personal relationships, like his passion for his profession, were manifestations of his large appetite for life.

BY ONE COUNT, Fuller was a member of eighteen professional associations and twelve private or social clubs and organizations.[92] The list encompasses the length and breadth of his business, technical, and private interests. His involvement with two associations attests to his dedication to service and the improvement of public health.

APHA annual and quarterly reports during 1894–1900 summarized Fuller's early participation in the business of the association and published the papers he presented at its annual conferences. His work also appeared in the quarterly issues of *Journal APHA*. Reports from the organization's annual conferences also listed Fuller's positions on volunteer committees. After he was elected to the association in 1892, no doubt under the sponsorship of William T. Sedgwick, he wasted no time in getting deeply involved in one of the association's committees.

The 1894 APHA conference in Montreal, Canada, was a watershed event for water bacteriology. Two reports were issued by the association's Committee on Pollution of Water Supplies. Fuller was a member of this committee and, along with seven senior members of the association, formed a subcommittee to organize the various approaches to isolating and identifying bacteria. Although an attempt was made to resolve the disparities by mail, it soon became apparent that the difficult task had to be accomplished face to face. The Convention of Bacteriologists was held in New York City, June 21–22, 1895.[93] The papers published from that conference and the discussions recorded by a stenographer showed a large divergence in methods used to analyze water samples for bacteria.[94]

One passage from the discussion at the 1894 conference provided a historical footnote to one of the most routine bacteriological procedures practiced today. A participant who was comparing the bacteria recoveries on gelatin and agar growth media described a problem with water condensation in the agar plate. Fuller's response provided a solution.

> "I would say in regard to the interference by water of condensation that we have noted that point, and have overcome the difficulty by inverting the plates after the agar has become solid. By this means the water is kept on the lower surface of the Petri dish, during the period of incubation while the glycerine agar is upon the upper surface. We have regularly adopted this custom for some time with satisfactory results."[95]

Although the 1894 conference provided a lot of input to the subcommittee, the attendees and the members of the subcommittee could not come to a final agreement on methodologies. Therefore, a committee of nine highly respected bacteriologists was appointed to come up with an approach that would be acceptable to all. Fuller was one of the committee members; he was 27 years old in 1895.[96]

The original intent for the convention and the nine-member committee was to publish routine methods for isolating and identifying bacteria in water. The committee's report, published in 1898, stated, "The report is not intended to be a complete treatise upon bacteriological technique. Its purpose is to make certain recommendations concerning methods to be pursued in the study of bacteria, with the view of securing greater uniformity and exactness in the determination and description of the characters of bacterial species."[97] In effect, the recommendations in this report, along with later improvements, led to the development of standard methods for enumerating indicator organisms such as B. coli (now called total coliforms) and total bacteria (now called the total plate count or heterotrophic plate count).

Fuller's contributions to the field of public health were noted in one of his obituaries: "In American Public Health Association affairs, he was one of the initiators and contributed largely to the development and widespread adoption of the *Standard Methods of Analyses for Water and Sewage*, sponsored by the A.P.H.A."[98] The first edition of *Standard Methods* was published in 1905. When the sixth edition was published in 1925, Fuller was still very much involved. He became president of APHA in 1929.

There are no records of Fuller's early history with the American Water Works Association (AWWA), but he undoubtedly did the same thing with AWWA that he did with APHA—joined committees, volunteered, and worked diligently to improve public health and advance the drinking water industry. He was appointed and elected to positions of increasing responsibility in AWWA until 1920, when he had a profound impact on the future of the association and the water industry.

"In 1920, at the Montreal Convention of the A.W.W.A., Fuller negotiated the organization of a committee to codify and standardize water works practice. The Association before that time had developed a few specifications documents, but its relation to the preparation of those documents was not that of leadership but rather of cooperative participation. The group under his leadership and chairmanship was first called the Standardization Council, later the Committee on Water Works Practice."[99]

More than any other individual, Fuller was responsible for the publication of the *Manual of Water Works Practice* in 1926.[100] Serving on the committee with him were George A. Johnson, George C. Whipple, and Abel Wolman.

Today, the work of AWWA's Standards Council touches every aspect of the safe and reliable distribution of drinking water to the public. Thousands of people devote long hours to refining voluntary standards for chemicals, pipes, meters, and machinery. In 1923, in recognition of his organization skills and long service, Fuller was elected president of AWWA.

During his life, Fuller received only one award. In 1903, the American Society of Civil Engineers awarded him the Thomas Fitch Rowland Prize for his paper describing the new mechanical filtration plant at Little Falls, New Jersey.[101] Given the contributions he made during his lifetime, it is astonishing that he did not receive more awards while he was alive. Thirty-seven years after his death, he was elected to AWWA's Water Utility Hall of Fame, a significant honor. But most practitioners in the water field know George Warren Fuller because one of the top annual awards bestowed on AWWA men and women is named after him.[102]

FULLER PUBLISHED INCESSANTLY. No complete listing of his publications exists, but if such an inventory were compiled, it would number well over 100 journal articles, reports, and books.

The bibliography of this book lists well over a dozen articles and reports for which he served as author or co-author. In his 1933 paper published in *Journal AWWA*, Fuller must have been proud to report that a total of thirty-one million people were receiving filtered water from filtration plants producing about 3,800 million gallons per day.[103]

Fuller is generally given credit for writing three books. *Sewage Disposal* was published in 1912, and an update, *Solving Sewage Problems*, was published in 1926 with his partner, James R. McClintock, as co-author.[104] The third book he is credited with is his report on the Louisville filtration studies.[105] Even though this publication is, in reality, a project report, it was published as a book in 1898 by Van Nostrand. Just as easily, his masterwork on the filtration investigations at Cincinnati could be considered a book.[106]

<hr />

1   Ensign Fuller is first mentioned in the town records of Dedham, Massachusetts, on November 25, 1642. On November 22, 1643, he married Hannah Flower at Dedham, Massachusetts; Ensign Thomas Fuller, ancestry.com; Fuller, "Descendants of Ensign Thomas Fuller," 157–8.

2   Ibid.

3   "George W. Fuller," *Journal AWWA*, 950; Not much is known about George Newell Fuller. He was born on the Fuller family property and was described simply as a farmer.

4   She attended Mount Holyoke College but did not graduate. Her final year at the institution was 1865.

5   Fuller kept in touch with his younger sister in later years. She married Carl W. DeVoe and moved to Jerome, Idaho, "George W. Fuller," *Journal AWWA*, 953; George Warren Fuller owned a ranch in Idaho and likely visited her there, Harding, "Personal Reminiscences of George Warren Fuller," 1526.

6   "George Warren Fuller Award," *Journal AWWA*, 284.

7   George must have felt he had lost both parents at the same time. Perhaps he was looking for a stable family life to replace the one he had lost when he married in 1888, only two years after leaving high school.

8   Lucy Fuller, Deaths Registered, 118; her gravestone at the Evergreen Cemetery in West Medway, Massachusetts states March 19, 1893, as her date of death; Lucy came from a family who immigrated to America from New Brunswick and Prince Edward Island, Canada. Her father was born about 1830 and listed his occupation as farmer. Her mother, Sarah, was born about 1845. The farming family had seven children, three boys and four girls. They must have moved to Boston from New Brunswick sometime between 1877 and 1880. The youngest boy, Harry, was born in New Brunswick about 1877, Census Record, Lucy Hunter, 1880; The 1880 U.S. Census showed that Lucy's family lived in Boston at 218 Bennington Street (Census Record, Lucy Hunter, 1880),

which is now near Boston's Logan International Airport but was located near culti-vated land in the late 1800s. The address is about three miles from the MIT campus, as the crow flies.

9   Draft Registration Card, Myron E. Fuller, June 5, 1918.

10   Passport Application, Fuller, May 2, 1890.

11   Lucy Fuller, Deaths Registered, 118.

12   Fuller, *Sewage Disposal*, vii.

13   Leonard, *Who's Who in Engineering*, 472.

14   Draft Registration Card, Myron E. Fuller, June 5, 1918.

15   Ibid.

16   Marquis, *Who's Who in America*, 758.

17   Passport Application, Fuller, May 31, 1900.

18   Census Record, 1920.

19   Passport Application, Fuller, July 11, 1923.

20   Census Record, George W. Fuller, 1910.

21   Harding, "Personal Reminiscences of George Warren Fuller," 1523.

22   Passport Application, Fuller, May 2, 1890.

23   "George Warren Fuller," *Successful American*, 101; "George W. Fuller," *Water Works and Sewerage*, 235.

24   Harding, "Personal Reminiscences of George Warren Fuller," 1523.

25   "George Warren Fuller Award," *Journal AWWA*, 284

26   Whipple, *State Sanitation*, 83.

27   Winslow, "They Were Giants," 15.

28   Ibid., 17.

29   Ibid., 18.

30   Ibid.

31   Jordan, Whipple, and Winslow, *Pioneer of Public Health;* Winslow, "They Were Giants," 19.

32   U.S. engineers' reliance on European experience for water treatment expertise oc-curred again more than 70 years after Fuller's time in Berlin. After the U.S. devel-opment of mechanical filtration and chlorination in the early part of the twentieth century, Americans showed little interest in studying what the Europeans were doing. European engineers insisted on using slow sand filters for decades into the twentieth century, and they never fully embraced the use of chlorine for disinfection. Because chlorine gas was used as a terror weapon during the Great War of 1914–18, its large-scale use in drinking water was simply not possible in Europe. No amount of insistence on the role of "nascent oxygen" in chlorine disinfection was going to fix European per-ceptions. In the late 1970s, U.S. water utilities became interested in granular activated carbon and ozone treatments, which had been developed into mature technologies in Europe. U.S. engineers once again made pilgrimages to European research centers to learn the latest advances, Miller and Rice, "European Water Treatment Practices."

33   Between Jersey City and Water Company, February 5, 1909, 5064.

34   Brock, *Robert Koch*, 199–214.

35   George Warren Fuller, *American National Biography*, 145; "George W. Fuller," *Transac-tions*, 1654; "Sad Milestone," *American Journal of Public Health*, 895; "George W. Fuller," *Water Works and Sewerage*, 235.

36   Sedgwick, "Data of Filtration, 69–75; "Arrangement and Working of Filter Beds," *Engineering Record,* 3.

37   "Remarkable Letter," *Journal American Medical Association,* 122.

38   "George W. Fuller," *Water Works and Sewerage,* 235; "George W. Fuller," *Journal AWWA,* 951; *New York Times.* June 16, 1934.

39   Passport Application, Fuller, May 2, 1890.

40   "George W. Fuller," *Water Works and Sewerage,* 235.

41   New York Passenger List, George Fuller, November 5, 1890.

42   Baker, *Quest,* 64, 80.

43   Whipple, "Biographical Memorandum of George W. Fuller," 1; this and other biographical memoranda in the Harvard Archives were written by Whipple as part of preparations to write his book *State Sanitation.*

44   Between Jersey City and Water Company, February 5, 1909, 5064.

45   Sedgwick, *Principles of Sanitary Science,* 139–43.

46   Ibid.

47   Whipple, *State Sanitation,* 80.

48   Ibid.

49   Sedgwick, *Principles of Sanitary Science,* 139–43.

50   "Report on Water-Supply and Sewerage," Minutes of Proceedings of the Institution of Civil Engineers, 380–2.

51   Fuller, "Special Biological Work," 620–33; Fuller, "Differentiation of the Bacillus," 637–44.

52   Fuller, "Special Biological Work," 620–33.

53   Fuller, "Differentiation of the Bacillus," 637–44.

54   Fuller, "Purification of Water by Sand Filtration," 449–538.

55   Ibid., 452; Fuller, "Progress in Sanitary Science," 73–4.

56   Fuller, "Progress in Sanitary Science," 73–4.

57   Sedgwick, "Data of Filtration," 72.

58   Fuller, "Purification of Sewage and Water," *State Board of Health,* 401.

59   Ibid., 453–5.

60   Hazen, *Filtration of Public Water-Supplies,* 96–100.

61   Baker, *Quest,* 231–3.

62   Fuller, *Purification of the Ohio River at Louisville,* 387.

63   Ibid., 440.

64   Ibid., 46.

65   Baker, *Quest,* 332–4.

66   Ibid.

67   Ibid.

68   "Purification of River-Water," *Lancet,* 331–2.

69   Ibid., 331.

70   Fuller, *Purification of the Ohio River at Louisville.*

71   Fuller, *Purification of the Ohio River for Cincinnati,* 377–402.

72   News from the Classes, *Technology Review,* 471.

73 Between Jersey City and Water Company, February 5, 1909, 5065.

74 Ibid., 5065–6.

75 "Rudolph Hering," *Journal AWWA*, 304–6.

76 "George W. Fuller," *Transactions*, 1654; "Sad Milestone," *American Journal of Public Health*, 896.

77 "George W. Fuller," *Journal AWWA*, 951.

78 Moore, *Purification of the Washington Water Supply*, xii–xvi.

79 *Report of the Filtration Commission of Pittsburgh*, 1–8.

80 Abplanalp, "Little Falls Treatment Plant," 1; Fuller, "Filtration Works at Little Falls," 156.

81 Abplanalp, "Little Falls Treatment Plant," 1–2.

82 Hendricks, *Water Treatment Unit Processes*, 537.

83 Ibid., 647–8.

84 "Sad Milestone," *American Journal of Public Health*, 895–6.

85 *New York Times*, July 19, 1918.

86 Harding, "Personal Reminiscences of George Warren Fuller," 1527.

87 Sackett, *New Jersey's First Citizens*, 197–8; Scannell, *New Jersey's First Citizen's and State Guide*, 171–2.

88 Census Record, George W. Fuller, 1910.

89 Wolman, "George Warren Fuller," 12.

90 Ibid. 3.

91 Harding, "Personal Reminiscences of George Warren Fuller," 1527.

92 For a complete list, consult the many biographies and obituaries for Fuller in the Bibliography.

93 *New York Times*, June 23, 1895.

94 American Public Health Association, 1895, 380–516.

95 "Discussion," American Public Health Association, 178.

96 The best analogy for Fuller's participation on this august committee would be for a high school basketball star to be called up to a National Basketball Association professional team. The high school star would then do spectacularly well and end up becoming a team leader.

97 Welch, "Introduction to the 1897 Report of the Bacteriological Committee," 10.

98 "Sad Milestone," *American Journal of Public Health*, 896.

99 "George Warren Fuller Award," *Journal AWWA*, 288.

100 "George W. Fuller," *Journal AWWA*, 952.

101 "George W. Fuller," *Transactions*, 1657.

102 "George Warren Fuller Award," *Journal AWWA*, 284–8.

103 Fuller, "Progress in Water Purification," 1575.

104 Fuller, *Sewage Disposal*; Fuller and McClintock, *Solving Sewage Problems*.

105 Fuller, *Purification of the Ohio River at Louisville*.

106 Fuller, *Purification of the Ohio River for Cincinnati*.

8

Jersey Water Wars

*"Water from the Boonton Reservoir was first deliv-
ered to Jersey City on May 23, 1904."*

The Dutch established a settlement on the site of Jersey City
in the seventeenth century. Along with Jamestown (Virginia),
Plymouth (Massachusetts), and New Amsterdam (New York City),
it was one of the earliest European settlements in North America.[1] Early settlers arrived in the area in 1612 or 1613, and Dutch
settlers built a few houses in the area now encompassing Jersey
City in 1633. In 1660, the "somewhat notorious" Peter Stuyvesant
founded the Town of Bergen on land purchased from the native
population.[2] Bergen consisted of a few square blocks containing
buildings and surrounded by palisades. In the center of the settlement was a public well (no doubt quite shallow) that some have
considered the first public water supply in New Jersey. The well
functioned for about 200 years before being abandoned and covered over. Wells and cisterns furnished water to the settlers until
the mid-nineteenth century.[3]

Jersey City is located on the western shore of the Hudson River
across from lower Manhattan (Chapter 5, Figure 5-1). Compared
with Paterson, the other major New Jersey city that is relevant
to the story of Dr. John L. Leal, Jersey City's rapid population
growth was delayed for a couple of decades. In 1850, Paterson had
a population of more than 11,000, whereas Jersey City's population was only about 7,000. Ten years later, in 1860, Jersey City's
population had ballooned to more than 29,000, besting Paterson
with slightly fewer than 20,000 people. During the nineteenth
century and early twentieth century, population growth in Jersey
City would dramatically exceed that of Paterson, reaching a peak
of more than 316,000 people in 1930.

Location was the key to Jersey City's growth. Throughout the
latter half of the 1800s, Jersey City became the terminus for many
of the major railroads coming from the Western United States.[4]
The rail terminals fed goods and raw materials into a ferry system
that transported cargo across the Hudson River to New York City.

Ships also loaded and unloaded cargo at the railroad terminals on the Jersey City wharf and distributed it throughout the northeastern United States and beyond. Jersey City became a major manufacturing center at the same time its influence as a transportation hub was expanding.

As the area's population increased during the early 1800s, it became apparent that local wells and cisterns were no longer capable of supplying water for both domestic and industrial use. In 1851, a Water Commission was established for Hoboken and Jersey City. The commission considered a number of water projects including:

- Building a 250,000-gallon reservoir that would be filled by local streams,
- Importing water from Rockland Lake in New York state—about 28 miles north of Bergen Square on the west bank of the Hudson River,[5]
- Building a dam on the Hackensack River above Newark, New Jersey,
- Purchasing water from the Passaic River at Paterson from the Society for Establishing Useful Manufactures,
- Purchasing water from the Passaic River at Dundee Lake below Paterson, and
- Taking water from the Morris Canal, which terminated in Jersey City.

All of the proposed projects were either too expensive, involved technological limitations, or suffered from water quality problems. Some of the proposals involved all three difficulties.[6]

The project that was ultimately chosen was a water diversion point on the Passaic River at Belleville—about eight miles from the center of Jersey City. The first water to be delivered from the Belleville project was sent to Jersey City on June 16, 1854. Water was pumped to distribution reservoirs and delivered by gravity to the city.[7] The quality of the water from Belleville was reported to be good in 1854, when the population of Paterson, the major upstream city, was only about 16,000. However, Paterson's exploding population growth soon resulted in large flows of sewage contaminating the Belleville water supply. Newark, too, was growing rapidly, and sewer construction to remove sewage was a priority there. Many of Newark's sewers discharged into the Passaic River, where their contents reached the Belleville intake through tidal action. The "organic matter" being discharged from Paterson was dismissed by some as not a problem because it was completely "oxidized" by the time it reached Belleville.[8] Of course, that was not the case, and the "organic matter" in the water was the least of the problems with the Belleville supply.

DURING ALL OF ITS EFFORTS to obtain a secure water supply, Jersey City had to deal with various entrepreneurs who wanted to export water from the Passaic River basin to quench the insatiable thirst of millions of people across the Hudson River. In addition to the scheme to transport water from New Jersey to New York City described in Chapter 5, there were other, less advertised efforts to siphon off New Jersey's water supplies. Buried in the 1894 minutes of the Jersey City Board of Public Works was a curious motion made by the board's president.

> "Whereas the city of New York desires to obtain a better water supply for certain portions of said city and whereas the city of Jersey city can by a slight increase of its pumping and conduit facilities furnish such supply, and whereas in order to convey the said water to New York legislative aid may become necessary therefore

> "Resolved, that this Board requests his Honor the Mayor to appoint a committee consisting of five property owners from each Aldermanic District of Jersey City, his Excellency the Governor of the State, the Mayor, the President of the Board of Finance and Taxation, the President of the Board of Aldermen, the Corporation Counsel and the Chief Engineer of the Board of Works to meet in connection with the Board of Works and take such action in the matter as may be necessary to obtain the sanction of the Legislature in the matter."[9]

No record exists of the governor's attitude about Jersey City's attempt to drag him into a water scheme, and there was no mention of the scheme's outcome in later minutes of the Board of Public Works. No doubt this resolution was shelved along with all of the other half-formed proposals for New York City to "steal" New Jersey's water.

Disease incidence. Pollution of the lower Passaic River continued to worsen as a result of ever-increasing sewage discharges from Paterson and Newark. Jersey City experienced excessively high numbers of typhoid fever deaths during the latter part of the nineteenth century up to 1895. The death rates were typically greater than 25 per 100,000, a figure considered "acceptable" for large cities during this period. The typhoid fever death rates ultimately forced city leaders to take seriously the need for an uncontaminated water supply. Clearly, Jersey City was in the grips of the Sewer Pipe–Water Pipe Death Spiral.[10]

Rockaway River Supplies

Between 1893 and 1895, the city entered into a contract with the East Jersey Water Company (EJWC), obligating the company to provide a temporary water supply that would protect the health of its citizens until the planned upland water supply on the Rockaway River was put into operation. On October 12, 1895, Jersey City abandoned the Belleville water supply and began receiving a temporary supply from the Little Falls water works owned and operated by EJWC.[11] The supply from Little Falls in 1895 was excess water available from the Pequannock pipeline originally built to provide water to Newark; it was not filtered or treated in any fashion.[12] Still, the bacteriological quality of the Little Falls supply was far better than the supply at Belleville. It would not be until 1902 that the mechanical filtration plant designed by George Warren Fuller began operation at Little Falls.

Because Jersey City was in debt to the full extent permitted by law, its officials turned to the private sector to secure a permanent water supply.[13] The city experienced many false starts in its search for a permanent supply, receiving bids from various individuals and water companies over the years. "Since November, eighteen hundred and ninety-two [until 1899], the city has advertised eight times for bids for a new water supply."[14] Most of the bids and proposals involved EJWC and a businessman named Patrick H. Flynn. The New Jersey Supreme Court threw out one bid. Once again, the actions of the Morris Canal and Banking Company complicated the water rights picture in New Jersey. The company was being prosecuted for some misdeeds, and this invalidated the bidding process and contract award.[15] At least one publication suggested that the bidding involved some nefarious dealings, including the charge that Patrick H. Flynn was merely a "cover" for EJWC.[16]

The final bidding was held on July 18, 1898. It was a frustrating and confusing process, and in the end there was only one man standing—Patrick H. Flynn. The bid actually contained two proposals for supplying water to the city. One involved constructing multiple reservoirs in the Rockaway River watershed and was not accepted by the Board of Street and Water Commissioners. On December 8, 1898, the board accepted Flynn's other proposal. Flynn was known to be a ruthless businessman, and it seems odd that Jersey City would choose to involve him in its water supply project.

The proposal that was accepted outlined a plan to build a reservoir at "Old Boonton" and to supply 50 million gallons per day (mgd) to the city, with the possibility of supplying 70 mgd sometime in the future. Stripped of all of the complicated (and important) contract language, Flynn's proposal was to build a dam, reservoir, and pipeline and to sell those facilities on completion to the city for the sum of $7,595,000, which translates to more than $175 million in equivalent buying power today.[17] The reservoir was to act as a "fine settling basin" that would purify the water.[18]

Before acceptance of the bid, the senior staff of the city's water department and the members of the Board of Street and Water Commissioners conducted "a very careful investigation of the water-sheds with a view of ascertaining what means would be taken to prevent pollution." In the board's minutes, these city officials admitted there was no need to redirect the pollution coming from Dover, Rockaway, and Boonton. "We believe that the means ["fine settling basin"] proposed by the contractor for this purpose will be efficient."[19] These extraordinary statements directly contradicted the city's entire case in the lawsuit it filed six years later.

Patrick H. Flynn signed a contract with Jersey City on February 28, 1899. This contract and its subsequent modifications became the subject of the two Jersey City trials held from 1906 to 1910.

On May 4, 1899, just two months after signing the contract, Flynn organized a private company, the Jersey City Water Supply Company (JCWSC). As expected, the company's officers were all friends of Flynn. "Although Mr. Flynn does not figure in the company, it is understood that he will have absolute control of the work of building the pipe lines and reservoirs."[20]

Nothing about the contract to build the Boonton water supply was simple. After about $1 million had been expended on construction of the dam and reservoir, the work was suspended for about one year. Clearly, Flynn was in over his head, and he tried to get out of his legal obligations by transferring the contract to JCWSC.[21] However, the company was insolvent, and it was taken over by the New Jersey General Security Company in 1902. All of the officers of the old company resigned, and new officers were installed, including a new president, Cornell University engineering professor Edmund L. Gardner.[22]

THE ROCKAWAY RIVER ORIGINATES in the mountains in Sussex County, New Jersey. It wends its way through Morris County until it meets the main stem of the Passaic River about 10 miles northwest of Newark (Figure 5-1). The total drainage area of the Rockaway River is 138 square miles, but, according to one source, only 121.5 square miles encompass the watershed above Boonton Dam.[23]

In the early 1900s, the watershed above the dam consisted of about 80 percent forested land, about 20 percent farmland, and miscellaneous urban areas. Although the watershed contained no large towns, several of its small to medium-sized towns and villages would become the central focus of the two trials.

In the early 1900s, populations in the towns of Dover, Boonton, Hibernia, and Rockaway, New Jersey, were 6,300, 4,000, 1,600, and about 1,000, respectively.[24] Wastes from the town of Rockaway were not a subject of concern during the two Jersey City trials. Concern for contamination of the Rockaway River above Boonton centered on Dover, Boonton, and Hibernia.

Industrial activities contributed additional sources of contamination. A rag mill discharged large amounts of industrial waste into the river not far above the reservoir. A mining operation near Hibernia dumped mine wastes into a creek tributary to the Rockaway River. Other industries were located on the banks of the river near Dover and Boonton.

Jersey City appointed Garwood Ferris, engineer for the Jersey City Board of Street and Water Commissioners, to be in charge of the new waterworks under construction at Boonton. He was responsible for supervising a team of inspectors who examined all aspects of the dam and pipeline construction.

A careful reading of the exhibits in Volume 7 of the Jersey City trial transcripts shows that before the lawsuits against Flynn and JCWSC were filed, documents and communications from Jersey City's engineer in charge indicated that the city was generally pleased with the quality of the water expected from the Boonton supply.

Dr. John L. Leal was interested in cleaning up sources of direct contamination on the Rockaway River during his employment as health officer for Paterson, New Jersey. As the sanitary adviser to JCWSC, he was responsible for ensuring that contamination sources above Boonton Reservoir were eliminated.

In early 1904, there was an exchange between Ferris and Leal regarding the pollution of the Rockaway watershed. In a letter, Leal assured Ferris that JCWSC would remove any identified source of pollution above the Boonton Dam. However, the city was obligated to bring any pollution source to the company's attention so that a cleanup could be initiated.[25]

Construction of Boonton dam, reservoir, and pipeline. Patrick H. Flynn selected the site for Boonton Dam shortly after signing the contract with Jersey City. As described in a newspaper article, the dam site was on the "Banta Property" near the town of Boonton, and "There were no engineering difficulties to interfere with the speedy execution of the work."[26] Engineering difficulties were never the problem with the project. Money and water quality were the major stumbling blocks.

When Boonton Dam was constructed, its design was one of the most innovative of its time. The dam extended for 3,150 feet, of which a 2,150-foot stretch was a masonry dam and the remaining 1,000-foot segment was an earthen dam with a concrete core wall. The dam's maximum height was 114 feet, and it was 77 feet wide at the base. The top of the dam was 310 feet above sea level, and the reservoir impounded behind it was, at most, 100 feet deep at its maximum water elevation. More than seven billion gallons of water were stored in the reservoir when the water level reached the elevation of the spillway.[27]

The pipeline connecting Boonton Reservoir with the distribution reservoirs in Jersey City was about 23 miles long. Seventeen miles of the aqueduct consisted of 6-foot-diameter riveted steel pipe, and reinforced concrete pipe formed another four miles. The remaining distance was serviced by an 8-foot-diameter tunnel lined with brick or concrete.[28]

The exhibits for the first trial show an early pattern of cooperation between the contractor and the city. As with all construction projects, every contingency could not be anticipated in the original contract, but the contractor and the city apparently were able to work out needed changes to the project during late 1899 and early 1900. After construction work resumed in 1902, the city and the contractor continued to cooperate to get the job done. Numerous communications from Garwood Ferris bear this out.[29]

In March 1900, construction of the pipeline to Jersey City was in progress. As of September 7, 1900, excavation for the dam was complete. Riveted steel pipe continued to be laid. Digging for the Watchung Mountain tunnel began in 1900 and was about one-third complete by May 1901. In May, however, work was suspended until Flynn could secure a loan.[30]

An article in the *Boonton Times* claimed that Flynn had agreed to build sewers for the town of Boonton to prevent pollution of the reservoir that was under construction. Although the article went into great detail on how the system would be built,[31] the story was pure fantasy.

In 1902, the new representatives of JCWSC signed two contracts with Jersey City. The first was an agreement to continue temporarily supplying water from Little Falls. The second allowed the newly constituted company to complete the Boonton Dam, reservoir, and transmission pipeline. In April 1902, a subcontract was executed with a construction company to complete Boonton Dam. Work on the dam site resumed shortly thereafter and continued through 1902 and 1903. In January 1904, a group of civil engineers from around the country visited the completed dam at Boonton. The level of water in the reservoir continued to rise throughout 1904.[32]

Water from the Boonton Reservoir was first delivered to Jersey City on May 23, 1904. On November 15, 1904, JCWSC gave the city notice that the project was complete and requested the agreed-on payment.[33] In April 1905, John L. Leal filed a $500,000 bond on behalf of JCWSC with Mayor Mark Fagan of Jersey City. Filing the bond was an essential step in the effort to get Jersey City to buy the Boonton waterworks from JCWSC.[34]

While payment for the entire waterworks was under consideration, Patrick H. Flynn and JCWSC began billing Jersey City for water deliveries. The bills were based on amounts agreed to in the original and subsequent contracts. Now the fun would begin.

IN LATE MAY 1904, after water from the Boonton Reservoir was first delivered to the city, Ferris and Leal conducted a detailed inspection of the watershed. Ferris's report was generally positive, but he identified several sanitary concerns in the Dover area. He referred to the bacteriological testing JCWSC was doing on the Rockaway River and its tributaries above the reservoir, and he suggested that the city hire its own bacteriologist if the results of JCWSC's testing were not provided to him.[35]

Apparently, Ferris's concerns about the sanitary quality of the Rockaway River were not satisfied. On September 19, 1904, Jersey City approved an agreement with George C. Whipple to conduct sanitary investigations on the new supply.

After Whipple was hired, he and engineering expert Emil Kuichling inspected the Rockaway River watershed. Based on the extensive litigation that erupted soon afterward, a newspaper report of their findings was surprising. On November 19, 1904, Whipple submitted a report to Jersey City in which he was said to have stated, "The report as to the condition of the water supply was most favorable . . . Expert Whipple sets forth that he had made daily analyses of the water for several months. Minor impurities were found from local pollutions caused by inflowing streams, but these impurities were eliminated long before the water reached the Jersey City reservoir." A water commissioner confirmed Whipple's findings, stating that ". . . the [report] of the experts was satisfactory and that the condition of the water was excellent."[36]

Eleven days later, in a long report to the Board of Street and Water Commissioners dated November 28, 1904, Ferris noted that JCWSC had given official notice that it had completed its project and that the facilities were ready to be purchased. Ferris then catalogued how the project met or did not meet the requirements of the contracts. Overall, he was pleased with the physical structures, even though the invert of the Watchung Tunnel did not meet specifications. "Taken as a whole, I deem the construction work of the plant as creditable in design and in execution."[37]

Such was not the case for the quality of the water, in Ferris's opinion. His report maintained that the contract requirement for the water to be pure and wholesome was to be met as the water entered the reservoir and not as the water was delivered to Jersey City.[38] This was a ridiculous assertion.

What happened during those intervening eleven days? One interpretation of the city's sudden concern about water quality is that the primary reason for the legal challenges was to reduce the amount of money it would have to pay JCWSC. In the rough-and-tumble world of obtaining water supplies in New Jersey, both contractors and cities resorted to court action to improve their business positions and financial obligations. It is likely that the lawsuit Jersey City brought against Patrick H. Flynn and JCWSC was primarily a tool for obtaining a lower purchase price for the Boonton water supply.

In a report by the Board of Street and Water Commissioners dated January 27, 1905, Jersey City made it plain that its officials were getting cold feet over the imminent purchase of the Boonton waterworks and that the city would use the courts to improve its position regarding the purchase. "It is our purpose that before the works have been finally accepted, even remote cause for complaint shall have been remedied, meaning that to that end we will exhaust every legal power at our disposal."[39]

On September 25, 1905, a report by the Board of Street and Water Commissioners noted that the city solicitor had filed a bill of complaint against Patrick H. Flynn and JCWSC. The city alleged that the waterworks were not complete and asserted that if the facilities were accepted, payment would be less than the contract price of $7,595,000 because of deductions associated with actions necessary to complete the facilities, as specified in the contract.[40]

The trial began in February 1906 before Vice Chancellor Frederic W. Stevens (see Chapter 9). A short article in the *Boonton Times* gave an estimate for completion of the trial—"several days"[41]—but it took several years to resolve the battle between Jersey City and JCWSC.

1   "Jersey City Past and Present," New Jersey City University.

2   Ibid.

3   Harrison, "The Public Water Supplies of Hudson Co.," 1–2.

4   On the wall of the Jersey City Library, there is an extraordinary photo taken where Jersey City met the New York Harbor shoreline showing dozens of terminals for hundreds of railroad tracks.

5   Rockland Lake is about five miles north of the existing Tappan Zee Bridge.

6   Harrison, "The Public Water Supplies of Hudson Co.," 4–5.

7   Ibid., 5–7.

8   Manual of the Board of Public Works 1883–84, July 3, 1883, 39–40.

9   Manual of the Board of Public Works 1883–84, January 14, 1884, 128.

10 See Chapter 2.

11 "Water Supply of Jersey City," compiled by the Free Public Library of Jersey City, 2. The temporary contract was signed between Jersey City and EJWC on October 12, 1895. A supplemental contract, signed on April 22, 1897, extended the terms for providing a temporary supply to the city and allowed for water from the Passaic River at Little Falls to be included in the supply in addition to water from the Pequannock River. Between Jersey City and Water Company, Vol. 7—Exhibits, 3908–18.

12 Colby and Peck, *International Year Book-1899*, 858.

13 "Jersey City's Water Supply," *City Government*, 99.

14 Jersey City Board, "Water Question to the Taxpayers," 3–4; Between Jersey City and Water Company, Vol. 7—Exhibits, 3956.

15 Ibid.

16 "Jersey City's Water Supply," *City Government*, 100.

17 Bureau of Labor Statistics, CPI Inflation Calculator; the calculator provides an estimate of the current buying power of the original amount.

18 Jersey City Board, "Water Question to the Taxpayers," 4–5; Between Jersey City and Water Company, Vol. 7—Exhibits, 3958.

19 Jersey City Board, "Water Question to the Taxpayers," 5.

20 *Boonton Times*, May 4, 1899.

21 *Boonton Times*, December 14, 1899.

22 Between Jersey City and Water Company, April 18, 1907, 3034–5.

23 "Water Supply of Jersey City," compiled by the Free Public Library of Jersey City, 2; other figures that have been cited for the watershed areas above Boonton Dam are 118.2 and 118.4 square miles.

24 Winslow, "Water-Pollution and Water-Purification," 4.

25 Between Jersey City and Water Company, Vol. 7—Exhibits, 3983–4.

26 *New York Times*, October 25, 1899.

27 Harrison, "The Public Water Supplies of Hudson Co.," 11–2.

28 Ibid., 10–1.

29 Between Jersey City and Water Company, Vol. 7—Exhibits.

30 *Boonton Times*, May 16, 1901.

31 Ibid., July 11, 1901.

32 Ibid., January 28, 1904.

33 Between Jersey City and Water Company, Vol. 7—Exhibits, 3995.

34 *Evening Journal*, April 22, 1905.

35 Between Jersey City and Water Company, Vol. 7—Exhibits, 3989–90.

36 *Evening Journal*, November 19, 1904.

37 Between Jersey City and Water Company, Vol. 7—Exhibits, 3995–4002.

38 Ibid., 4002–3.

39 Ibid., 4004.

40 Ibid., 4010–1.

41 *Boonton Times*, February 23, 1906.

9

Bacteria on Trial—1906 to 1908

". . . the defendant company may . . . present other
plans or devices for maintaining the purity of the water. . ."
— Stevens, In Chancery of New Jersey,
Final Decree, 1908, 4163

Both Jersey City trials were held before the Chancery Court of New Jersey, which had jurisdiction over equity cases, including property disputes. The Chancery Court also held jurisdiction over divorce cases and delegated jurisdiction over lunacy proceedings. Historically, a chancery court (otherwise known as an equity court or court of equity) was authorized to apply principles of equity, as opposed to law, to cases brought before it. In English history and under English law, the decisions of equity courts were not always consistent and were generally not precedent-setting. Chancery Courts are so-named because the concept for this system under English law began with petitions to the Lord Chancellor of England.

Transcripts from the Jersey City trials show that the rules of the Chancery Court for expert witnesses in the early twentieth century were pretty much the same as they are today. Fact witnesses could testify only to what they experienced directly; they were not allowed to give their opinions on some of the larger issues of the trial. Expert witnesses had to be qualified to testify about issues specifically related to their background and expertise. A chemist was usually not allowed to provide an opinion on engineering matters. Typically, at the beginning of an expert witness's direct testimony, the attorney would elicit the education, training, and experience that qualified him (and all of the experts in both trials were men) to provide his opinions.

Although fact witnesses played an important role in these lawsuits, the largest portion of the testimony came from expert witnesses. The experts from both sides were asked for their opinions on the meaning of the voluminous data that had been collected. This testimony was particularly important for the bacteriological data, which were being collected with methods that were constantly being revised and improved during the time of the trials.

One attorney in the Jersey City trials apparently did not have much respect for expert witnesses in general and in particular for those provided by the defendants in the first trial. "James B. Vredenburgh gave the experts a shot. He intimated that he did not take much stock in their testimony in law suits [sic]. He said they constructed theories."[1]

George C. Whipple, who was an expert witness for the plaintiffs in both Jersey City trials, was discouraged with the tasks he was required to perform as an expert. Later, in his personal journal, he referred to his testimony in a Passaic River pollution case in 1914:

> "My examination by an old lawyer named Boggs was a farce. He had no idea of what I was to testify to and asked more fool questions than I have ever heard in the same space of time. I hate this kind of thing. The whole system of expert testimony is wrong. I am strongly tempted never to take another case."[2]

Charles-Edward A. Winslow, another expert witness for the plaintiffs, also held a dispirited view of expert testimony. In a 1916 letter to William T. Sedgwick, his former professor at the Massachusetts Institute of Technology, Winslow referred to another lawsuit and the excesses fostered by experts. "I do not know of anything which the Lederle Laboratories or Professor Prescott have done for which they could be legitimately criticized, although of course we are all apt to be prejudiced and take somewhat extreme stands when we get into a court case."[3]

No matter what the personal opinions of both sides' experts were, the outcome of the case would rise and fall on their credibility.

IN THE TWO JERSEY CITY TRIALS, both sides were permitted to present opening statements to alert the judge about what they intended to prove during the court proceedings. Testimony generally began with the plaintiffs being required to prove they had an actionable case so the trial could proceed. The defendants' case followed, with witnesses providing evidence for why their point of view should be accepted. Briefs, motions, and stipulations on a variety of issues were filed with the judge at the beginning of and throughout the trials. Many of these filings did not make their way into the transcripts. Some of the judge's decisions were referred to in statements by the judge or in arguments by the attorneys.

Both sides made summary statements at key points during the trials to assist the judge in weighing both positions. In his rulings during the trials, the judge often summarized the direction in which he thought the facts and the expert opinions were leading. Fact and expert witnesses were called by both sides, and exhibits were submitted and explained as part of the witnesses' testimony.

After the fundamental components of each side's case were established, the opposing side was allowed to offer rebuttal witnesses, who were called to give expert opinions that rebutted or countered the contentions of the other side's experts. Testimony went on until either the judge or the attorneys felt the topics had been adequately covered. The judge would rule on whether the fact and expert testimonies were admissible. In the second trial, Special Master William J. Magie interpreted what was admissible rather liberally. He allowed much more evidence into the trial than he excluded.

After the trial, the judges issued their rulings. In the first case, Vice Chancellor Stevens issued an opinion on May 1, 1908, and a final decree several weeks later on June 4. In the second case, Special Master Magie issued his ruling on May 9, 1910, after 38 days of testimony but only a few days after the last brief had been filed.

Both sides then filed appeals, and the courts announced various rulings. In these cases, the judges' rulings were appealed all the way to the New Jersey Supreme Court, which published its ruling many months after the conclusion of the trials.

From the time Jersey City filed the original complaint to the final ruling issued by the New Jersey Supreme Court, more than five years transpired. Although this case didn't last a record length of time for a legal case in New Jersey, its resolution took a significant amount of time by anyone's reckoning.

IN NEW JERSEY IN 1906–1908, when the first trial took place, there was a great deal of blurring of the lines separating attorneys, judges, legislators who made the law, and politicians and businessmen who were supposed to abide by the law. This is true today, of course, with attorneys serving in legislatures, running businesses, and then moving to a final stage of their careers as judges. But in the early 1900s, and especially in New Jersey, the legal system comprised a much smaller group of men who knew one another, were members of the same clubs, and sometimes worked with one another on the same side of one issue and on different sides of another issue. The Jersey City lawsuit paired two antagonists who knew each other well. James B. Vredenburgh and William H. Corbin were principals of the two leading law firms in Jersey City in the early 1900s.

What was indisputably one of the most important water quality cases of the twentieth century was argued by outstanding attorneys before distinguished judges. All of the main participants attended Ivy League colleges (Princeton, Columbia, Cornell) and read for the law with distinguished attorneys or went to a first-rate law school (Columbia). The transcript for both trials comprises more than 6,800 pages and can be surprisingly interesting because the attorneys plotted their strategies carefully and the judges kept them on track to address the critical issues at hand.

Many of the expert witnesses in the first trial also appeared as experts in the second trial. In the first trial, the plaintiffs' experts included George C. Whipple, Charles-Edward A. Winslow, William T. Sedgwick, and Emil Kuichling. For the defendants, the main expert witness was Dr. John L. Leal, but significant testimony was also provided by Dr. George E. McLaughlin, John H. Cook, Rudolph Hering, Dr. Herman C.H. Herold, and George A. Johnson.

ON AUGUST 1, 1905, JERSEY CITY filed an injunction against Patrick H. Flynn and the Jersey City Water Supply Company (JCWSC). The city wanted to prevent Flynn and his company from suing over payment for the water delivered from the Boonton supply. At the end of the month, on August 29, Vice Chancellor J. J. Bergen issued his ruling. He found that grounds for an injunction existed but ordered Jersey City to pay the defendants 4 percent annual interest on $7,595,000 for the period May 24–August 24, 1905. The interest payment was to be made by October 2, 1905, but deductions, based on a variety of criteria, were allowed. A modification of Judge Bergen's ruling was made on October 1, 1907, and attested to by then-Chancellor William J. Magie. The modified ruling required the city to pay the defendants $97,943.75 each quarter, beginning with a payment on August 29, 1907, and continuing every quarter thereafter.[4]

In October 1905, Patrick H. Flynn assigned all of his rights under various contracts to JCWSC.[5] Flynn had been out of the picture for some time, but now it was official. Although his name was still included in the legal proceedings over the purchase price of the waterworks at Boonton, he had nothing to do with the trials.

All of these legal maneuverings, plus plenty more to come, resulted in the first Jersey City trial presided over by Vice Chancellor Frederic W. Stevens. The trial began on February 20, 1906. The court proceedings and depositions lasted until May 29, 1907, covering more than 40 days of testimony.

Vice Chancellor Stevens issued his opinion on May 1, 1908. Because of the complex nature of the facts of the case, the opinion was more than 100 pages long.[6]

At its heart, the first trial dealt with a contract dispute. Stevens's opinion was brutally frank about the complexity of the case.

> "The case involves a variety of questions, nearly all of them depending upon the proper construction of the contract of February 28, 1899, and of three other contracts supplemental thereto. Some of these questions are by no means easy of solutions."[7]

A few of the questions Stevens needed to resolve included:
- What was meant by the original contract?
- How did the subsequent contracts amend that meaning?
- What verbal assurances were given or extra-contract letters were written to amend the meaning of the contract?
- How much should Jersey City pay for the waterworks?
- How much should Jersey City pay for the water delivered by JCWSC after completion of the Boonton waterworks?
- What interest did Jersey City owe for not paying the contract price when required?
- How much did JCWSC owe in liquidated damages?

After examining the basic contract provisions and the arguments and counter-arguments as to what the contract meant, Stevens arrived at the following conclusions:[8]
- Patrick H. Flynn and JCWSC were obligated to build a dam, reservoir, and 23-mile pipeline to Jersey City to provide a water supply of up to 50 million gallons per day.
- Once the facilities were finished, JCWSC notified the city that they were complete.
- Jersey City completed its legal obligations to complete the purchase, and the facilities passed inspection.
- The city was obligated to pay the company $7,595,000, minus deductions for contract provisions that were not met.
- Jersey City had to pay JCWSC for the water delivered at the rate specified in the contract until the city purchased the facilities.

The contract specified deadlines when the contract provisions had to be completed. Sorting through the arcane contract language, Stevens was certain that the contract laid out a clear plan. "It is hard to imagine how language could have been more explicit."[9]

A key deadline in the original contract was the completion date for the dam, reservoir, and pipeline—August 28, 1901. A supplemental contract extended this deadline to December 25, 1903. Because water deliveries from the Boonton waterworks did not begin until several months beyond the deadline, Stevens ruled that liquidated damages applied at the rate of $500 per day, for a total of $72,500, to be deducted from the purchase price. Again he considered the contract language clear: "It would be quite impossible for the parties to have expressed themselves with greater clearness."[10]

At one point during the proceedings, representatives of the city claimed they had been "injured" by receiving water from the Little Falls source. In blunt language, Stevens dismissed this claim: "The injury on this head is hardly more than fanciful."[11]

In writing about the issue of liquidated damages, Stevens was actually commenting on the overall contracting process and the competence of the parties involved.

> "The parties were dealing at arms length. They had competent advice. They were peculiarly well informed in respect of the matters they were contracting about, and they were dealing with a subject incapable of being reduced to a certainty by any legal rule for the assessment of damages. No oppression, no unconscionable circumstances are shown; no inequality such as to shock the conscience of the Chancellor. . . . Under these circumstances, it seems to me that it is the duty of a court of equity to specifically enforce the contract and not to nullify it."[12]

THE CONTRACT WAS EXPLICIT in requiring JCWSC to provide water that was "pure and wholesome."

> "'The water proposed to be furnished is pure and wholesome. The plan has been prepared so as to prevent all contamination thereof from any source in accordance with the specifications.'

The specifications provided [are] as follows:

'The water to be furnished must be pure and wholesome for drinking and domestic purposes.'"[13]

Stevens made clear what he understood the phrase "pure and wholesome" to mean. His definition provides a distinct judicial opinion on just how pure "pure" should be.

"Again, the requirement that the water must be pure and wholesome does not mean that it shall be absolutely pure—of such purity as could be obtained in a labora-tory—all that is required is that it be 'free from pollution deleterious for drinking and domestic purposes.'"[14]

Stevens summed up the contract requirements regarding the company's obligation to provide "pure and wholesome" water to Jersey City under all likely conditions.

"The thing [waterworks] to be delivered [by JCWSC] is a plant capable of preventing contamination from any source, at any time, under any conditions likely to occur, and not a plant that may be effective under favorable conditions for a part of the year but ineffective at other times. As this is a very important part of the case, I may be permitted to illustrate further. Drought in summer is no uncommon occurrence; heavy rain following drought is no uncommon occurrence; high winds accompanying rain [are] no uncommon occurrence. If the plant be capa-ble of delivering pure and wholesome water in ordinary weather but not on the happening of the occurrences mentioned, either separately or together, then I take it that the plant would not be so completed as to meet the requirements of the contract."[15]

Once Stevens made this finding, hundreds of pages of testi-mony came into perspective. The defendants had argued that, on average, the water as delivered to Jersey City had low bacteria counts. The plaintiffs had argued that averages obscured the fact that several times per year, under various meteorological and operational conditions, high bacteria counts and many samples with positive B. coli analyses showed that the water served to Jer-sey City was contaminated.[16]

Stevens noted the groundbreaking changes that had occurred over the previous 30 years in relation to protecting water supplies from disease-causing germs. He noted agreement among the experts on both sides (including Leal) that pathogenic agents could survive for weeks once they were introduced into a water supply. Quiescent water, as provided for by Boonton Reservoir, reduced the numbers of these pathogens through a variety of mechanisms, including the disinfecting effect of sunlight on the upper layers of the water, deposition in the sediments, and predation by other organisms in the water. The question then became how long was the water subjected to the purifying attributes of Boonton Reservoir and was the storage sufficient to purify it under all likely conditions?[17]

Purification resulting from storage was crucial because both sides agreed that the water passing Boonton, New Jersey, and entering Boonton Reservoir ". . . was little better than an open sewer."[18]

Leal made the following points during part of his testimony at the first trial:

- As the sanitary adviser for JCWSC, he was responsible for protecting the Boonton Reservoir watershed. He diligently inspected the watershed himself, and he oversaw a group of men who regularly patrolled the watershed looking for illegal discharges of sewage or other contaminants. He resolved hundreds of individual sources of contamination—such as overhanging privies and leaking cesspools—and brought a number of complaints to trial to abate the nuisances created by contamination of the Rockaway River.

- He presented data demonstrating that total bacteria concentrations and incidences of B. coli were very low, on average, as a result of the "treatment" provided by Boonton Reservoir.

- He worked with physician and consulting bacteriologist Dr. George E. McLaughlin, who analyzed the bacteriological quality of water as delivered to Jersey City, 23 miles from Boonton Reservoir.

Leal's testimony and his theory of water treatment by means of the storage capacity of Boonton Reservoir were supported by a number of studies during this period. One study in particular stands out. Dr. Alexander Cruikshank Houston was director of water examinations for the Metropolitan Water Board in London, England. This is the same person who in 1905 had initiated the chlorine feed to the water supply at Lincoln, England, during a typhoid epidemic (see Chapter 4). In 1908, *Engineering News* published Dr. Houston's report on a study investigating the effect of storage on the survival of typhoid bacilli in water.[19] He found that a storage period of one week resulted in a 99.9 percent reduction in typhoid bacilli. He favored storage as a pretreatment for filtration—either slow sand or mechanical filtration. In summary he stated:

> "The advantages accruing from adequate storage of water are of a general character and are not confined to the elimination of danger from typhoid fever."[20]

An editorial in the same issue of *Engineering News* explored the possibility of using large storage reservoirs to replace expensive filtration plants.[21] Water professionals were searching for alternatives to doing nothing to control epidemic typhoid fever during this period. Coincidentally, the chloride of lime feed facility was under construction at the Boonton Reservoir when Houston's report was published.

The plaintiffs' attorney presented credible witnesses who convinced Stevens that during storm events, water bypassed a large area of the reservoir, allowing higher-than-acceptable concentrations of bacteria to remain in the water delivered to Jersey City. Figure 9-1 is a copy of the diagram from Stevens's opinion showing the relatively short distances from the reservoir influent (in the northwestern portion of the reservoir) to the spillway and gatehouse (on the northeastern side of the reservoir), from which water was delivered to Jersey City.[22]

The plaintiffs conducted several "float" studies showing that the main course of flow was across the narrow northern portion of the reservoir toward the gatehouse. Stevens believed that during significant rainfall and runoff events, water from the contaminated river would short-circuit a large area of the reservoir. He summed up his understanding in one sentence: "There can be no question that in two or three days some of the inflowing water [from the contaminated river] would have reached Jersey City."[23]

Bacteriological data were available to back up the hydrology and expected purification of water in the reservoir, but interpretation of the data was in the eye of the beholder. Most of the data quoted in Stevens's opinion related to total bacteria count, and it appeared that results from gelatin plate incubations were used most frequently if not exclusively. The influence of Dr. Robert Koch's original method for counting total bacteria in water

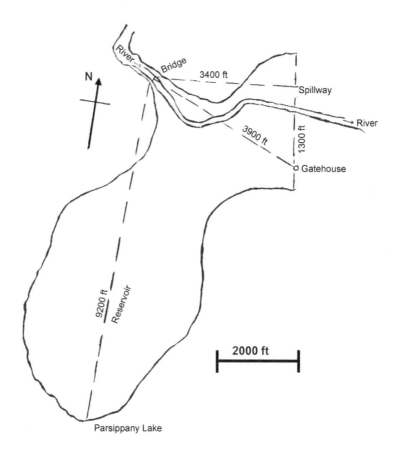

Figure 9-1   Boonton Reservoir—relative locations of inlet and outlets

extended through the 1880s and 1890s and into the early twentieth century with the Jersey City trial testimony.

Stevens's interpretation of the data from both plaintiffs and defendants was succinct. "These analyses show that when the water reaches Jersey City it contains many more bacteria on some days than it does on others."[24] In fact, despite some problems with sampling locations in Jersey City, Stevens was convinced that high concentrations of bacteria at the taps in Jersey City were connected with some of the high flow events sampled on the Rockaway River.[25]

After an extensive discussion of the total bacteria count results, the hydrology of the Rockaway River and watershed, and the purification processes in the reservoir, Stevens stated his conclusion.

"I am on the whole, obliged to conclude that all the evidence favors the theory that water, under certain combinations of circumstances occurring perhaps on an average two or three times a year, will pass from the mouth of the river to the Jersey City reservoirs in two or three days. . . . Now, this appears to be the difficulty with the defendants' case on this branch of it. *The reservoir does its purifying work imperfectly at the time when that work is most needed.*"[26] (emphasis added)

Stevens then attacked the company's use of percentage removals of bacteria through the reservoir and averages of bacteriological data to prove that the reservoir was doing an efficient job of removing pathogens. "It is very easy to see that by taking averages, very good water—water above the standard—delivered at one time may be used to cover up the faults of polluted water delivered at another time."[27]

Stevens's opinion shows that he listened carefully to the bacteriological data and to the expert opinions on guidelines for determining what constituted safe drinking water. As guidance for understanding the importance of total bacteria count, Stevens used a statement by Leal dealing with the operation of the mechanical filtration plant at Little Falls. "'We don't care so much about the rate of efficiency, but we want less than 100 bacteria [per milliliter].'"[28]

"This, no doubt, is a perfectly accurate statement of the matter. It agrees with what is testified to by all of the experts; but so far from bearing out Dr. Leal's theory of the reservoir as a purification basin, it shows very clearly that the important thing is not so much [a] high percentage of efficiency as it is absolute results."[29]

Stevens observed that none of the company's experts suggested a guideline for good quality water based on B. coli test results. Stevens noted that B. coli removal efficiency data were not included in the treatment plant data supplied by the defendants. He also observed that it was, therefore, not possible to compare the B. coli removal results with B. coli data for treated water delivered to Jersey City.

Thus, Stevens was left with only the B. coli rule for water safety espoused by Whipple, whom he quoted in his opinion.

> "'If the water regularly shows the presence of b. coli in ten cubic centimeters and not in one cubic centimeter it may be safe to use. If it contains b. coli in one cubic centimeter and not in one-tenth cubic centimeter it may be considered as of doubtful quality. If it contains b. coli in one-tenth cubic centimeter but not in one-hundreths [sic] cubic centimeter it may be considered as much too polluted to be safely used. If it contains b. coli in one-hundreths [sic] cubic centimeter the water is quite certain to be seriously polluted.'"[30]

Stevens applied the rule to B. coli concentrations in the water delivered to Jersey City customers. He put significant weight on the B. coli data because he was aware that these data were more indicative of undesirable pollution than total bacteria counts. He found that during the years 1904, 1905, and 1906, the water delivered to Jersey City was of doubtful quality for 25.5, 18, and 29 days, respectively.

Several experts for the company presented evidence showing that after sedimentation treatment was implemented in Boonton Reservoir, the typhoid fever death rate in Jersey City was much lower than the typhoid death rates in other cities. Stevens acknowledged the improvement in Jersey City's typhoid statistics but dismissed them by falling back onto the crucial contract language: "In the case in hand Jersey City bargained, not for water less polluted than that of some other cities, but for pure and wholesome water."[31]

Stevens was respectful of the opinions of the company's experts—Leal, McLaughlin, Johnson, and Hering—who claimed that the water from Boonton Reservoir was safe. But he was not impressed that no typhoid epidemic had occurred while water from the reservoir had been delivered. He agreed with the plaintiffs' experts that because no epidemic of typhoid fever had occurred in the watershed, the purification potential of the reservoir had not been put to the test for protecting the residents of Jersey City against typhoid fever bacteria.

Stevens agreed with the testimony of Whipple, Winslow, and Sedgwick, who claimed that the water from Boonton Reservoir was ". . . at times highly impure and unwholesome." Summarizing this point, Stevens wrote: "I think the weight of the evidence is that while much has been done toward securing the end in view, the works are not yet 'so prepared as to prevent all contamination from any source.'"[32]

THE ORIGINAL CONTRACT between Jersey City and Patrick H. Flynn allowed for a $500,000 deduction from the water supply's purchase price if Flynn (or his successor, JCWSC) did not obtain the water rights held by the Morris Canal and Banking Company on the Rockaway River upstream of Boonton Reservoir. This was a considerable amount of money at the time and it dwarfed some of the costs associated with building sewers and sewage disposal facilities.

By the time of construction of the Boonton Reservoir and the two trials, the Morris Canal was little used. Railroads had taken over the transport of goods throughout New Jersey, New York, and Pennsylvania. However, the Morris Canal and Banking Company had the right to use the waters of the state of New Jersey to operate the canal for a period of 99 years, dating from the inception of the canal's charter in 1824. The proviso was that the water had to be used to maintain navigation of the canal.[33] The defendants in the Jersey City trials had made an arrangement with the Morris Canal and Banking Company and the Lehigh Valley Railroad Company (the lessee) to use the Morris Canal as the destination of waste flows from the town of Boonton. It was known that water from the Rockaway River watershed was being put into the canal even as late as the time of the trials (1906–1910). However, Vice Chancellor Stevens observed that the contract between the two parties depended on continued navigation of the canal.[34]

Stevens found that the provisions of the Morris Canal contract were still viable and that JCWSC had not obtained the rights to use the canal for waste removal.[35] Thus the court upheld the terms of the canal's charter and ordered $500,000 to be deducted from the purchase price of Jersey City's water supply. In terms of buying power, the deducted amount is equivalent to more than $11 million in today's dollars.[36]

Vice Chancellor Stevens ended his 100-page opinion after a discussion of the Morris Canal and a few "minor objections."[37] He waited several weeks for appeals and other motions to be filed before issuing his final decree.

On June 4, 1908, the final decree summarized Stevens's decision and set the stage for the next trial. In large part, the defendants won. Stevens ordered the company to deliver the waterworks to Jersey City, and he ordered the city to pay the defendants $7,595,000, minus various deductions. The question was how big would the deductions be?

On the other hand, the plaintiffs also won a critical victory. Stevens's ruling stated unequivocally that the water supply as delivered to Jersey City was not pure and wholesome.

> "And the court being also of the opinion and adjudging that Patrick H. Flynn and the Jersey City Water Supply Company have not complied with those provisions of the said contract which stipulate that the supply of water delivered to Jersey City shall at all times be free from pollution; it appearing that the works, as a mechanism for the purification of the supply, are not at all times adequate and reliable. . . ."[38]

Final resolution of the amounts the city would be required to pay the water company is discussed at the end of Chapter 11.

Stevens's ruling required the water company to mitigate the significant sources of contamination in the Rockaway River watershed and permitted the costs of sewers and sewage disposal facilities to be deducted from the purchase price. Stevens then made history by allowing for an alternative to the expensive sewers and sewage treatment plant.

> "In lieu of and as a substitute for all or any of the sewers and sewage disposal works . . ., the defendant company may, within ninety days from the date hereof [June 4, 1908], present *other plans or devices* for maintaining the purity of the water delivered by the company to the city throughout the year. . . ."[39] (emphasis added)

The defendants asked the Vice Chancellor to put this wording into his final decree. Why he complied with their request has never been explained in documents that survive this period of New Jersey history.

Stevens then turned over to Special Master William J. Magie the responsibility to determine the cost of the sewers and sewage disposal plants and ". . . to investigate and report upon other plans and devices as alternative remedies presented by the defendant company for delivering water to the city in a pure and wholesome condition throughout the year."[40]

1   *Evening Journal*, April 27, 1907.

2   Whipple, "Journal, January 8, 1914."

3   Winslow to Sedgwick, Charles-Edward Amory Winslow papers, May 23, 1916.

4   Bergen and Magie, "Order of Injunction and Modifying Order," Vol. 7 Exhibits, 4043–6.

5   Jersey City Board, "Water Question to the Taxpayers," 6.

6   Stevens, In Chancery of New Jersey, Opinion, 1908, 4047–148. In my copy of Volume 7 of the 12-volume set of both Jersey City trial transcripts found in the museum of the Passaic Valley Water Commission, marginal notes were made throughout Stevens's opinion. The notes are obviously in the handwriting of William H. Corbin. On the front cover of the volume is written in pencil "WHC." A rubber stamp on the inside of the front cover and the first blank page identifies the book as belonging to Corbin—'Return to William H. Corbin Jersey City [the location of his office]'. The notes in the margins are clearly in the same hand and were written with the same or a similar pencil.

7   Stevens, In Chancery of New Jersey, Opinion, 1908, 4049.

8   The wording of the summaries is the responsibility of this author.

9   Stevens, In Chancery of New Jersey, Opinion, 1908, 4055.

10 Ibid., 4068.

11 Ibid., 4072.

12 Ibid., 4074–5.

13 Ibid., 4078.

14 Ibid., 4080.

15 Ibid., 4082; no notes penciled by Corbin in the margin disagree with this critical finding.

16 In marginal notes, Corbin noted what he thought was an inconsistency in Stevens's opinion. On page 4083, he wrote a question mark and highlighted Stevens's statement that ". . . in general, the reservoir does act effectively as a sedimentation basin. . . ." On page 4103 is the oft-quoted passage that ". . . two or three times a year . . ." contaminated water was delivered to Jersey City. These two passages are not inconsistent and may be an example of Corbin searching for a basis on which to justify his appeal.

17 Stevens, In Chancery of New Jersey, Opinion, 1908, 4083–8.

18 Ibid., 4081.

19 "Laboratory Tests of the Effect of Storage," *Engineering News*. Report by A.C. Houston, 247–8.

20 Ibid., 248.

21 Editorial, *Engineering News*, 257.

22 Stevens, In Chancery of New Jersey, Opinion, 1908, 4091.

23 Ibid., 4093–4.

24 Ibid., 4098.

25 Ibid.

26 Ibid., 4103–4.

27 Ibid., 4109.

28 Ibid., 4107; As noted in Chapter 3, Robert Koch first cited this limit defining good quality water produced by a slow sand filter.

29 Ibid., 4107.

30 Ibid., 4111.

31 Ibid., 1908, 4115.

32 Ibid., 4119–21.

33 Trowbridge, *Reports on the Water Power*; Sullivan, *Paterson Manufactories*.

34 Stevens, In Chancery of New Jersey, Opinion, 1908, 4145–7.

35 Ibid., 4134.

36 Bureau of Labor Statistics, CPI Inflation Calculator; the calculator provides an estimate of the current buying power of the original amount.

37 Regarding the removal of the rag mill at Powerville, Stevens stated on pages 4136–7 of his opinion: "It conclusively appears that from Mr. Corbin's testimony that the Water Company had notice of it [a letter signed by Flynn and Edwards that was transmitted during the signature session of the original contract between Flynn and Jersey City]." In the margin, Corbin scribbled "absolutely not."

38 Stevens, In Chancery of New Jersey, Final Decree, 1908, 4160.

39 Ibid., 4163.

40 Ibid., 4165.

10

Chlorination Plant at Boonton

". . . it is very probable that the process will, in a few
years, be very extensively used. . . ."
—Harrison, "The Public Water Supplies of Hudson Co.," 13

Between the two Jersey City trials, the revolutionary water treatment process using chloride of lime was conceived, designed, and implemented. How could Leal have been permitted to add chlorine to an existing water supply when there were no long-term examples of this treatment in the United States? Adding chemicals to drinking water was serious business in 1908 because of the public's chemophobia, but no New Jersey Department of Health regulations were in place to stop him or give him permission.

Still, this was no junior engineer experimenting with an unknown technology. This was a practicing physician, who had been the health officer for a major city in New Jersey. When Leal engaged George Warren Fuller and his firm to design and oversee construction of the chloride of lime facility, this was equivalent of the United States assembling a "dream team" of the country's finest basketball players for the 1992 Olympics. No hacks. No politician–engineers. Simply the best of the best brought together to protect public health.

The use of chlorine in Jersey City was not due solely to the fact that Leal was a physician. At the turn of the twentieth century, three groups of experts were engaged in protecting drinking water and public health:

- Doctors, who treated diseases that we now know originated in contaminated drinking water.
- Public health experts, who were responsible for collecting and analyzing disease statistics and making sure infectious diseases were under control. These experts included bacteriologists and epidemiologists who worked in academia or for state boards of health.

- Engineers, who were responsible for designing and building public works projects whose ultimate aim was to protect public health—sewer systems, potable water distribution systems, and treatment systems for sewage and drinking water.

Leal was an expert in the first two areas of proficiency, and he knew what to do about the third. He hired the best sanitary engineer in the United States to provide the engineering assistance.

On their own, engineers would never have been allowed to add chlorine to drinking water. Public health experts, though aware of the general disinfection capability of chloride of lime, would not have had the practical expertise to design and build such a system. Academic experts in bacteriology were competent at testing and investigating, but they would not have made the leap to applying a disinfectant to drinking water; they would have been too concerned that they did not know everything about the chemistry and microbiology of disinfection.

Leal did not need to know everything. On the basis of his education and experience, he knew enough to make the leap. As a public health expert, he understood that the Sewer Pipe–Water Pipe Death Spiral had to be broken by means of technology that did not rely on expensive civil works like sewers, sewage treatment plants, and filtration plants. He did not have training as an engineer, so he was not a slave to building things. He certainly knew that moving sewage to a river or lake that served as someone's water supply was not the solution. There had to be a way of killing the disease-causing organisms, even if the method was not perfect and did not kill all bacteria. That method was disinfection using chloride of lime.

What Leal could not do was design a foolproof process that could demonstrate how the concept worked. Although he had tested chlorine in the laboratory, he had not designed a system to feed a disinfectant to drinking water. The electrolytic chlorine generator he had worked with on the lab bench made chlorine gas just fine, but a later attempt to apply chlorine gas at full scale was unsuccessful. If Leal had attempted to use chlorine gas at full scale at Boonton on his own, the undertaking would have been a failure, and chlorination of drinking water would have been set back decades.

The key step in the design of the Boonton chlorination system was setting the chlorine dosage, and Leal maintained control of that critical aspect. Once the dosage was established, all other components of the design would follow—the size of the tanks, the sizes of the orifice plate openings, the concentration of the chloride of lime solutions, the amount of power needed to run the mixers, and the size of the building to house all of the equipment. The detailed design of tanks and pipes would have been far different for chlorine dosages of less than 1 part per million (ppm) compared with dosages similar to that of 4,200 ppm used in Maidstone, England in 1897.

ALTHOUGH FULLER WAS TRAINED as a chemist and was a student and practitioner of bacteriology, he had been operating as a sanitary engineer since 1890. In 1908, could he have come up with the idea to add chlorine to the Boonton Reservoir water? It is highly unlikely that he would have hit upon that solution. He knew how to build water filtration plants, sewers, and sewage treatment plants, but he was a conservative engineer who implemented what his clients wanted. If Leal had traveled to Fuller's office on June 19, 1908, and asked him to design sewers and sewage treatment plants, Fuller would have been glad to do so. He would never have said, "Don't build sewers. That will not solve the problem. What you really want to do is add a poison to the water supply."

Leal asked Fuller to do something quite different from simply building a physical facility. He asked Fuller to take a radical departure from the norm. Fuller must have known that he was taking a personal risk. If the chlorine addition had turned into a disaster, Fuller would have been in as much trouble as Leal. Fuller had to be able to assess for himself that adding low doses of chlorine to the Jersey City water supply had a good chance of working. Up to 1908, he had not published anything on disinfection of drinking water, but he must have been well aware of the literature. With his background as a chemist and bacteriologist, he surely had an idea that Leal's proposal could work. Because he and Leal were both bacteriologists, they spoke the common language of the microscope, gelatin plate, and fermentation tube.

Fuller must have been energized about the chance to build such a plant. He knew about previous tests in which chlorine in one form or another had been added to drinking water, but even during typhoid fever outbreaks, chemical disinfectants were added only as a last resort and for limited periods. The project for Jersey City would break new ground and establish technological leadership in the United States. Because of the lawsuit and Leal's control over the water supply at Boonton, Fuller knew he could get the work done without having to obtain permission from bureaucrats.

At the same time, Fuller must have had some reservations about the project. He was undoubtedly aware of the paper Whipple had presented at an American Water Works Association (AWWA) conference in Boston just two years before Leal's visit to his offices. Whipple had not recommended adding a chemical disinfectant to drinking water supplies. His presentation simply explored the alternatives for disinfection, but the audience attacked him for even considering such nonsense. Because the fear of adding chemicals to water was rampant in the United States, Fuller also would have realized that professional censure could follow his involvement in such a scheme. He must have pondered the situation and considered all the angles. One comforting thought had to be that he knew he could make the concept work. He knew how to design chemical feed systems, and similar systems had already performed reliably at full scale.

Whatever his thought process, Fuller agreed to collaborate in Leal's bold plan, and Fuller's staff immediately began to create the practical apparatus that would set the chlorine revolution in motion.

BY 1908, NUMEROUS PUBLICATIONS had reported on bench-scale chlorination studies and their effects on bacteria. Chlorine killed bacteria and was particularly effective against pathogenic bacteria. Chlorination's utility had been known for decades, but its practicality and reliability had not been verified on a continuous basis at full scale in the U.S.

Leal did not have time for a pilot study. He certainly did not have time to build a demonstration-scale facility to test the new technology. How would the chloride of lime be fed? Would it clog the feed mechanism? How would the chlorine dosage be controlled? Leal understood that the key to success was keeping the dosage significantly below 1 ppm to avoid consumer complaints about the chemical's taste and odor at higher dosages. The public was familiar with the smell of chloride of lime because it had been used for decades as a disinfectant to clean filthy streets and for more than a century as a bleaching chemical to whiten laundry. If the chloride of lime feed system lost control of the amount of chemical being fed and a slug of high chlorine residual was delivered to Jersey City, Leal knew that would define the failure of the process.

Incredibly, Leal and Fuller jumped from bench scale to full scale—at 40 million gallons per day. This was a leap of faith for both of them. No one had done anything like this before.

Multiple sources of information provide details on the design of the chloride of lime feed system at Boonton. On June 8, 1909, Leal, Fuller, and George A. Johnson gave presentations on the plant's design and operation at AWWA's annual conference in Milwaukee, Wisconsin. Testimony at the second Jersey City trial described the plant facilities in some detail, and later publications gave an overview of the facilities along with selected design details.[1]

Figure 10-1 is a schematic of the chloride of lime feed facility at Boonton. According to Fuller's testimony, he made only nine engineering design drawings to guide the contractor during construction of the plant.[2] For an equivalent facility today, dozens of drawings would be required.

The chloride of lime facility was housed in a one-story wooden building that was constructed adjacent to the gate house located at the foot of Boonton Dam.[3] In addition to all of the mechanical equipment required to feed chloride of lime, the building housed a small laboratory used to perform simple chemical tests and to conduct bacteriological examinations.

The concentrated chloride of lime powder was put into dissolving tanks along with dilution water from the reservoir (Figure 10-1). Typically, the bleaching powder contained 35 percent available chlorine. A highly concentrated solution of chloride of lime was made in the dissolving tanks and then fed by gravity into the solution tanks. More dilution water was added to the solution tanks to create the desired strength for the chloride of lime mixture. Triplicate pairs of dissolving and solution tanks allowed the operator to produce large batches (about 10,000 gallons each) of 0.5–1 percent dilute solutions.

A belt-driven turbine pump[4] (in duplicate) moved the dilute solution up to one of the two orifice tanks. The orifice tanks were positioned at a relatively high elevation, enabling them to feed chlorine solution by gravity into the chamber below. The chamber was downstream of the 48-inch pipelines connecting the outlet tower of the dam to the pipeline delivering water to Jersey City. Duplicate orifice tanks were a critical design factor because chloride of lime in 0.5–1 percent solutions tended to build up solid deposits on the sides of the orifice plate and obstruct the opening.

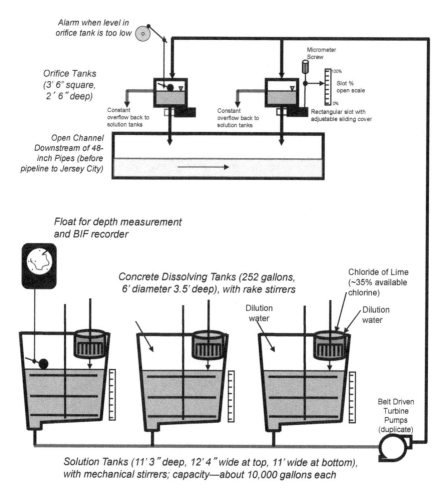

Figure 10-1   Schematic of chloride of lime feed system at Boonton Reservoir, 1908

If one of the orifice tanks became plugged, the other one could be used while the plugged one was cleaned.

Flow of the chloride of lime solution was set and maintained by delivering a slight excess flow from the turbine pumps to the orifice tanks and allowing the excess solution to overflow out of the tanks. This arrangement ensured that a constant head was always maintained above the orifices at the orifice tank outlet. At the bottom of each orifice tank was a plate held in position by bolts. Each plate had a fixed opening made of hard rubber that was

0.75 inches wide and 4 inches long. A movable cover was mounted over the fixed orifice. The cover could be moved backward or forward to set the desired length of the opening. The cover was moved by a micrometer screw device connected to a pointer that indicated on a scale how far the orifice was open.

The quantity of dilute chloride of lime solution that was fed was based on calibration of the orifice device. For a set orifice opening, the amount of liquid fed in a specific time period was noted. If the pump became blocked and the level of solution in the orifice tank dropped, a gong would sound to alert the operator. The level of liquid fed from the solution tank was monitored with a float that rested on the solution surface and was connected to a depth recorder manufactured by Builders Iron Foundry Company (Figure 10-1).[5]

The concentration of the dilute chloride of lime solution was determined by the Penot titration method.[6] No reliable method of measuring chlorine residual in treated water was available to Leal, Fuller, and Johnson, but quantifying the amount of chloride of lime solution flowing into the mixing chamber as well as the flow of water through the chamber made it possible to determine the amount of chloride of lime fed into the water in parts per million. The flow of water in the 48-inch pipes feeding the chamber was determined by the setting of the gates that fed the outlet valves and pipelines at the base of the dam.[7]

Dilute chloride of lime solution was fed through perforated pipes into the mixing chamber from the reservoir. After this, the water took several turns, traversed flanges and valves, and was "very much in commotion." In other words, the water flow was turbulent, and the dilute solution of chloride of lime was thoroughly mixed before the water began its 23-mile journey to Jersey City.

In 1908, the chloride of lime powder was delivered by rail and truck in sheet-iron drums containing 750 pounds of bleaching powder. Dissolving a known amount of the powder in water and determining the percentage of available chlorine by means of the Penot method determined the strength of the chloride of lime material. In the basement of the chloride of lime building was enough storage space to house 120 drums, for a total storage capacity of 90,000 pounds—at least one year's supply of the chemical.

Figure 10-2 shows a schematic of the alum feed facility at the Little Falls treatment plant, which had been designed by Fuller and began operation in 1902.[8] The similarities between the two

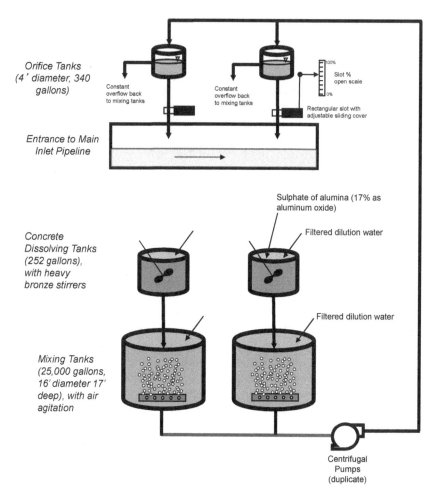

Figure 10-2   Schematic of alum feed system at Little Falls treatment plant, 1902[9]

chemical feed designs are unmistakable. Fuller clearly used the successful alum feed system at the Little Falls treatment plant as the model for the chloride of lime facility at Boonton. He was able to design and build the successful Boonton facility in 99 days only because he had worked out most of the kinks with a similar chemical feed system six years before.

The two designs also had some differences. Fuller wisely replaced the air agitation system used at Little Falls with mechanical stirrers at Boonton. In the chloride of lime system, the biggest problem with air agitation would have been the release of chlorine gas from the concentrated solution of chloride of lime. Such a release would have damaged the mechanical and electrical equipment in the building and created a fatally toxic atmosphere for the operators.

The 25,000-gallon mixing tanks at Little Falls were 2.5 times larger than the tanks at Boonton. Because the alum dosages at Little Falls (8–34 ppm) were significantly higher than the chloride of lime dosages (0.2–0.35 ppm) at Boonton, more solution needed to be fed at the Little Falls plant, and a day's worth or more of chemical needed to be stored there to avoid frequent operator involvement.

Three mixing tanks (or solution tanks) were constructed at Boonton, whereas only two were built at Little Falls. This is an excellent example of a conservative engineering design in the face of uncertainty. Fuller designed three solution tanks at Boonton undoubtedly because he was not sure how the chloride of lime solutions would be made or handled in actual practice. He also must have been concerned that if a batch of the chloride of lime solution was bad (too dilute or too degraded), it would have to be dumped, which would have complicated operations with only two tanks available. All in all, the tank design and arrangement at Boonton provided elegant solutions to design problems and achieved important improvements to an already proven chemical feed system operating at the Little Falls plant.

THE CHLORIDE OF LIME PLANT began operation on September 26, 1908. George A. Johnson operated the plant from the start until the end of that year, December 31. Leal took over operations after that date. Although testimony from the second Jersey City trial and the papers that Leal, Fuller, and Johnson published show that all Johnson did was operate the plant for three months, his role with the Boonton plant became exaggerated over subsequent years.

According to Johnson, only one person—in addition to the staff required to run the hydraulic facilities at the dam—was needed to operate the chloride of lime plant and perform the necessary testing. The plant operator's duties included making up the chemical solutions, keeping records of the amount of chloride of lime applied to the water, collecting samples, testing samples for chemical and bacteriological parameters, and preparing daily reports.

Also according to Johnson, very little trouble occurred with clogged orifices. However, accounts from operators of ensuing chloride of lime facilities cited clogging of the measuring and feeding devices as a major drawback of the process. The chlorine dosages during the first months of operation at Boonton were typically 0.2 or 0.35 ppm of available chlorine.

Six large tables in Johnson's 1909 paper summarized the operational, chemical, and bacteriological testing data for October through December 1908.[10] The results of these tests were discussed during the second Jersey City trial.

After Leal took over operation of the plant on January 1, 1909, the chlorine dosage was maintained at 0.35 ppm. In February, March, and the first part of April, the dosage was reduced to 0.2 ppm. On April 24, 1909, the dosage was raised to 0.35 ppm, where it remained until Leal testified at the second trial on October 14, 1909.[11]

AT THE SECOND TRIAL, the Secretary-Treasurer of the Jersey City Water Supply Company (JCWSC) testified that he had kept track of all the costs associated with construction of the chlorination facility at Boonton beginning on July 28, 1908. The total cost—including fees paid to Hering and Fuller for design and construction supervision—was $20,545.64. This sum has an equivalent buying power in today's dollars of about $470,000.[12]

Johnson's paper stated that the plant's operational costs were only $0.14 per million gallons of water treated. These costs included the operator's salary, bleaching powder, coal for heating the building, and miscellaneous costs for laboratory supplies. No power costs were incurred because water power from the dam was used to run the mechanical equipment.[13]

How much were Hering and Fuller paid for designing the plant and supervising its construction? Fuller had a reputation for being good at the business of consulting. "Fuller had remarkable ability when it came to making money. He could pretty well size up what a municipality was prepared to spend for a particular project in the way of engineering fees and then talk them into spending a substantially larger sum."[14] The apparent arrangement for this project was that JCWSC would cover Hering and Fuller's costs plus an engineering fee.

An exhibit provided by the company's Secretary-Treasurer during the second trial shows that the amount paid to Hering and Fuller for all of the design and construction management work was $5,297.64.[15] In today's dollars that works out to more than $120,000.[16] The payment seems fairly modest for what was delivered—the first permanent chlorination facility in the world. The fee certainly was not excessive. No engineering consulting firm today would agree to take the kind of risks Fuller took or to deliver a project so fast for that amount of money.

1   Leal, "Sterilization Plant of the Jersey City Water Supply Company," 100–9; Fuller, "Description of the Process and Plant of the Jersey City Water Supply Company," 110–34; Johnson, "Description of Methods of Operation of the Sterilization Plant," 135–47; Between Jersey City and Water Company, February 5, 1909, 5081–91. Unfortunately, differences in the dimensions of some of the tanks and facilities were noted in the various publications. The most likely dimensions are shown on Figure 10-1.

2   Between Jersey City and Water Company, February 5, 1909, 5081.

3   The building was 33 by 67 feet or about 220 square feet in area. The various pieces of equipment were housed either on the main floor or in a basement. Fuller's paper gives details of the equipment locations.

4   Nominal flow rate of 25 gallons per minute.

5   The BIF recorder has been the workhorse of water treatment monitoring for more than 100 years.

6   Fresenius, *Quantitative Chemical Analysis*, 505-6.

7   Between Jersey City and Water Company, February 4, 1909, 4992.

8   Fuller, "Filtration Works at Little Falls," 168–72.

9   Ibid.

10   Johnson, "Description of Methods of Operation of the Sterilization Plant."

11   Between Jersey City and Water Company, October 14, 1909, 6742–3.

12   Bureau of Labor Statistics, CPI Inflation Calculator; the calculator only goes back to 1913 but provides a reasonable estimate of the current purchasing value of the original cost of the plant's construction.

13   Johnson, "Description of Methods of Operation of the Sterilization Plant," 145.

14   Harding, "Personal Reminiscences of George Warren Fuller," 1525.

15   Between Jersey City and Water Company, February 23, 1909, 5598–5600.

16   Bureau of Labor Statistics, CPI Inflation Calculator; the calculator provides an estimate of the current buying power of the original amount; the estimate is sufficient for assessing how reasonable Hering and Fuller's charges were in 1909.

11

Chlorine on Trial—1908 to 1910

*"I do therefore find and report that this device is
capable of rendering the water delivered to Jersey City,
pure and wholesome. . . ."*
— Magie, In Chancery of New Jersey, 13

The principal attorneys in the second Jersey City trial were the same—William H. Corbin for the Jersey City Water Supply Company and James B. Vredenburgh for Jersey City. The special master for the trial was William J. Magie, an outstanding jurist who played a leading role in New Jersey judicial history. William D. Edwards, who was present as a member of the firm Collins & Corbin, conducted some of the questioning of witnesses for both the plaintiffs and the defendants. George L. Record, corporation counsel for Jersey City, was listed as an attorney for the plaintiffs, but he did not conduct any questioning during the two trials.

The second trial consisted of 38 trial days over a 14-month period. A total of 73 witnesses were called, hundreds of exhibits were submitted and examined, and more than 2,700 pages of testimony were taken and transcribed. Hundreds of pages of testimony covered routine information such as who took which samples and who did the analyses.

The conduct of the second trial was not linear. Witnesses were called out of order, then recalled, and then called again much later to rebut testimony from the other side. The trial's progress was interrupted for months for a variety of reasons.

Topics covered during the trial included:
- Arguments between the attorneys regarding the purpose of the trial and whether Justice Magie should consider chlorine as an alternative to sewers
- Detailed grilling of Dr. John L. Leal by both attorneys on what he did, why he did it, how he did it, and what adding chlorine to water meant
- Testimony from the plaintiffs' experts on the costs of sewers and sewage treatment

- Testimony from the plaintiffs' experts on the effect of sewers and sewage treatment on water quality
- Testimony from the defendants' experts criticizing the cost figures of the plaintiffs' experts and presenting their own sewer cost estimates
- Testimony from the defendants' experts on design and construction of the chloride of lime plant; operators testified about how they ran the plant
- War of Experts
  - Famous experts for the defendants who claimed that chlorine was good and was a major advance in protecting public health and who said sewers would not improve the bacteriological quality of water from Boonton Reservoir
  - Famous experts for the plaintiffs who claimed that chlorine was bad and should not be used in drinking water and who said the best way to provide safe drinking water was to protect the source
- Testimony by experts on both sides about the chemistry, analysis, and mode of action of chlorine in water and about the issue of whether nascent oxygen or chlorine itself accomplished the disinfection
- Testimony by experts for the defendants and plaintiffs on the bacteriological tests and how they were conducted
- Miscellaneous testimony from both sides that had little to do with the decision on the suitability of chlorine use—for example, rainfall records, time of travel tests on the Rockaway River and Boonton Reservoir, evidence of sewage contamination in the Rockaway watershed, a census of privies and cesspools, and details of sample collection and analysis
- Testimony on other matters that were not related to the central discussion of chlorine treatment of drinking water—for example, riparian rights along the Rockaway River

One thing was consistent between the two Jersey City trials: Dr. John L. Leal sat at Corbin's side through both trials and in some cases helped formulate questions for the plaintiffs' expert witnesses. One newspaper account of the second trial noted, "Mr. Corbin, with Dr. Leal to prompt him, fired all sorts of technical questions at the city's high priced expert."[1]

IN CONCERT WITH VICE CHANCELLOR STEVENS'S RULING, the defendants filed a report with the Chancery Court on September 3, 1908, describing in general terms the alternative plans or devices that they claimed would maintain the purity of the water to be delivered from Boonton Reservoir. The report reviewed Vice Chancellor Stevens's ruling and emphasized that the contract between the parties did not call for a filtration plant. The report made it clear that pollution from agricultural and urban runoff would continue to contaminate the water entering Boonton Reservoir even if sewers were built.[2]

The plan devised by the defendants was described only as a system of "works" that needed to be built by the short deadline ordered by the court. The engineering firm of Hering and Fuller—"the most eminent sanitary engineers in the country"—was named as being in charge of the design and supervision of construction. The report noted that delays in completion of the "works" resulted from a variety of circumstances including the "slowness of contractors." Given the incredible speed with which the plant was built, a complaint about slow contractors seems almost churlish.

In a somewhat disingenuous tactic, the only treatment process mentioned in the report was the addition of oxygen to the water. The report claimed that liberated oxygen rendered the treated water "practically sterile." No mention was made of using chloride of lime. The only description of the "works" was that they consisted of some tanks in a building. The opinion expressed in the report was that the works would:

> ". . . enable the company to literally fulfill its contract
> as interpreted by the court, namely: 'to supply to Jersey
> City, every day in the year, water pure and wholesome
> for drinking and domestic purposes'. . . ."[3]

A cost estimate for construction of the plant was included in the report. The estimate—$25,325—was based on what was known at the time, but the plant was not finished when the report was submitted.[4]

The second trial started on September 29, 1908. The first order of business on the opening day was a request by the defendants to postpone everything. Corbin made a long statement in which he, once again, summarized the opinion and decree by Vice Chancellor Stevens. He also described in general terms the "other plans or devices" that the company was installing at the Boonton

Reservoir site as ". . . an experimental plant for the [introduction] of oxygen into the flow of water as it comes from the dam."[5]

Corbin stated that the experimental plant was put into use "last Saturday," which would have been September 26, 1908. He noted that Vice Chancellor Stevens desired daily bacteriological analyses during the first trial but that the company had not gathered these data that frequently. Corbin said the company had been taking daily bacteriological samples over the summer and wanted to continue the sampling through the next few months in order to catch rainfall and significant runoff events. He also wanted more time to operate the "works" to demonstrate conclusively that the water delivered to Jersey City would be "pure and wholesome." He requested a three-month adjournment.

Vredenburgh acknowledged the need for a delay but stated that two months would be sufficient. His position was that if the water was of doubtful quality, the risk of Jersey City's population contracting waterborne diseases was too high and no extended delay in finding a solution should be allowed. He was particularly concerned that a typhoid fever carrier could contaminate the water above Boonton Reservoir. He expressed concern about high death rates from childhood diarrhea and said this problem was related to the quality of the drinking water.

Vredenburgh also complained that Jersey City was paying the company for water delivered from Boonton Reservoir and that it would be significantly cheaper for the city to purchase the dam, reservoir, and "works" rather than continuing to pay the water delivery charge.

Vredenburgh stated his understanding that the treatment to be applied to the water consisted of passing electricity through air to produce ozone, which would then be introduced into the water. There is no mention in the trial transcripts, exhibits, or reports of the company that it was testing ozone or proposing its use. The company's insistence that its staff would be adding oxygen to the water to sterilize it may have given Vredenburgh and the city the impression that ozone was the treatment method selected.

Given his questions and comments, Special Master William J. Magie clearly understood both counsels' arguments for adjournment. Even though he was not up to speed on all aspects of the case, he could rely on the vice chancellor's opinion, which required him to carefully examine the "other plans or devices." He agreed to a three-month adjournment and scheduled the second day of trial for January 5, 1909.

ON THAT SECOND DAY OF TRIAL, Vredenburgh opened the case for the plaintiffs by calling one of his expert witnesses, Nicholas S. Hill. Vredenburgh did not deliver an opening statement for the plaintiffs. Presentation of the plaintiffs' case lasted nine days and occupied more than 650 pages of the trial transcript.[6]

Hill's testimony focused on the costs of sewers and sewage disposal facilities. An experienced engineer, he was chief engineer of the Water Department of New York City with responsibilities for the water supply in the borough of Manhattan.

He gave detailed testimony on the costs of individual sewers, intercepting sewers, and various sewage disposal plant designs. He also presented some data on the bacteriological contamination of the Rockaway River. The cost estimates changed over time as both sides debated the location of sewers and the need for sewage treatment. Interestingly, Hill discussed the possibility of disinfecting the treated sewage with chloride of lime.[7]

Final cost estimates for sewage disposal at Boonton and Dover are shown in Table 11-1, along with descriptions of two proposals for eliminating sewage contamination from Hibernia. These proposals were included in Magie's decision.[8] The plans for Boonton and Dover were expensive, ranging in cost from $600,000 to $800,000 ($14 million to $18 million in equivalent buying power today).[9]

Plan 2 was much more expensive than Plan 1 because of the extremely long sewer needed to move Dover's sewage 13 miles to the Boonton Dam area for treatment and eventual discharge into the Rockaway River 200 feet below the dam. The two treatment options offered for Hibernia seemed absurd. Suggesting the use of a "pail system" instead of privies did not make sense, given the likely inattention to removal of waste contents. In some ways, the proposed pail system was equivalent to London's early 1800s "night soil man" system, which was subject to abuses and midnight dumping in unapproved river sites.

After Hill's presentation on costs, Vredenburgh presented evidence that significant sewage contamination from Dover, Boonton, and Hibernia was entering the Rockaway River above Boonton Reservoir. Corbin objected to the extensive discussion of contamination presented by the plaintiffs, arguing that this issue had been settled in the first trial. Vredenburgh responded that he had to show the extent of sewage contamination of the Rockaway River to justify the extensive nature and costs of the collection and treatment systems proposed for all three towns. Magie noted

Table 11-1 Description and costs of proposed sewage collection and treatment systems above Boonton Reservoir

City	Project	Population Served	Average Daily Flow million gallons per day	Treatment Process	Capital Cost	Capital Cost per Capita	Effluent Discharge Point
Dover	Sewers	7,000	0.95	Primary sedimentation, primary contact beds, secondary sedimentation basin, addition of sterilizing medium (chloride of lime or sulphate of copper)	$212,407.00		
	Treatment plant				$115,633.00		Mill Brook, 12 miles above Boonton Reservoir
	Force main, pumping plant, land				$22,832.00		
	Subtotal Plan 1				**$350,872.00**	**$50.12**	
Boonton	Sewers	4,333	0.632		$193,299.00		
	Treatment plant			Grit chamber, primary sedimentation, primary contact beds, secondary sedimentation basin, sand filters, pump station	$55,085.00		Rockaway River, 200 feet below Boonton Dam
	Force main, pumping plant, land				$5,095.00		
	Subtotal Plan 1				**$253,479.00**	**$58.58**	
Total Plan 1					$604,351.00	$53.33	
Dover-Boonton	Sewers	11,333	1.673		$645,101.55		
	Treatment plant			No process details provided	$150,000.00		Rockaway River, 200 feet below Boonton Dam
	Force main, pumping plant, land				$5,095.00		
Total Plan 2					$800,196.55	$70.61	
Hibernia	Plan 1 – Pail system in town and sand filters for mine effluent				$19,245.00		
	Plan 3 – Stormwater sewers and drains with cut-off embank-ment along Hibernia brook. Effluent to be treated on sand filter				$78,152.00		

that the issue might have been decided in the first trial but that he would hear the plaintiffs' testimony on contamination anyway.[10]

Vredenburgh's main expert for presenting the sewage contamination data was Charles Edward North, a bacteriologist at the Lederle Laboratories in New York City. North and several assistants reported on extensive sampling of the Rockaway River and sewage flows from Dover, Boonton, and Hibernia.

Another expert witness for the plaintiffs presented testimony on a census of privies and cesspools and their deteriorated conditions in the vicinities of the three towns. North testified that he had observed a substantial flow from the Rockaway River entering the Morris Canal in the vicinity of Dover.[11] The relevance of the Morris Canal to the case was noted in Stevens's judgment at the end of the first trial (see Chapter 9).

Figure 11-1 depicts a portion of the Rockaway River watershed and indicates several relevant geographic relationships. Because the town of Boonton was immediately upstream of the reservoir, any contamination from that town would definitely have had adverse effects on the quality of the water entering the reservoir. Less obvious was the potential for contamination from waste discharges from Dover, which was about 13 miles upstream. More remote than these two towns was the village of Hibernia and the mining operation that discharged waste into the local stream. Hibernia was located on a minor tributary to the Rockaway River more than 15 miles upstream of the reservoir. A time-of-travel study conducted by Hill showed that water from Dover took almost 21 hours to reach the influent to Boonton Reservoir (at the Rockaway River's minimum flow of about 6 million gallons per day [mgd]).[12]

In addition to their proximity to the reservoir, these three sites were potential sources of contamination because of the amount of waste they produced. Dover, with a population of 7,000, could generate a waste flow of 0.95 mgd, which could significantly pollute the Rockaway River, depending on the flow in the river. The sewage flow from Hibernia was quite small, and the plaintiffs invested a lot of effort trying to convince Magie that its waste had to be treated or curtailed in some fashion.

ON FEBRUARY 3, WILLIAM H. CORBIN summarized the case for the Jersey City Water Supply Company. Included in his opening statement was some news that the plaintiffs claimed was a shocker. Through nine pages of the trial transcripts, Corbin summarized

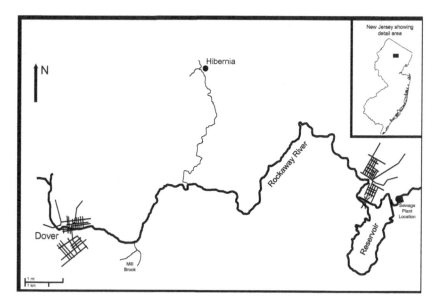

Figure 11-1   Map of Boonton Reservoir Watershed showing areas where
sewage collection and treatment were proposed (traced from
Google Maps)

the findings of the first trial and laid out the approach he planned
to take in the second trial. The most important finding from the
first trial was that ". . . two or three times per year . . ." especially
after large rainfalls, river water moved quickly across Boonton
Reservoir and the purification process normally provided
by the reservoir was not able to protect the water supply from
bacterial contamination. Corbin went into detail about Dr. Leal's
extraordinary efforts since his employment in 1899 to remove
pollution sources from the Rockaway River above Boonton. More
than 500 privies that were discharging directly into the river or its
tributaries had been removed.

Corbin also cited the low incidence of typhoid fever in Jersey
City as proof that Boonton Reservoir was providing the purification function that it had, in part, been designed to perform. His
justification of the "low" rate of typhoid fever in Jersey City is
astonishing when viewed from the present. "It was proved in the
case, and substantially agreed on [by] both sides, that in a great
city, if the death rate from typhoid fever was not more than about
twenty-five per hundred thousand population per year, there is no
ground to suspect that the water is impure."[13] Although a typhoid

fever death rate this high is beyond comprehension today, the threshold death rate of 25 per 100,000 was a well-known metric in its day and was cited by Allen Hazen in his classic book on filtration.[14] Corbin summarized recent typhoid fever statistics that were not heard in the first trial. "The death rate has fallen off to only thirteen per hundred thousand of population in 1907—a rate so low that it practically placed Jersey City at the head of all the large cities in the United States, and in 1908 it was a still lower amount—about ten."[15]

Corbin then presented a crucial argument of the defendants' case: building sewers to capture the sanitary wastes from a few upstream towns would not eliminate the elevated concentrations of bacteria in the water delivered to Jersey City. He contended that bacteria entered the Rockaway River in large measure by what are now called nonpoint sources—runoff from agricultural lands, urban landscapes, and streets. "That means of pollution would continue, and the count of b. coli and the count of bacteria would remain practically as it is—certainly not materially changed."[16]

Continuing his opening statement, Corbin said the defendants' strategy to avoid the expense of installing sewers in the watershed began after the opinion by Vice Chancellor Stevens on May 1, 1908, and before the decree was entered on June 4, 1908. "Being advised as we were, that the building of these sewers would not protect the water and certainly would not give us any relief [from high bacteria counts], after very careful consideration *we asked the court to put in an alternative reference* to your Honor, of a method which we should suggest . . . would make the water pure. . . ."[17] (emphasis added) Therefore, there is no mystery as to why the language potentially allowing "other plans or devices" was included in the final decree. The defendants asked for it. It is amazing that the defendants were able to influence the final decree to such an extent.

Corbin then described in general what Leal was to describe in detail—plans and devices had been prepared and operated to render the water ". . . substantially sterile, without bacteria . . . there are a few."[18] At this point, Corbin stated that the other plans or devices involved the addition of small amounts of chloride of lime to the water. In the final remarks of his opening statement, Corbin observed that the plaintiffs' attorney was obviously "disturbed" about the use of chloride of lime. Corbin was evidently reacting to the body language of the plaintiffs' attorney, but Vredenburgh held his fire until the next day.

On February 4, Vredenburgh strongly objected to the special master that chloride of lime be considered as the alternative to sewers. In effect, he claimed he was blindsided because Jersey City was never officially informed that this work was going on at Boonton. Despite his use of the polite legal jargon of the day, it is clear from the trial transcript that Vredenburgh and his client, Jersey City, were quite upset. Obviously, this was a matter for the court to rule on.

> "The Court—I conceive that the opening of Mr. Corbin is in the line of the direction of my order of reference. It requires me to investigate and report upon other plans and devices as alternative remedies presented by the defendant company for delivering water in a pure and wholesome condition. I think this [chloride of lime] is a plan and device, so far as I am able to see from the opening, that may be said to come within the direction to me in investigate and report on. Proceed."[19]

There is no mention in the transcript of a huge sigh of relief coming from the defendants' table, but Corbin must have been pleased. The court's ruling was a huge win for the defendants and allowed them to proceed with presenting mountains of evidence supporting their alternative over the next 28 days of testimony, which stretched over more than nine months.

Vredenburgh was not happy. He filed an exception to the ruling, which was an unusual move in this trial. Although the court noted his protest for the record,[20] and Vredenburgh renewed his objection many times throughout the trial, Justice Magie did not flinch.

CORBIN'S FIRST ORDER OF BUSINESS was to put Leal on the stand and lay out the defendants' main case for the special master. He then brought George Warren Fuller and George A. Johnson into the proceedings to establish how the plant was built, how it fed chloride of lime, and how effective it was at killing bacteria.

Leal's testimony began with a recitation of his background and qualifications (recounted in Chapter 6). Next he was questioned about his role in protecting the Rockaway River watershed, which had been covered in detail in the first trial. Leal then stated in his own words why an alternative to building sewers in the watershed was needed.

"Believing firmly as I do that the building of such sewerage systems and sewage disposal works would not accomplish the purpose but would leave the company in exactly the same condition in the future as in the past, I at once began to consider various means for the accomplishment of the desired ends. . . . if sewage systems or sewage disposal works were constructed and put in operation, in times of heavy rain, severe wash of the surface, or high water in the river, an examination of samples of water taken from the point of delivery in Jersey City will show too high a count [of] b. coli present in too small portions of water. This would be due to the wash of the surface. The majority of b. coli found at such times in my opinion do not come from human beings but from animals—from the washings of manured fields, stables and barnyards, streets, roads, etc., and this condition would exist just the same after the construction of those works [the sewers] as before."[21]

Corbin then asked Leal what he had done to develop the other plans and devices.

"Influenced in the way I have already described, I at once began to devise means to fulfil [sic] the contract in the way required by the Vice Chancellor's decision. I consulted with very many very eminent gentlemen on this line, and I had had personally a good deal of experience on this line in connection with the purification of the Little Falls water in 1898, 1899, 1900 and 1901, and I decided that the process indicated was the practical sterilization or disinfection, which is the more proper word, of the water coming from the dam, in this way removing the excess number of bacteria of which the Vice Chancellor has complained, and practically eliminating the b. coli from the water. This seemed to me to fulfill the Vice Chancellor's requirements, and I then proceeded to take action at once. On the 19th day of June I engaged the firm of Herring [sic] & Fuller to build a plant."[22]

Later, under cross-examination, Leal confirmed that he had been experimenting with disinfection of drinking water using chloride of lime since 1898.[23] He then described the treatment plant that added chloride of lime to the water before it was sent to Jersey City (see Chapter 10 for details).

Next, Leal explained how bacteria were removed when chloride of lime was added to water.

"Q. Now, doctor, describe to the court what this process is, and what the active agent is which is used for the accomplishing of this result, that is, the sterilization of the water, or rather the destruction of the bacteria in the water?

A. Why, the agent, and the only agent which actually does the work and accomplishes the purpose is oxygen, which we have called potential oxygen as a matter of convenience. It is really atomic oxygen. This is given off in various ways, from various substances, and at once attacks organic matter, and to speak plainly . . . it is a process of combustion practically, and the organic matter is burned up, if you want to put it that way, oxidized, and is returned to its ultimate elements. . . . we oxidize the organic matter, seeming to attack the bacteria first."[24]

Leal was explaining in detail the theory of disinfection that Corbin had alluded to in his opening statement. However, modern-day water treatment professionals might wonder what Dr. Leal was talking about. Oxygen killed bacteria? Leal was describing one view of chlorine chemistry that was believed at the time and that involved the liberation of what was generally called "nascent oxygen." The concept of nascent oxygen originated with James Watt in the early part of the nineteenth century. [25]

Leal can be forgiven for pressing the nascent oxygen theory of disinfection in 1909 because the theory was widely believed at that time. Nevertheless, declaring that oxygen was disinfecting the water was a huge advantage for the defendants. Oxygen was undoubtedly considered benign by the public as well as by the special master—after all, oxygen made up 21 percent of the atmosphere and everyone needed it to breathe—but most people in the early 1900s feared chemicals in water, food, and medicines. Chlorine had a poor reputation with the public as just another chemical and one that really smelled bad at that. Chlorine even burned the skin and eyes at high concentrations.

In later testimony, experts for the defendants and plaintiffs lined up on opposite sides of the nascent oxygen theory. Chlorine as a poison was discussed in some detail. Victory in the battle of the experts over whether oxygen or chlorine was killing the bugs would have a critical impact on the special master's final opinion.

Corbin needed to nail down when treatment of the water from Boonton Reservoir began:

"Q. Have you been using this [chlorine] continuously or only during heavy rains?

A. We have used it continuously.

Q. Night and day?

A. Yes, sir.

Q. From the time you began in September until the present time?

A. From the 26th day of September, every minute."[26]

Leal then explained that on the day of his testimony, February 4, 1909, they were testing the generation of chlorine by using an electrolytic salt solution instead of feeding chloride of lime.

Corbin asked Leal to describe how the amount of chlorine to be added was calculated, and this led to an unexpected discussion of who was in charge of making decisions at the Boonton plant.

"Q. (from Corbin): You might generally describe the method by which you determined that.

Mr. Vredenburgh [interrupting]: Would it not be more accurate to have the man who does it?

Mr. Corbin: Well, here is the man that directed it done. I am trying to give a general idea of what this plan is, to the court."[27]

Even though Leal never acknowledged the presence of chlorine in the water, it is remarkable that he would add a substance for which there was no method to measure its residual. His testimony was all about nascent oxygen being released by the added chlorine, and no analytical method for nascent oxygen existed. However, there were methods for measuring the effect of adding a disinfectant to water containing bacteria.

Leal was asked what monitoring was done at the plant and where the water delivered to Jersey City was sampled to determine the effectiveness of adding chloride of lime. Leal stated that bacteria were monitored daily at the plant by Mr. Johnson and his assistants. Water delivered to Jersey City was monitored every day (except Sunday) by Professor William Hallock Park of New York University and the New York Board of Health Laboratories and by Dr. George McLaughlin, the bacteriologist who had testified in the first trial. The defendants made it clear at this point in the second trial that they would be using bacterial test results as a primary piece of evidence proving that the addition of chlorine solved the bacterial problem raised by Vice Chancellor Stevens.

Corbin asked Leal to summarize the bacterial test results.

> "With the exception of one or two occasions, which probably were occasioned by something in the pipe, the water has been practically sterile and when I say that I mean it contains very few bacteria, not more than is very apt to go into distilled water unless you are very careful and the b. coli are, I think I am justified in saying, eliminated."[28]

Hundreds of pages of testimony from both sides recorded the exploration of this summary statement.

Next, Corbin asked Leal to put in perspective the importance of adding chloride of lime as a water treatment method.

> "I do believe . . . that this method is going to be . . . the fourth great advance in water purification; I should say, [slow] sand filtration was the first; the knowledge of what sand filtration meant . . . marks the second great advance. Then, the discovery, or at least the practical applications of coagulation, with the consequent rapid mechanical filtration I place third; . . . and this method, which we can call the disinfection of water supplies, I regard as the fourth great advance; and I think that the fourth is every whit as important as the rest."[29]

Leal then proposed that the usefulness of disinfection would come about by combining it with filtration to improve the bacteriological quality of the water. Though he did not state it explicitly, Leal was espousing the multiple-barrier concept, which would ultimately become a basic tenet of sanitary engineering. Leal went on to say that the chemical quality of the water was not changed in any material way except for a slight improvement in the color and a slight increase in hardness.[30]

Corbin then asked Leal the most important question of all. Every expert witness on both sides of the case was asked a form of this key question.

"Q. Doctor, in your opinion is this [adding chloride of lime] a safe and proper method of rendering the water pure and wholesome?

A. It think that it is the safest, the easiest, and the cheapest and best method for rendering this water pure every day in the year and every minute of every hour."[31]

Almost as important as the question "Is it safe?" was the question "Can we rely on it?" The reliability of chlorine disinfection was a central issue in the case, and testimony addressing this key point was taken from experts on both sides.

"Q. Is it reliable?

A. And the most reliable method. I believe the water supply of Jersey City today is the safest water supply in the world."[32]

As further support for the advisability of using chlorine in drinking water, Leal briefly described the recent example of drinking water disinfection in Lincoln, England (discussed in Chapter 4). Again, Leal stated his conviction that the use of chlorine was safe.

"Q. Any ill effect on the health of the people there?

A. Not the slightest."[33]

Corbin asked Leal if he drank the chlorinated water currently being served in Jersey City. Every person responsible for the quality and safety of a water supply since that time has been asked the same question. It is the ultimate test of whether a drinking water professional believes in what he or she is doing.

"Q. Do you drink this water [in Jersey City]?

A. Yes, sir.

Q. Habitually?

A. Yes, sir."[34]

An unfortunate exchange followed, showing that the questioning had moved beyond the professional into the personal.

"Q. [by Corbin]: Would you have any hesitation about giving it to your wife and family?

A. I believe it is the safest water in the world.

Mr. Vredenburgh: I ask that the answer be stricken out as to whether he would give it to his wife or children; I don't believe he is married; he would give it evidently to other people's children."[35]

Leal's wife, Amy Lubeck Arrowsmith, had died on June 1, 1903, six years before this exchange. No record exists of any further exchange along this line of questioning by Mr. Vredenburgh.

THE TRANSCRIPT FROM FEBRUARY 5 reveals that Leal had also installed a chloride of lime feed system at the filtration plant at Little Falls. He stated that he had experimented with chloride of lime addition some months before and that he was now using it daily. Thus, the trial transcript provides the first written evidence of the second use of chlorine to disinfect a drinking water supply.[36] This was also the first time chlorine was used in conjunction with mechanical filtration. Chlorine was tested at Poughkeepsie, New York, in early February 1909, but the permanent application of chlorine at Poughkeepsie did not begin until March 17, 1909.[37] Therefore, the Poughkeepsie water supply became the third example of chlorine disinfection, and this was the first time chlorine was used as an adjunct to slow sand filtration.

According to Leal, total bacteria concentrations in the Passaic River at Little Falls were 30,000–40,000 per milliliter. Prior to the use of chlorine, Leal had operated the filtration facility with the goal of producing water that contained fewer than 100 bacteria per milliliter; however, during some parts of the year, he was not able to meet that goal. Combining chlorine and filtration resulted in treated water that contained 6–8 bacteria per milliliter.[38] Leal then stated that filtration plants in Washington, D.C., and Albany, New York (that were not using chlorine) were also unable to produce filtered water containing less than a few hundred bacteria per milliliter.

Vredenburgh immediately confronted Leal on his citing Lincoln, England, as a successful example of using chlorine in a water supply. Vredenburgh specifically wanted to know if the population of Lincoln had been warned by the water company that "Chloros" was going to be used. It was Leal's understanding that the population received no warning. However, the detailed report of the Lincoln episode made it clear that chlorine was added with the population's knowledge.[39]

Vredenburgh raised the issue of public protests about the use of chlorine in the water supply and noted that 30 to 40 physicians had complained that chlorine in the water delivered to Lincoln caused eczema, colic, and diarrhea. The questioning from Vredenburgh became sharp at this point.

"Q. Answer the question, did the doctors do that? Do you know anything about it?

A. I believe there was some action taken."[40]

In fact, an account of the meeting with the doctors in Lincoln resulted in no changes to treatment, as indicated by the following excerpt.

"On February 24th the members of the medical profession practising in Lincoln, some 30 to 40 gentlemen, were invited to meet Drs. Houston and McGowan and myself [Dr. R. J. Reece] in conference. At this meeting I explained the difficulties that had to be dealt with, and the action adopted. It was stated by certain of the medical men present that the water as treated had caused colic and diarrhea among their patients and eczema, skin irritation, and conjunctivitis. Others had noticed nothing of the sort. This irritant action was considered by a few to be due to 'chlorination' of the water, but it was pointed out that chemical tests capable of detecting free chlorine in one part per two million failed to show any trace of chlorine in the water—and it was admitted by certain of these practitioners that they treated their enteric fever patients with chlorine mixture. The medical profession was divided in opinion as to the quality of the water supplied by the Corporation Waterworks, but they were unable to suggest any alternative treatment."[41]

Chlorine use was not immediately discontinued at Lincoln. The addition of Chloros continued until 1911, when a new water supply was acquired for the city.[42]

Vredenburgh's cross-examination of Leal focused on key issues that the attorney would continue to explore throughout the rest of the trial.

"Q. Doctor, what other places in the world can you mention in which this experiment has been tried of putting this bleaching powder in this same way in the drinking water of a city of 200,000 inhabitants?

A. Two hundred thousand inhabitants? There is no such place in the world, it never has been tried.

Q. It never has been?

A. Not under such conditions or under such circumstances but it will be used many times in the future, however.

Q. Jersey City is the first one?

A. The first to profit by it.

Q. Jersey City is the first one used to prove whether your experiment is good or bad?

A. No, sir—to profit by it. The experiment is over."[43]

The next question and answer focused a laser beam on the risk Leal took by adding chlorine to the water.

"Q. Did you notify the city that you were going to try this experiment?

A. I did not."[44]

Thus, Leal admitted under oath that he had no permission to add chlorine to the water. Surely Leal's decision qualifies as one of the boldest, most courageous acts in the history of drinking water treatment. Some might call it foolhardy or even irresponsible. However, because his actions had such a positive impact on the conquest of waterborne disease, there is no evidence that his peers ever criticized him for not asking permission.

Vredenburgh's questions got personal again, and Leal rose to his own defense.

"Q. And if the experiment turned out well, why, you made a fortune, and if it turned out badly—(interrupted)

A. I don't know where the fortune comes in; it is all the same to me. Why, Mr. Vredenburgh, we knew before we actually did it that it would serve our purpose."[45]

Leal never tried to patent his application of chloride of lime to drinking water. Others in the drinking water field obtained patents for filtration systems and other disinfection methods—for example, Ferrochlore in Belgium and Chloros in England.[46] John

L. Leal lived modestly both before and after his revolutionary breakthrough. He created the process for the benefit of the company he was working for and for all drinking water consumers. His dedication as a physician and public health advocate trumped any thoughts of personal profit.

Examination of George Warren Fuller. After Fuller described his education and experience, Corbin asked him to explain the nature of the treatment process. Fuller kept to the script: "This process which we designed a plant for, to be used at the Boonton dam, is an oxidation process pertaining especially to the destruction of objectionable bacteria."[47]

Fuller went on to describe the chemistry of how chlorine was changed when it was added to water and how "potential oxygen" or "atomic oxygen" was liberated when the chemistry proceeded to its natural conclusion. The oxygen that was released then oxidized and killed the bacteria.[48]

He stated that either chloride of lime or the electrolytic process could have been used to liberate the oxygen but that the amount of electric power available was not sufficient to run the electrolytic apparatus at Boonton Dam. Enough was known about chlorine chemistry at the time for Fuller to explain chlorine's reaction with naturally occurring organic compounds, usually described by the parameter of natural "color." He correctly described the competition of the oxidation reactions between the chlorine (or, in his parlance, the oxygen) and the competing targets—natural organic compounds and bacteria. Fuller also emphasized that one of the major findings from the work at Boonton was that only a fraction of a part per million was needed to kill the target bacteria.[49]

Corbin then moved on to asking about plant details. To engineers accustomed to water treatment plant design and construction taking two to four years, the time period Fuller described must seem little short of miraculous. He started the design on the day Leal visited his offices, June 19, 1908, and he was almost finished with the plans at the end of July—six weeks of design time. Part of July, all of August and most of September were devoted to construction.[50] As recorded in the trial transcript, the plant went into full and continuous operation on September 26.

Fuller described the processes and the bacteriological monitoring program at the plant in general terms, leaving the specifics to George A. Johnson, who would testify next. Corbin then asked the key questions and got the right answers. Fuller believed the oxidizing process was effective, safe, and reliable and that it produced pure and wholesome water.[51]

Fuller was one of the few scientist/engineers in the country who could competently answer the next question Corbin asked him: if the sewers and disposal facilities were installed as proposed by the plaintiffs, would the bacterial count of the raw water at the Boonton Dam be any different than it was without the facilities? Fuller answered that there would be no difference in the bacteria counts even after the expenditure of hundreds of thousands of dollars because the proposed sewer systems would not capture bacteria washed into the Rockaway River from land areas where farm animals were being raised and from human activities.

The answer to this question was the key to the entire second trial. Not only did the chloride of lime process have to reliably destroy the bacteria that made it through the reservoir, but it also had to give better results than installing sewers. Vredenburgh objected every time Corbin asked one of his experts this question, and every time Justice Magie allowed the question and answer into the record.

Moving on to the bottom line regarding sanitary improvements in water supplies, Corbin asked Fuller if Jersey City's typhoid death rate of 10 per 100,000 was indicative of a good public water supply. Fuller noted that this was an extremely low death rate for a large urban area like Jersey City. "I should say those data indicated Jersey City had an unusually good public water supply so far as its hygienic character is concerned."[52]

As Vredenburgh cross-examined Fuller, the lawyer tried to shake the witness's commitment to the answers he had given during direct testimony, but Fuller would not budge. Vredenburgh did elicit from Fuller a correct guess as to one method by which chlorine inactivated bacteria—cell membrane destruction. Fuller also stated that the last time he had actually performed bacteriological tests in a laboratory was in 1904 (not surprising, given the expansion of his other responsibilities). Fuller acknowledged that in the early 1890s, he personally had conducted the Lawrence Experiment Station studies that identified the natural rate of die-off for the typhoid bacillus in different kinds of water.[53]

Examination of George A. Johnson. After recounting his training and experience, Johnson described the plant's operations in detail, including the method of preparing chloride of lime solutions, examination of the available chlorine in the chloride of lime powder as delivered to the site, calibration of the discharge orifices, preparation and review of daily reports, and compilation of the bacteriological results from water samples collected after chloride of lime treatment.

Johnson then summarized the results of total bacteria analyses of the treated water during October, November, and December 1908. Samples were analyzed daily on gelatin plates incubated at 20 degrees centigrade, and at no time did total counts exceed 30 per milliliter. Johnson noted that these were very low bacteria concentrations and that it would be difficult for any filtration plant to achieve such low numbers. He also said the total bacteria counts were well below the drinking water standard of 100 total bacteria per milliliter set by the Imperial Board of Health of Germany.[54]

Treated water B. coli results were completely negative, according to the analytical method Johnson used, although this method was quite different from the total coliform method in use today.[55] Analyses of treated water in later months before the end of the trial detected a couple of positive B. coli samples, possibly caused by an upset in the chloride of lime feed process or the influx of a high number of bacteria.

Vredenburgh's cross-examination of Johnson was intense. The plaintiffs' attorney pounced on the inconsistencies in Johnson's terminology and the mistakes in his calculations.

Other experts for the defense. During the main part of the defendants' case, Corbin presented testimony from a number of expert witnesses in addition to Leal, Fuller, and Johnson. Leal either knew the experts personally or knew their work, and he had asked each of them to perform laboratory experiments using chloride of lime. He had also obtained their agreement to testify in the case. All of the experts were in communication with each other. Most knew one another personally or professionally, and certainly they read each other's reports stating their individual opinions in the case. Experts for the defendants included:

- Henry B. Cornwall, professor of chemistry at Princeton University;
- Franklin C. Robinson, professor of chemistry at Bowdoin College and at the State of Maine Medical School;
- Leonard P. Kinnicutt, professor of chemistry and director of the chemical laboratory at Worcester Polytechnic Institute;

- William P. Mason, professor of chemistry at Rensselaer Polytechnic Institute in Troy, New York; and
- Frank F. Wesbrook, professor of pathology and bacteriology at the University of Minnesota and director of the state Board of Health Laboratories of Minnesota.

The direct testimony of these witnesses proceeded along similar lines. Corbin wanted the witnesses to provide basic information and to answer key questions that supported the defendants' points of view:

- Is there anything in the water that could harm the consumer?
- What is your opinion of the propriety of using chloride of lime for the purification of water?
- Is the addition of chloride of lime to water safe?
- Is the addition of chloride of lime to water reliable?
- Was the chemistry of the water changed as a result of the chloride of lime addition?
- Is chlorine a poison?
- Was any free chlorine left in the water after chloride of lime was added?
- How was the bacteriological quality of Boonton Reservoir water changed with the addition of chloride of lime?
- What chemical reactions does chloride of lime undergo when it is added to water?
- What chemical agent is responsible for killing the bacteria?
- Did you perform analyses (chemical or bacteriological) on the water?
- Did you analyze the strength of the chloride of lime powder as delivered to Boonton?
- Is the water as treated and delivered to Jersey City suitable for a potable water supply?

The defendants' experts were unanimous in their opinions that the water treated with chloride of lime was safe, reliable, and appropriate and that no harm would come to people who consumed the water. The experts confirmed that chlorine was a poison in high concentrations, but they emphasized that no chlorine was present in the water treated with chloride of lime. Instead, nascent oxygen resulted from the chemical reactions, and oxygen was the agent responsible for the "oxidation" of bacteria.

Five-gallon demijohns of water from Boonton Reservoir were sent to the defendants' experts during October through December

1908, along with samples of the chloride of lime being used at the Boonton plant. Leal's instructions to the experts were simple: Apply chloride of lime to the Boonton water or any other water you choose, and let me know if it is effective in killing bacteria. Leal also asked them to conduct any tests necessary to convince themselves that the use of this form of chlorine was safe. Finally, he had asked most of them to visit the plant at Boonton so they could give opinions on the treatment's reliability.

The experts conducted chemical or bacteriological (or both) tests on the raw water shipped to them and on water they treated with chloride of lime. Because state-of-the-art water analysis for bacteriological quality was still evolving at this time, different investigators used variations of the available methods. In general, the bacteriologists tested the raw and treated water for total bacteria count and for the presence of B. coli. One researcher even dosed a specific raw water sample with typhoid bacteria and noted chlorine's effectiveness at killing this organism. The importance of the laboratory and full-scale disinfection results obtained by Leal and confirmed by the experts in their own laboratories was that only a fraction of a part per million of chlorine was needed to eliminate all indicators of sewage contamination (B. coli) and to reduce the total bacteria count to very low concentrations.

Several analyses of the bleaching powder shipped to the investigators showed the strength of available chlorine to be about 33–37 percent. Almost all of the experts relied on dosing calculations to determine the amount of chlorine added to the test solutions.

During the trial, virtually all of the experts discussed the chemical reactions that occurred when chloride of lime was added to water. The defendants' experts all testified that the substance doing the work of disinfection was not a form of chlorine but some form of "available oxygen," "potential atomic oxygen," or just "oxygen." Henry B. Cornwall referenced a number of other researchers' publications supporting his opinion that oxygen was the active chemical accomplishing disinfection and bleaching.[56]

The experts were asked if Leal's process of adding only a small amount of bleach to the already clean water from Boonton Reservoir produced a water supply suitable for a city such as Jersey City. They all said yes, with one of the more concise answers coming from Frank F. Wesbrook: "I think that it is safe and suitable as a water supply, a pure supply."[57]

Corbin also wanted some expert comments of a more general nature on the suitability of the water produced by Leal's process. Leonard P. Kinnicutt responded by saying ". . . I am of the opinion that the treated Boonton water contains nothing that is injurious, and is a suitable water for domestic and manufacturing processes."[58] The rest of the experts concurred with this view.

Henry B. Cornwall noted that the slight increase in the water's hardness after chloride of lime treatment was only 1.5–2 parts per million (ppm) in a supply that was already soft. He said there was no material negative effect from this slight rise in hardness.[59]

The plaintiffs' experts testified that the water treated with chloride of lime resulted in a blue color when tested with the potassium iodide starch test. Franklin C. Robinson, an expert for the defense, tested 75 *raw* water supplies and found that, over time, the potassium iodide starch test caused a blue color in water to which no chloride of lime had been added. He concluded that a naturally occurring substance in raw water had the same oxidizing effect as chloride of lime.[60]

Leonard P. Kinnicutt tested the ability of chloride of lime to kill bacteria from Boonton Reservoir water and from water seeded with intestinal bacteria.[61] In order to determine the bacteriological effects of adding chloride of lime, Frank F. Wesbrook applied the chemical to other water supplies, including water from the Mississippi River near Minneapolis and the lake supply for Toronto, Canada.

William P. Mason testified that over time he had changed his mind with respect to the advisability of using chemical disinfection to kill pathogenic microorganisms in contaminated water supplies. In his 1907 book on water treatment, Mason had stated, "To 'disinfect' a water by use of hypochlorites does not appeal to one as a suitable means for increasing its potability."[62] Also, in a published discussion of the paper on disinfection given by George C. Whipple at the 1906 annual conference of the American Water Works Association (AWWA) in Minneapolis, Mason had made it clear that he did not favor chemical disinfection and emphasized that water supplies should always be taken from unpolluted sources or filtered if they were subject to contamination. Mason stated that he had successfully tested chloride of lime in samples of the water supply of Troy, New York, dramatically reducing the bacteria content of that supply.

AFTER PRESENTING THE MAIN PORTION of his case, Corbin asked Magie on March 1 if the defendants would need to produce testimony countering the sewer and sewage treatment cost information presented by the plaintiffs. Corbin believed that if Magie ruled that the addition of chloride of lime was acceptable in meeting the requirement of "other plans or devices" specified by Vice Chancellor Stevens, it would not be necessary to delay the proceedings by presenting extensive testimony about costs from the defendants' perspective. Magie was concerned that if he ruled immediately that the chloride of lime treatment was an acceptable alternative to sewers, he ran the risk of being reversed on appeal. Magie stated that the defendants' position on the cost question should be presented, and he allowed a delay of about one month for the defendants' experts to prepare their estimates. Corbin noted that a month's delay would also allow additional time to gather operational data on the chloride of lime plant.[63]

Returning to court on April 5, 1909, Corbin made it clear in his opening remarks that the defendants disagreed with the plan for sewers and sewage treatment presented by the plaintiffs. Corbin interpreted Vice Chancellor Stevens's ruling as requiring that the company be responsible only for constructing a trunk sewer and that the collecting sewers and any sewage treatment were beyond the scope of the company's responsibilities. Corbin emphasized that the trunk sewer would extend only from Dover to below Boonton Dam and would be operated under gravity.[64]

Several witnesses testified over three days about the defendants' alternative sewer plan. Experts for this portion of the defendants' case included familiar witnesses such as George Warren Fuller and Rudolph Hering. The trunk sewer proposed by the defendants was sized to capture the sewage coming down the small creek from Hibernia (Figure 11-1). Because the Morris Canal was located in this area of the Rockaway River and the canal was designed to operate under gravity, it is not surprising that the trunk sewer proposed by the defendants ran alongside the canal for most of the distance to Boonton.

The cost of the trunk sewer was estimated to be $175,243.37,[65] hundreds of thousands of dollars (in 1909 dollars) less than the sewers and disposal facilities proposed by plaintiffs (Table 11-1). Fuller also presented costs for a disposal plant ($39,519—far less expensive than the plant proposed by the plaintiffs) to be located below Boonton Dam, but Corbin made it clear to the court that he did not believe a sewage disposal plant was part of the plan envisioned by Vice Chancellor Stevens.[66]

VREDENBURGH OPENED THE REBUTTAL CASE for the plaintiffs on April 28, the twenty-fifth day of 38 days of testimony in the second Jersey City trial. Vredenburgh told the court he would be presenting testimony on the cost of sewers and a sewage disposal plant, the quality of the raw water in the Rockaway River and its tributaries, the unsuitability of the chloride of lime treatment to protect the public from disease, and the inadequate time of testing for the new process.[67]

The plaintiffs' expert witnesses were equally as senior and accomplished in their fields as the defendants' experts. Vredenburgh drew heavily from the ranks of professors and former students associated with the Massachusetts Institute of Technology (MIT). There was like thinking at that institution on the matter of water contamination and public health threats. However, the first rebuttal witness called was Joseph A. Deghuee, a chemist at the Lederle Laboratories in New York City. Deghuee had performed many of the chemical analyses on water samples taken from the Boonton source of supply and the Jersey City distribution system. Dughuee's testimony included the surprising information that he had measured a "free chlorine" residual in the water distributed to Jersey City by using a new analytical method suggested by Earle B. Phelps, who would testify later.[68]

Other experts for the plaintiffs included:

- William T. Sedgwick, professor of biology at MIT;
- Charles-Edward A. Winslow, assistant professor of sanitary biology at MIT and biologist in charge of the Sanitary Research Laboratory and Sewage Experiment Station at Lawrence, Massachusetts;
- Earle B. Phelps, assistant professor of chemistry at MIT;
- George C. Whipple, consultant working with the firm Hazen and Whipple (two years after his testimony, he was hired as a professor at Harvard University);
- Henry Goodnough, chief engineer of the Massachusetts Board of Health;
- Ernst J. Lederle, chemist as well as president and principal owner of Lederle Laboratories in New York City; and
- Charles E. North, a bacteriologist with Lederle Laboratories.

Interchanges between the lawyers and the plaintiffs' expert witnesses were not as consistent as those involving the defendants' experts, but the testimony of the plaintiffs' experts had many points in common.

- The danger of polluted water came not only from the presence of pathogenic bacteria but also from the presence of other harmful substances. Adding chlorine to the water only removed living germs and did not remove other kinds of pollution.
- Boonton, Dover, and Hibernia contributed significant amounts of pollution to the Rockaway River.
- Installing sewers to remove waste from significant sources upstream of Boonton Reservoir would reduce the danger of infection and pollution of the water distributed to Jersey City residents.
- Preventing pollution by protecting the source of supply from contamination was the proper approach to providing potable water.
- Small amounts of disinfectant did not remove particles of feces or industrial pollution.
- The chloride of lime plant was unable to remove bacteria during periods when the reservoir was an imperfect barrier to bacterial contamination of the Jersey City water supply.
- In addition to the two days (January 28 and March 4, 1909) when high total bacteria counts and B. coli were found in the treated water, two to four days following October 26, 1908, also had high total bacteria counts and positive B. coli results. The expert witnesses attributed the high bacteria concentrations in October to heavy rainfall or "pipe moss" (what is now called biofilm that sloughs off of the interior of distribution pipes).[69]
- Variable amounts of organic matter in the source water would have varying effects on the ability of chloride of lime treatment to produce potable water.
- In order to set the proper dosage of chloride of lime, continuous bottle experiments would be necessary to account for the changes in raw water quality.
- Because the addition of chloride of lime to drinking water was a novel and experimental process, more experience was needed before it could be fully accepted.
- Operation of the plant for only seven months (September 26, 1908—May 1, 1909) was not sufficient time to test the treatment process.
- Chloride of lime addition in the small quantities used at Boonton was not effective at curing odor problems in the water.

- Disinfection of water cannot be considered equivalent to prevention of pollution.
- The treatment process at Boonton was not designed to deliver pure and wholesome water.

Given his testimony and publications, Sedgwick clearly knew that chlorine killed bacteria when it was added to contaminated water. Nevertheless, through a series of redundant questions, Vredenburgh tried to get Sedgwick to put a negative spin on disinfection.

"Q. Does, in your opinion, the disinfection of polluted water maintain the purity of it?

A. It certainly promotes the purity of it, but I should say it would help to avoid maintaining the impurity of it. No, it does not necessarily purify it."[70]

Responding to a question about the necessity of providing water from a pure source, Sedgwick took a leap backward in the common understanding of waterborne disease transmission.

". . . it is an interesting question today whether mere pollution apart from infection with specific disease germs is in itself detrimental to the public health. We used to say it was not. So long as we avoided infection we were all right, but there is some evidence today, and a good deal of evidence, to show that *mere filth itself does somehow or other affect the public health*. . . ."[71] (emphasis added)

Phelps's testimony included a description of a new analytical method for detecting chlorine in water. Phelps testified that he had determined that orthotolidine, an organic compound derived from coal tar, was specific for determining the presence of free chlorine. Orthotolidine in a solution of water was normally pink in color unless it was in the presence of free chlorine or hypochlorous acid, when it became golden yellow.[72]

Phelps emphasized that he had only recently developed the orthotolidine test and that it was known to only two or three other people.[73] Not surprisingly, counsel for the defendants questioned Phelps closely on how widely his test was accepted. Even though the test was in its infancy at the time of the trial, with modifications it became the standard method for measuring chlorine residuals in drinking water for many decades. Subsequent research showed that orthotolidine reacted with other oxidizing substances, and the method was modified to eliminate interferences. In the 1970s and 1980s, the DPD[74] test replaced the orthotolidine test for free and total chlorine, particularly after orthotolidine was found in animal experiments to be carcinogenic.[75]

Dr. Charles E. North had already testified in the second trial, providing general testimony on bacteriological tests on samples from the watershed. The scope of his testimony expanded significantly when he was asked to introduce an additional piece of the plaintiffs' case. On May 5, he gave an astonishing answer to a question from Vredenburgh.

> "Q. Suppose that a water has added to it some chemical disinfectant which kills most of the bacteria, how would that affect the value of the bacteriological analyses of that water?
>
> A. Well it would depreciate the value very much, because it would utterly destroy the inference which was always drawn from the bacteriological analyses regarding the other pollutions which are present in the water, that is, any artificial removal of the bacteria or destruction of the bacteria."[76]

In effect, North said that killing bacteria, especially the bacteria that caused disease, would make it harder to determine if sewage contamination had occurred. This convoluted question and answer was designed to set up an important part of the plaintiffs' case—the filtering cloths. With only 12 days remaining in the testimony phase of the trial, Vredenburgh must have had plenty of signals from Justice Magie that Leal's arguments that chlorine effectively killed bacteria and removed the threat of disease were making headway. Vredenburgh resorted to playing on fears of disease transmission as old as miasmas and filth. North described filtering 150 to 450 gallons of water from a pipe in the Jersey City distribution system through cotton flannel cloth, which captured sediment from the water.

North also presented results from two analyses showing positive results for B. coli in water treated with chloride of lime. The defendants had already shown that with their analytical method, B. coli was absent except in two instances. In effect, the disagreement was a conflict over the analytical methods used by the defendants' and plaintiffs' laboratories,[77] and Magie would sort out the methodological differences in his final ruling.

For several days, Vredenburgh presented witnesses who claimed that when chlorine-treated water from Boonton Reservoir was run through filters they designed, the water produced filter cloths that were soiled. The defendants' witnesses countered by saying essentially, "It ain't so, and even if it was—so what?" These arguments were recounted in hundreds of pages of trial transcript. Even though the testimony on particulates removed by filter cloths was not supported by good science, it made good newspaper copy and resulted in at least three stories in the local Jersey City paper. One of the articles reported that the introduction of the filter cloth evidence caused a "sensation" in the hearing room.[78]

One witness who supported the significance of the soiled filter cloths was Daniel D. Jackson, director of the laboratories for New York City's Department of Water Supply, Gas and Electricity. Jackson's testimony was mostly devoted to describing how particles trapped in filter cloths demonstrated the presence of filth in the water treated with chloride of lime. Thirty cubic feet of water from various parts of the Jersey City distribution system was filtered through cotton flannel, the sediment trapped on the cloths was washed off with one liter of water, and the resulting black material was placed in glass bottles and brought into the hearing room as exhibits. Because water from Boonton Reservoir was not treated with slow sand or mechanical filtration, it is not surprising that the plaintiffs' experts were able to demonstrate that particles were present in the water. What was surprising was how the experts tried to connect the particulates to filth and thus to a public health risk.

By presenting an analogy, George C. Whipple attempted to connect the material captured on the filter cloths with sewer gas and health problems, stating:

> "To what extent that is a real danger the sanitary statistics do not show, or to what extent the drinking of contaminated—or I would say decomposing organic matter predisposes to other diseases cannot be stated very definitely. But we have an analogous case in that of sewer gas, **where the breathing of sewer gas has been shown to be deleterious by way of predisposing to other diseases.** It is believed that that is the chief danger of the sewer gas, that people do not get infectious germs from sewer gas so much as they get their systems run down by breathing air that is not of the best quality.

Cross-Examination by Mr. Edwards:

Q. Then the learning on this subject is practically in embryo, is it not?

A. I believe it is, because I think the future studies are going to show that there is some very important relation between filth and the public health, even though the filth is in water and may not be infectious."[79] (emphasis added)

And thus, the courtroom was rocketed back decades to the Great Stink of London and belief in the miasma theory alleging that filth and miasma caused disease.

For reasons not stated on the record, the trial was adjourned from late May to October 4, 1909. About a month before the lengthy adjournment, an editorial cartoon in the Jersey City *Evening Journal* showed two old codgers looking over a newspaper, with one noting that the Boonton water case had been postponed again. The other old-timer reflected that his father knew about this case as a boy. The title of the cartoon was "In 1998."[80] Finalizing the case did not take 99 years, but the trial might have seemed that long to the participants.

IN THE MIDDLE OF THE SECOND TRIAL, Leal, Fuller, and Johnson traveled to Milwaukee, Wisconsin, for AWWA's annual conference. Three papers included in the conference program described in detail an overview of the addition of chloride of lime at Boonton (Leal), design and construction of the plant (Fuller), and operation of the facilities (Johnson).[81]

A newspaper account at the time alerted the populace of Jersey City that Dr. Leal was going to present a paper on the chloride of lime process, and the article identified two representatives of the Jersey City water board who would attend the meeting as some sort of "truth squad."[82] City officials feared that Leal would get conference attendees to entertain a motion to approve the use of chlorine at Boonton and that this would complicate the city's lawsuit. Apparently, the *Evening Journal* also sent a reporter to Milwaukee, because a story appeared in the newspaper the same day Leal presented his paper—June 8, 1909. The article quoted Leal verbatim because the newspaper account was virtually identical to parts of the paper Leal eventually published.[83]

No such motion was presented to the full conference, but the process's approval by most of the discussants of the three papers was helpful to the defendants. All three papers and the discussions following them were published in the conference proceedings and are now considered classic publications in the field of drinking water treatment.

Because it was common for a discussion to follow the presentation of important papers at AWWA's annual conferences, a stenographer took down the comments and interchanges between members of the audience and the presenters. Often the people making the comments were allowed to edit their remarks before they were published in the proceedings. The discussions of the papers presented by Leal, Fuller, and Johnson produced some memorable exchanges.[84] Positive comments were made by Henry B. Cornwall, J. J. Flather, William P. Mason, Rudolph Hering, R. E. Milligan, Edward Bartow, and George Warren Fuller.

Cornwall, an expert witness for the defendants, had been one of Leal's professors at Princeton. His comments during the discussion emphasized that oxygen was the agent responsible for killing bacteria and not chlorine. He also noted that chloride of lime was able to perform its disinfection function with only a fraction of a part per million. Flather's comments focused on his visits to the Boonton plant and to other treatment facilities that were using chloride of lime disinfection. (The use of chloride of lime for disinfection was spreading even before the trial ended.) Rudolph Hering spoke of the "debt of gratitude" owed Leal, Fuller, and Johnson for developing the chloride of lime process.

The biggest impact of the discussion undoubtedly stemmed from the comments by William P. Mason who, at the time of the conference, was president of AWWA. Dr. Mason had testified as an expert witness for the water company three months before the conference—a fact he did not mention in his discussant remarks.

During the discussion of Whipple's paper at the 1906 AWWA annual conference in Minneapolis, Mason had stated that adding chemicals to kill pathogens was not a good idea. Three years later, his wholehearted support for the use of chlorine to disinfect water must have made an impression on the audience in Milwaukee and on anyone who read the discussion after it was published. Here was the president of AWWA, the most important professional body in the drinking water field, and a former skeptic with respect to chemical disinfection, praising the use of chlorine to all who would hear.

Another discussant at the 1909 conference, Nicholas S. Hill, raised concerns about the Boonton chloride of lime process. He identified himself as having been one of the plaintiffs' expert witnesses in opposition to chloride of lime treatment. He was evidently uncomfortable being one of the few naysayers in the discussion. Searching for support, he named some of the other witnesses for the plaintiffs—Sedgwick, Winslow, Phelps, Whipple, Goodnough, and Jackson. His remarks were general, claiming only that the process was neither sufficient nor complete.

LEONARD P. KINNICUTT WAS RECALLED to the stand on October 4, 1909, to discuss the orthotolidine test described by Phelps. Kinnicutt had found that orthotolidine reacted with oxidants other than chlorine. During his testimony, Kinnicutt conducted an experiment in which he added lead peroxide to an orthotolidine solution and observed the yellow color characteristic of a positive reaction. It was Kinnicutt's opinion that if a test using orthotolidine resulted in the yellow color, it merely indicated the presence of an oxidizing substance. He had demonstrated that the orthotolidine test was not specific for chlorine.[85]

George A. Heulett, a chemistry professor from Princeton University, was called as a defendants' witness to discuss the acid–base chemistry of water and explain why it was not possible for Boonton Reservoir water treated with chloride of lime to contain free chlorine. Heulett presented more chemistry equations than all of the other witness combined. Magie must have struggled to understand how all of the detailed testimony on chemistry fit into the case. Henry B. Cornwall was called to testify again about filter cloths, and George A. Johnson testified again about bacteriological methods. At this time, he also testified about his experience with the Bubbly Creek plant near Chicago.[86]

George Warren Fuller was recalled twice during the defendants' rebuttal case. He presented revised cost figures for the proposed sewage disposal facility below Boonton Dam and noted for the record that he believed the water treated with chloride of lime and delivered to Jersey City was pure and wholesome. He covered a wide range of topics supporting the testimony of the defendants' experts.[87]

At the end of the defendants' rebuttal case, Leal was called to the stand and asked again to summarize his medical career to lay the foundation for a critical question.

"Q. Were you accustomed to use chlorine in your practice, as a medicine?

A. Yes, I used it very extensively. At the time that I started to practice medicine we did not know nearly as much as we do now about the cause of preventable disease, and in diphtheria, scarlet fever, smallpox and diarrhoeal diseases of infants *chlorine water* was a favorite remedy and one in common use. I have used it *thousands of times*, not only in the case of adults but in the case of infants; I would state it was especially used with infants."[88] (emphasis added)

Leal added that the treatment was not being used currently because it had been replaced by other therapies. However, he stated that when it was used, the strength of the chlorine water was 0.4 percent free chlorine, which translated to 4,000 ppm, and that the typical medical dose was 4 drams (about 15 milliliters). The original sources describing the preparation of chlorine water (Chapter 2) made it clear that the active substance in chlorine water was actually chlorine dioxide and that the concentration of chlorine dioxide was about 3,000 ppm.

After laying this foundation, Leal described how little chlorine theoretically remained in the water after the chlorination treatment at Boonton, explaining that it was infinitesimally small compared with the medicinal dose of chlorine used in the past. He quoted a figure of how many gallons of the minuscule amount of chlorine would have to be drunk to reach the medicinal dose of 4,000 ppm in 4 drams of liquid.[89]

Several pages of Leal's testimony show that he reassured Special Master Magie that the theoretically small amounts of chlorine in treated water would cause no adverse health effects. Leal maintained the opinion that no free chlorine was present in the treated water but that comparing the amount of chlorine the plaintiffs' witnesses said was present in the water and the medicinal dose of chlorine was a valid evaluation.[90]

Pages of testimony recorded the condition of filter cloths after water had passed through them. Water from Jersey City's distribution system and samples from many other raw and filtered supplies were filtered through cotton flannel cloths for comparison purposes.[91] Leal's testimony on October 14 brought the filter cloth comparisons up to date.

"Q. Have you formed any opinion as to the effect on the health of consumers of water, of organic matter in such quantities as are found in the Boonton water?

A. Why, *there is no evidence whatever to show that organic matter, per se, has any deleterious effect upon the health* whatever. The only advocates of such theory are laymen who have neither had a medical education or practical training, and who are not the ones to investigate a subject of this sort."[92] (emphasis added)

Leal noted that food is made of organic matter and that it should not be harmful in itself. Then he dived into the ptomaine pool.

". . . It is true that in concentrated organic matters you may get decomposition advanced to such an extent, and certain forms of decomposition especially, that you will have the development of ptomaines, and if you get into the system some of these ptomaines it will cause disease—it will cause trouble, it will not cause specific disease, but it acts as a poison; but there is no case on record where there has ever been sufficient ptomaine developed in a water supply as to cause any poison, and it is impossible. It is too dilute, there is not enough organic matter in any water which the eye or the nose would allow you to drink, which could by any possibility cause ptomaine poison. . . ."[93]

Leal then took on the theory of vital resistance.

". . . Now as to the very vague theory that though it does not cause disease it may lower the vital resistance and so interfere with the general health and make the person using such water more liable to disease—I will state again *that is a theory formulated by laymen*. If they know what vital resistance is, there is not a physician in the world [who] does. I would be very glad to be taught what it is by some of these lay-gentlemen that speak so glibly about it." (emphasis added)

Leal was directly attacking the testimony of Sedgwick, Winslow, and Whipple, calling them "lay-gentlemen." They must have been furious when Leal directed his attack on vital resistance at them. Next, Leal took on the outmoded theory of sewer gas as a corollary to organic matter in water and its impact on vital resistance. His language showed that he felt strongly about the subject.

> "Q. Is there any amount of organic matter in this Boonton water which would affect or in any way deteriorate the vital resistance whatever, that is, of the people drinking it?

> A. There is none. I will state the same thing that has been threshed out before, with respect to sewer gas. At first sewer gas was supposed to be a cause of disease. Oh! It caused diphtheria, and it caused scarlet fever and it caused typhoid fever, and it even caused smallpox, and I don't know what it didn't cause. Well, gradually that was exploded. Then . . . gradually a theory was formulated that even if it did not cause this specific disease it lowered the vital tone and vital resistance, and rendered people more liable to those diseases, ***and now that is gradually being discarded, just the same as this other theory about organic matter will be discarded.***"[94] (emphasis added)

Sedgwick stated in his testimony that between two and a dozen cities were adding chloride of lime to their water supplies but that they were doing so in a "quiet way" so the public would not be alarmed.[95]

At the end of his testimony on October 14, Leal made a startling statement. He knew positively that 27 cities were using chloride of lime disinfection in their water supplies, most of them without notifying their customers. He speculated that hundreds more were testing the treatment method. Although he had been pledged to secrecy by many of the water supply operators, he listed a dozen cities that were using chloride of lime, including Cincinnati, Ohio, and Harrisburg, Pennsylvania.[96] The chlorine revolution had begun.

WILLIAM J. MAGIE SUBMITTED his Special Master's Report on May 9, 1910.[97] The introduction noted that he held the first session under his assignment on September 29, 1908. He wrote that from that date, hearings were held for "forty days or more" and that he had not received the last pleadings and briefs from the litigants until April 19, 1910. Thus, he issued his findings only 20 days after the proceedings were completed but 588 days after the first day of the second trial.

Magie's report consisted of three topics: (1) the cost of sewers and sewage disposal plants intended to prevent contamination of the Rockaway River, (2) other plans and devices for producing pure and wholesome water, and (3) the cost of obtaining water rights from riparian owners downstream from Boonton Dam.

Magie found that the cost of constructing sewers (including obtaining the attendant right of way) and sewage disposal facilities to prevent contamination of the Rockaway River from the Boonton and Dover communities would total $438,189.11. However, he agreed with Dr. Leal that such facilities would not adequately protect the Jersey City water supply.

> "I do further report that if the *sewers* contemplated by the plan and scheme presented and the cost of which I have considered, shall be built, they *will not be capable of substantially preventing the contamination of the Rockaway river above the Boonton Reservoir*, from the sewage of Dover and Boonton, without other additional works. They will not intercept the surface water carrying the wash of fields and roads, and possibly at times contamination from neglected and overflowing privies. They will not intercept any seepage from neglected privies, through permeable soil. They can only be made available by the exercise of power, to compel the connection of all the privies within the district, conceived to be threatening to the purity of the Rockaway river, with some means of discharging the sewage into the connecting sewers and so to the trunk sewer."[98] (emphasis added)

One of Magie's findings was of critical importance to the defendants because he laid to rest the concern that chlorine was a poison that would harm members of the public who consumed the water.

> "Upon the proofs before me, I also find that the solution described *leaves no deleterious substance in the water.* It does produce a slight increase of hardness, but the increase is so slight as in my judgment to be negligible."[99] (emphasis added)

The Special Master's Report then delivered the finding the defendants had been waiting for:

"I do therefore find and report that this device is capable of rendering the water delivered to Jersey City, pure and wholesome, for the purposes for which it is intended, and is effective in removing from the water those dangerous germs which were deemed by the decree to possibly exist therein at certain times."[100] (emphasis added)

Magie's finding, summarized in this one sentence, anointed the use of chlorine in drinking water with the imprimatur of judicial legitimacy.

Magie found that the capital cost of the chloride of lime feed facilities was $20,545.64 and the annual operating and maintenance cost was $2,100. Interestingly, Magie also found that the addition of sodium hypochlorite was equivalent to the use of chloride of lime, thus opening the door for the use of this alternative technology once the reliability problems could be solved.

At no point in his ruling did Magie give any credence to the filter cloth residues or the lowering of vital resistance by the ingestion of organic material present in Boonton Reservoir water. In this regard, Magie's ruling was a complete victory for the defendants.

JERSEY CITY FILED EXCEPTIONS to the Special Master's Report on May 20, 1910. Not surprisingly, the plaintiffs disagreed with Magie's finding that no deleterious substances remained in the water after the use of chloride of lime.

"Whereas the Master should have found that chloride of lime is a poison and will leave a deleterious substance in the water, if used in sufficient quantities. That the use of this solution is an experiment, the effect of the continual use of which for a series of years, as an addition to water intended for drinking and domestic purposes, is unknown and now unknowable, and that its use, depending as it does, on the skill and care of employes [sic], is too dangerous to be adopted by the complainant."[101]

The plaintiffs also took issue with the finding that adding chloride of lime was an effective process for killing bacteria, including pathogens. They attacked this finding by focusing on the inability of chlorine to kill spore-forming bacteria and the fact that filth still remained in the water.[102]

Even as early as 1909, it was known that chlorine did not kill spores—specifically the spores of *Bacillus subtilis*. If the court had accepted this exception, it would have significantly limited the use of chlorine and its subsequent efficacy in controlling typhoid and other waterborne diseases. Fortunately, most of the bacteria that cause disease do not form protective spores.[103]

Jersey City was particularly concerned about Magie's finding that the use of chloride of lime rendered the water "pure and wholesome." The city repeated its contention that building sewers was the only suitable option.[104]

THE HONORABLE FREDERIC W. STEVENS issued his final ruling in the case on November 15, 1910.[105] With one exception, Vice Chancellor Stevens supported all of the findings in the Special Master's Report.

> ". . . the Chancellor doth, by virtue of the power and authority in him vested, ORDER, ADJUDGE AND DECREE, that the said exceptions to the Master's Report be overruled . . . [and] that the said Master's Report . . . be in all things approved, ratified and confirmed."[106]

One of the exceptions dealt with use of the Morris Canal as the recipient of flow from sewers and drains from the city of Boonton. Rather than deduct the cost of building sewers in Boonton to bypass the Morris Canal, Vice Chancellor Stevens allowed the defendants, if they chose to do so, to post a bond for the construction price of $58,300. Table 11-2 summarizes the deductions from the contract payment specified by Stevens.[107]

Finally, Stevens specified the amount Jersey City owed Jersey City Water Supply Company (Table 11-3).[108]

As with all litigation regarding water during this period in New Jersey, this was not the end of the story.

Jersey City appealed Stevens's decision. Given the care with which both judges conducted the two cases and the exalted positions of these two men in the New Jersey judiciary, it would have been surprising if their rulings and judgments had been reversed by higher courts.

Indeed, the New Jersey Court of Errors and Appeals and the New Jersey Supreme Court affirmed Stevens's decision.[109]

ONLY ONE ARTICLE SUMMARIZING AND ANALYZING the Jersey City trials was published right after the second trial, and it was written by C.-E. A. Winslow. Winslow summarized the importance of the case: "This case will, I think, take its place as one of the

Table 11-2 Authorized deductions from contract payment

Deduction Number	Reason for Deduction	Amount of Deduction
1	Defects in 72-inch pipe under the Hackensack River	$5,000
2	Liquidated damages for delay in contract completion	$72,500
3	Defects in invert of the Watchung Tunnel	$18,500
4	Obtaining diversion rights from riparian owners of Rockaway River rights downstream of Boonton Dam	$7,000
5	Cost of bond for construction of Boonton sewers	$58,300
6	From initial judgment, as authorized under contract dated July 8, 1901	$500,000
	Total Deductions	**$661,300**

Table 11-3 Final calculation of contract payment

Description	Amount
Original contract amount	$7,595,000
Total deductions	$661,300
Amount to be paid to Jersey City Water Supply Company	**$6,933,700**

historical law-suits in connection with the subject of water-supply . . ." In Winslow's opinion, the case had two main results: (1) it provided a thorough discussion of the uses and limitations of self-purification of wastes discharged into rivers and lakes, and (2) it successfully demonstrated the tremendous advantages of using chloride of lime to disinfect contaminated water supplies.[110]

Winslow agreed with the evidence presented in the second trial that chloride of lime eliminated bacteria, including B. coli and disease-causing organisms. He also noted the existence of bacteria that were resistant to disinfection by chlorine. Specifically, he was aware that anthrax bacillus and *Bacillus sporogenes* formed spores that were highly resistant to chlorine. He also expressed a concern about the tubercle bacillus and its waxy coating, but this organism was not a concern in drinking water.[111]

He cautioned that chlorine was not a panacea: "I do not believe that the bleaching-powder process will work miracles, and that it will be possible, by its use, to turn sewage into drinking-water. . . ." He went on to declare that chlorination was proven as a major tool for water treatment but added ". . . I do believe that it adds a third to our two recognized methods of water-purification, storage and filtration."[112] Winslow's comments are especially interesting, given that he testified as an expert for the plaintiffs in the second Jersey City trial.

THE DEATH RATE FROM TYPHOID FEVER in Jersey City dropped significantly from 1880 to 1925, as shown in Figure 11-2.[113] The 45-year period puts into clear perspective the effects of water supply changes and chlorination on waterborne disease in the city. Moving the water supply from Belleville to Little Falls resulted in a huge reduction in the typhoid fever death rate. It is not clear whether the institution of filtration at Little Falls had any measurable effect on Jersey City's typhoid death rate. Given the variation in the yearly death statistics, it appears that the change from a filtered water supply at Little Falls to the Boonton supply with sedimentation only in 1904–1907 resulted in no degradation of quality. However, the institution of chlorination at Boonton Reservoir had an immediate and substantial effect; it lowered the death rate to single digits and ultimately to low single digits.

For the entire period shown on Figure 11-2 and for most of the 104 years of continuous chlorination at Boonton Reservoir, free chlorine was used. In the late 1930s, Jersey City introduced the use of chloramines, which required the installation of an anhydrous ammonia feed system at Boonton.[114] Chloramines became popular at the time because they were known to have fewer taste and odor problems compared with free chlorine. Most utilities switched back to free chlorine during World War II because of ammonia gas shortages. Chloramines did not become popular again until after the discovery of disinfection by-products—specifically, trihalomethanes—in 1974.[115]

ONE OF THE BIGGEST FIGHTS in the second Jersey City trial was about sewers in the Rockaway River watershed. It would take 15 years, but the city eventually built the facilities its representatives described during the trial. A trunk sewer intercepting the wastes from Dover, Boonton, and smaller habitations in the Rockaway River watershed was completed in 1925. Continuing to confound and delay water and sewer development in the watershed, the Morris Canal figured into the final plan for sewers. The proposed route for the intercepting sewer included part of the right-of-way for the Morris Canal, and the

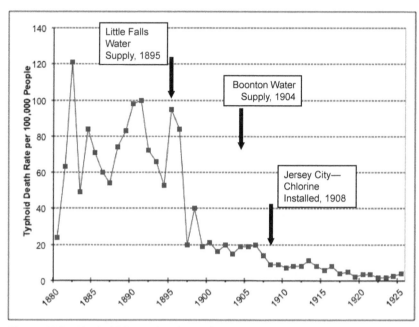

Figure 11-2   Typhoid fever death rate in Jersey City, 1880–1925

canal had to be abandoned before the sewer could be completed. At about the same time, a sewage treatment plant at Boonton was finished. The plant employed the trickling filter method of sewage treatment followed by sand filtration and chlorination of the plant's effluent.[116] Because of a number of delays, the sewage treatment plant was not put into operation until December 14, 1928.[117]

1   *Evening Journal*, May 8, 1909.

2   Between Jersey City and Water Company, Filed September 3, 1908, 4187–92.

3   Ibid., 4191.

4   Ibid., 4192.

5   Between Jersey City and Water Company, September 29, 1908, 4195–6.

6   Between Jersey City and Water Company, January 5 to February 3, 1909, 4215–4892.

7   Further discussion about chlorination of the Dover sewage disposal plant effluent is contained in Leal's testimony, Between Jersey City and Water Company, February 5, 1909, 5018–20.

8   Magie, In Chancery of New Jersey, 1–15.

9   Bureau of Labor Statistics, CPI Inflation Calculator.

10   Between Jersey City and Water Company, January 19, 1909, 4454–66.

11   Between Jersey City and Water Company, January 20, 1909, 4571–3.

12   Between Jersey City and Water Company, January 6, 1909, 4294; January 21, 1909, 4620–7, 4670.

13   Between Jersey City and Water Company, February 3, 1909, 4910–1.

14   Hazen, *Filtration of Public Water-Supplies*, 129.

15   Between Jersey City and Water Company, February 3, 1909, 4915.

16   Ibid., 4914.

17   Ibid.

18   Ibid., 4915.

19   Between Jersey City and Water Company, February 4, 1909, 4921–2.

20   Ibid., 4922.

21   Ibid., 4929–30.

22   Ibid., 4932–3.

23   Ibid., 4993.

24   Ibid., 4935–6.

25   Race, *Chlorination of Water*, 15.

26   Between Jersey City and Water Company, February 4, 1909, 4937.

27   Ibid., 4940.

28   Ibid., 4943.

29   Ibid., 4949–50.

30   Ibid., 4950–1.

31   Ibid., 4964–5.

32   Ibid., 4965.

33   Ibid., 4967.

34   Ibid., 4971.

35   Ibid.

36   Ibid., 4972.

37   Baker, *Quest*, 153.

38   Between Jersey City and Water Company, February 4, 1909, 4972.

39   Reece, "Epidemic of Enteric Fever in Lincoln," 117.

40   Between Jersey City and Water Company, February 4, 1909, 4977.

41   Reece, "Epidemic of Enteric Fever in Lincoln," 117–8.

42   Houston, "B. Welchii, Gastro-Enteritis and Water Supply," 484.

43   Between Jersey City and Water Company, February 4, 1909, 4978.

44   Ibid.

45   Ibid.

46   Baker, *Quest*, 327–8. Baker summarized a number of patents granted regarding chemical oxidizing agents added to water. The first American patent (1888) on chlorination of water was held by Albert R. Leeds, a professor of chemistry at Stevens Institute of Technology.

47   Between Jersey City and Water Company, February 5, 1909, 5069.

48   Ibid., 5074; Between Jersey City and Water Company, February 8, 1909, 5093.

49 Between Jersey City and Water Company, February 5, 1909, 5077–8.

50 Ibid., 5080.

51 Between Jersey City and Water Company, February 8, 1909, 5097.

52 Ibid., 5104.

53 Ibid., 5122.

54 Between Jersey City and Water Company, February 10, 1909, 5180.

55 Ibid., 5180–1.

56 Between Jersey City and Water Company, February 18, 1909, 5473; see also 5478.

57 Between Jersey City and Water Company, March 1, 1909, 5676.

58 Between Jersey City and Water Company, February 26, 1909, 5616.

59 Between Jersey City and Water Company, February 17, 1909, 5429.

60 Between Jersey City and Water Company, February 23, 1909, 5554–5.

61 Between Jersey City and Water Company, February 26, 1909, 5607–10.

62 Mason, *Water Supply*, 182.

63 Between Jersey City and Water Company, March 1, 1909, 5690–2.

64 Between Jersey City and Water Company, April 5, 1909, 5693–4.

65 Ibid., 5705.

66 Between Jersey City and Water Company, May 6, 1909, 5750–94.

67 Between Jersey City and Water Company, April 28, 1909, 5989–92.

68 Ibid., 5992–6008.

69 Between Jersey City and Water Company, May 7, 1909, 6234.

70 Ibid., 6218.

71 Ibid., 6227.

72 Between Jersey City and Water Company, May 11, 1909, 6320, 6329.

73 Ibid.

74 N,N-diethyl-p-phenylenediamine

75 Palin, "The Determination of Free and Combined Chlorine in Water"; Palin, "Current DPD Methods for Residual Halogen Compounds; Disinfection Committee AWWA, "Committee Report," 222.

76 Between Jersey City and Water Company, May 5, 1909, 6098–9.

77 Ibid., 6095.

78 *Evening Journal*, May 12, 1909; *Evening Journal*. May 13, 1909; *Evening Journal*. May 20, 1909.

79 Between Jersey City and Water Company, May 21, 1909, 6509.

80 *Evening Journal*, April 30, 1909.

81 Leal, "Sterilization Plant of the Jersey City Water Supply Company", 100–9. Fuller, "Description of the Process and Plant of the Jersey City Water Supply Company," 110–34; Johnson, "Description of Methods of Operation of the Sterilization Plant," 135–47. Histories of drinking water disinfection seldom mention that another paper on the application of chloride of lime was given at the same conference and published in the same proceedings. A.E. Walden described disinfection studies on the Baltimore water supply, "Application of Hypochlorite of Lime," 26–38; he also described the use of orifice boxes as part of a chemical delivery system that was clearly influenced by Fuller's design of the Little Falls Treatment Plant. Leal published a similar paper on

the chloride of lime feed facility at Boonton Reservoir in the *Journal of the Engineers Society of Pennsylvania*.

82 *Evening Journal*, June 2, 1909.

83 *Evening Journal*, June 8, 1909.

84 "Discussion" of three sterilization plant papers, *Proceedings AWWA*, 147–62.

85 Between Jersey City and Water Company, October 4, 1909, 6539–68.

86 Between Jersey City and Water Company, October 4-5, 1909, 6569–642, 6666–74.

87 Between Jersey City and Water Company, October 6, 1909, 6675–707, 6725–7.

88 Between Jersey City and Water Company, October 14, 1909, 6731.

89 Ibid., 6731–808.

90 Ibid., 6731–3. The comparison between the infinitesimal amount of chlorine in treated drinking water compared with the "chlorine water" treatment for typhoid fever was used in many public relations pieces in the subsequent years, including one in 1925, "Jersey City's Water Supply Improved," *Aquafax*, 4.

91 Years after chlorine was in place but before completion of a filtration plant in 1979, water from Boonton Reservoir was being filtered through "finely woven cheesecloth screens" at two locations prior to its distribution to consumers. No useful bacteriological function was served by the cloth filters, and they must have been a maintenance headache for the Jersey City staff. The insistence of the plaintiffs' experts on filth being removed by filter cloths had this unintended consequence, which to the author's knowledge was not practiced by any other water utility in 1939; Ohland, "The Jersey City Water Supply," 34–5.

92 Between Jersey City and Water Company, October 14, 1909, 6766–7.

93 Ibid.

94 Ibid., 6768–9.

95 Between Jersey City and Water Company, May 7, 1909, 6221–2.

96 Between Jersey City and Water Company, October 14, 1909, 6769–71.

97 Magie, In Chancery of New Jersey, 1–15.

98 Ibid., 7.

99 Ibid., 13.

100 Ibid., 13.

101 Between Jersey City and Water Company, Exception to Master's Report, Filed May 20, 1910, 2.

102 Ibid., 3.

103 Chlorine's inability to inactivate spores is similar to its inability to kill *Cryptosporidium* in the oocyst stage, as evidenced by the epidemic of cryptosporidiosis in Milwaukee, Wisconsin, in April 1993.

104 Between Jersey City and Water Company, Exception to Master's Report, Filed May 20, 1910, 4.

105 Stevens, In Chancery of New Jersey, Decree Confirming Report, November 15, 1910, 1–10.

106 Ibid., 3.

107 Ibid., 8–10.

108 Ibid., 9.

109 Baker, *Quest*, 339; "Mayor & Aldermen of Jersey City v. Jersey City Water Supply Co. Court of Errors and Appeals. November 21, 1911.

110 Winslow, "Water-Pollution and Water-Purification," 16; his paper was presented at a conference of the Western Society of Engineers held in Chicago on March 16, 1910.

111 Ibid., 15.

112 Ibid., 16–7.

113 Fuller and McClintock, *Solving Sewage Problems*, 106–7.

114 Ohland, "The Jersey City Water Supply," 34.

115 McGuire, "Eight Revolutions," 131–2.

116 Potts, "Sewage Treatment Plant at Boonton."

117 "Scraps of Information," January 1929.

12

Revolution and Conquest

*"There can be no doubt that the development of a
practical method of water disinfection during the last two
years marks an epoch in the art of water purification."*
— Winslow, "Field for Water Disinfection," 1

Chlorine use exploded after the success at Boonton Reservoir, and a number of reports catalogued the growth of its use. Francis F. Longley, a consulting engineer working for Allen Hazen and George C. Whipple, conducted a census of water utilities thought to be using chlorine in 1914.[1] Questionnaires were sent to 220 utilities around the country, and half of them were returned. Figure 12-1 shows the rapid rise in chlorine use in the United States from 1908 to 1914. Jersey City was the only utility using chlorine in 1908, but by 1914, more than 21 million people were receiving water from chlorinated municipal supplies.[2]

The population served by municipal water supplies in 1914 was just over 40 million people (estimate based on interpolation of population data from 1910 and 1920 and other U.S. Census records).[3] Fully 53 percent, or 21 million, of those 40 million people were known to be receiving water that had been disinfected with chlorine. And this number does not include the customers of utilities that were using chlorine but did not respond to Longley's questionnaire.[4]

In 1918, it was estimated that 3,000 million gallons per day (mgd) were being treated with chlorine in more than 1,000 North American cities.[5] If the same estimating factors are used, 3,000 mgd would have been the amount needed to serve a population of 33.3 million people, or 75 percent of the population served by municipal supplies. In 1922, another estimate pegged the number of water systems using chlorine at 3,000, and they were serving about 30 million customers.[6] An estimate in 1926 placed the amount of water treated with chlorine every day at 4,000 mgd in about 3,200 U.S. communities. At that time, 4,000 mgd would have been enough to serve about 44 million people.[7]

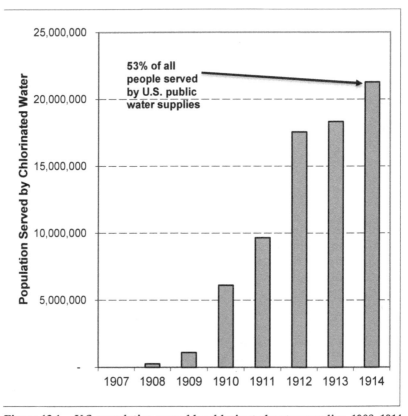

53% of all people served by U.S. public water supplies

Figure 12-1   U.S. population served by chlorinated water supplies, 1908–1914

Case Histories

A few case histories will illustrate how some cities were able to initiate chlorination quickly and how others took a while to institute this advance.

PATERSON AND NEARBY CITIES. The application of chloride of lime at the Little Falls Treatment Plant was mentioned in the transcript of the second Jersey City trial. Dr. John L. Leal began feeding chloride of lime at the Little Falls plant on or before February 4, 1909.[8] The chlorine dosage he specified is not known, but, on the basis of Leal's usual practice, we can assume it was a fraction of a part per million (ppm), probably less than 0.5 ppm.

As noted in Chapter 11, this was the first application of chlorine in a plant using mechanical filtration.

POUGHKEEPSIE, NEW YORK. Surprisingly, George C. Whipple, who had opposed chlorination during the second Jersey City trial, suggested the third application of chlorine to a water supply. Poughkeepsie is a medium-sized city located along the Hudson River about 70 miles north of New York City. In a report to the city of Poughkeepsie, Whipple recommended that the coagulant preceding slow sand filtration be replaced with chloride of lime. This protocol began as a test on February 1, 1909. On March 17, Poughkeepsie initiated continuous chlorination with the use of a permanent chemical feed apparatus.[9] Whipple also recommended installation of a chloride of lime feed at a mechanical filtration plant in Watertown, New York.[10]

CHICAGO, ILLINOIS. Chlorination of the Chicago water supply started in 1911 (see Chapter 2). From 1911 to 1916, chlorine feed systems were installed in nine pumping stations that withdrew water from Lake Michigan, according to one author.[11]

A more detailed account of circumstance in Chicago showed that the first attempts at chlorination of water from the lake used chloride of lime solution tanks. From 1912 to 1913, chloride of lime feed systems were installed in three intake cribs, But the cribs' location several miles offshore created operational problems. Cold weather caused the chloride of lime solutions to freeze in the tanks and pipes. In addition, bad weather and equipment problems interrupted offshore delivery of the chloride of lime material. It is also likely that the chlorination system operators had problems consistently feeding a low dose of chlorine because many consumers complained about the chlorine taste of the water.[12]

Cylinders of compressed chlorine gas were just becoming available about the time Chicago was installing chlorination. Using both homemade and purchased materials, the water department staff built pressurized chlorine gas feed systems at all nine intakes and had the systems installed by October 6, 1916. Because this was pioneering work, Chicago discovered all of the problems related to materials that came in contact with chlorine gas.[13] When vacuum chlorinators became available, Chicago abandoned the dangerous pressurized systems.

SEYMOUR, INDIANA. THE RECORD-TIME APPLICATION of chloride of lime in a small system in Indiana serves as another example of why the use of chlorine in the form of chloride of lime exploded after the experience at Boonton Reservoir.[14] The town of Seymour is located in south-central Indiana, and in 1910 it had a population of approximately 6,000. At the beginning of that year, the city was using alum as a coagulant for its surface water supply followed by sedimentation in a too-small basin and subsequent filtration. However, the quality of the product water was unsatisfactory—500–600 total bacteria per milliliter and the frequent presence of B. coli. Huge alum doses, as high as 43 ppm, were being used.

Shortly after news of the decision by the New Jersey courts, Seymour installed two small galvanized iron tanks piped to feed a solution of chloride of lime. The cost was a mere $30. A chloride of lime dosage of 6 pounds per million gallons was used. With 35 percent available chlorine in chloride of lime, the dosage translated to approximately 0.25 ppm. Bacterial results during January 1910 showed that half of the time, no bacteria were detected in the treated water and the other half of the time, total bacteria concentrations were very low, 1–20 per milliliter.

DANVILLE, ILLINOIS. The same report that recounted the experience of Seymour, Indiana, also described a successful application of chloride of lime at Danville, a coal mining center in 1910 with a population of about 28,000. At Danville, a chlorine dosage of 0.6 ppm yielded satisfactory results.[15]

QUINCY, ILLINOIS. The application of chloride of lime at Quincy in 1910 was an example of what not to do. The treatment plant personnel had no idea what was needed and had not read the papers published by Leal, Fuller, and Johnson. The stated purpose of the chloride of lime treatment at Quincy was to improve operation of the existing mechanical filtration plant, which had several physical defects. Thus, the chlorination system was operated only intermittently. The operators filled an oil drum (with a capacity of 40–50 gallons) with water and then added chloride of lime by hand to achieve the desired concentration for 1–2 hours of operation. A valve on a pipe at the bottom of the drum regulated the flow of the chemical into the water. Of course, the head continually decreased as the solution was used, and this reduced the flow past the valve.[16]

The flow of the chemical had to be adjusted constantly. Not surprisingly, the water company received a lot of consumer complaints about a "medicinal" flavor in the water. Not only was there no control over the amount of chemical fed, but the initial target dosage was 4.3 ppm—far above the odor and flavor thresholds for free chlorine. After a few days of complaints, the dosage was reduced to 2.4 ppm—still significantly above the odor and flavor thresholds.[17] It was clear that W.R. Gelston, superintendent of the Citizen's Water Company of Quincy, was unaware that he needed to keep the chlorine dosage far below 1 ppm or risk the ire of his customers.

TRENTON, NEW JERSEY. Trenton, the capital of New Jersey, was home to about 97,000 citizens in 1911. The city's water source was the Delaware River, which had been grossly contaminated with sewage for decades. Typhoid fever was ever-present in the city, and occasionally epidemics broke out, causing much higher death rates. The typhoid fever death rate during 1902–1911 ranged from 26.2 to 84.3 per 100,000 people, with an average of 49.7 per 100,000.[18]

Despite the water supply's wholesale killing of Trenton's citizens, there was tremendous opposition to installing filtration or any other kind of effective treatment. Outstanding treatment experts such as Allen Hazen and George Warren Fuller prepared two separate designs for filtration plants, both of which languished without being implemented. Finally, the New Jersey Board of Health had had enough. In early 1910, the board issued a "compulsory order" for Trenton to treat its water supply and made the order effective shortly thereafter, on June 15.[19] The Trenton Water Board began to install a chloride of lime feed system, but, incredibly, the local health board vetoed the plan. Wasting no time, the New Jersey Board of Health filed a lawsuit shortly after the June 15 deadline to compel the city to move forward with its plans.[20]

Even the lawsuit did not fix the problem, because there was still much opposition to the plan. The city took more than a year to hire a consultant and build a plant. The chloride of lime feed system was completed in early November 1911, and testing of the process with a dosage of 0.4 ppm began on November 9. At almost the same time, the state Board of Health became aware that a typhoid epidemic had broken out in Trenton. The board "summoned" Mayor Frederick W. Donnelly and told him more effort had to be made to curtail the epidemic. The mayor reluctantly agreed. The chlorine dosage was raised to 0.8 ppm on December 3, and the rate of typhoid fever cases quickly decreased.[21] The mayor's reluctance stemmed from his certainty that the higher dosage would result in more taste and odor complaints from citizens who did not want the chemical added to their water in the first place. The local newspaper printed a statement describing the need for the higher dosage.[22]

The design of the chloride of lime feed system consisted of two small mixing tanks, two solution tanks of 1,500 gallons each, and only a single constant-head orifice feed tank. George A. Johnson was apparently in charge of the design for the firm of Hering and Fuller.[23] No reason was given for the lack of redundancy in the greatest potential failure point in the process—the single constant-head orifice feed tank.

A newspaper article on December 21 reprinted a letter from George A. Johnson, who tried to reassure the public. Johnson noted that many cities had already installed chloride of lime feed systems, and he quoted the Magie decision in the second Jersey City trial, which declared the process to be effective at destroying pathogens. Johnson also disputed claims that chlorine was poisonous and dragged out the assertion that oxygen in an active state was responsible for the disinfection process. In short, the citizens of Trenton raised the same objections as the plaintiffs in the Jersey City case. Fortunately, Johnson was able to present the facts plainly because of the precedents of other cities' experiences and the legal rulings affirming that the recommended treatment was appropriate.[24]

In the case of Trenton, Johnson went further than he did in the Jersey City case and recommended filtration. A mechanical filtration plant went into service on Trenton's water supply in 1914.[25] In 1920, the typhoid death rate in Trenton had decreased to 8.3 people per 100,000.[26]

NASHVILLE, TENNESSEE. In 1910, Nashville withdrew its water supply from the Cumberland River, which was extremely turbid and known to be contaminated with high concentrations of bacteria. Average daily water use in Nashville in 1909 was approximately 14 mgd for a population of about 118,000. Raw water from the river was pumped into a 51-million-gallon reservoir for a 3.5-day detention time. Alum (as aluminum sulfate) was added to the water entering the reservoir at a dosage of 17 ppm.

Significant removal of turbidity and bacteria took place in the settling reservoir.

Beginning in August 1909, only two months after the three papers on chlorination at Boonton were presented at the June conference of the American Water Works Association (AWWA) but before the Jersey City judicial decision, the water company in Nashville added a chlorine dose of approximately 0.6 ppm to water leaving the reservoir. With the addition of chloride of lime, the overall treatment resulted in product water containing no B. coli and very few total bacteria. The cost of the chemical feed system was only $800, and the cost of chloride of lime treatment was $1.05 per million gallons.[27]

MINNEAPOLIS, MINNESOTA. The water source for Minneapolis in 1910 was the Mississippi River. At that time, the city had a population of about 380,000, and average daily water use was about 20 mgd. Water was pumped from the river into two rectangular basins with a total capacity of 97 million gallons. Plagued with outbreaks of typhoid fever, the city had considered several treatment options including the installation of slow sand filters. However, none of these plans came to fruition because of the resistance of the electorate and the costs of the projects. On February 25, 1910, a chloride of lime treatment system was put into operation after an alarming increase in typhoid cases and deaths in the city.[28]

CINCINNATI, OHIO. The application of chlorine to Cincinnati's water supply was chronicled decades later by Colleen K. O'Toole in a Ph.D. dissertation that examined this new technology in detail.[29]

Joseph W. Ellms, the treatment plant superintendent in Cincinnati, had worked with George Warren Fuller on the Louisville filtration study from 1895 to 1897 and on the Cincinnati filtration study during the two years following the pioneering work in Louisville. Ellms had also written a book on water treatment. Published at the beginning of the twentieth century, the book described state-of-the-art water treatment at that time.[30]

Despite this experience, for a period of eight years Ellms chose to treat Ohio River water with chlorine intermittently. O'Toole speculated on the reasoning behind Ellms's decision.

"In summary, Joseph W. Ellms began experimenting with chlorine disinfection at the Cincinnati Water Works

in 1909. It is understandable that he might have continued to experiment with the new process, using it only
intermittently for several years. Presumably because of
seasonal difficulties in application in conjunction with
concerns for cost-effectiveness, however, Ellms decided
to only intermittently chlorinate Cincinnati's water supply through 1917. Ellms'[s] chlorination practices certainly did not interfere with the continuing decline in
the city's typhoid fever mortality rate. Cincinnati's reputation for having a pure and wholesome water supply
and one of the most advanced water works in the country
remained high throughout Ellms'[s] tenure."[31]

Indeed, the historical record shows erratic implementation of
chlorination in Cincinnati from 1909 to 1917, including intermittent use primarily in the winter and spring seasons (Figure 12-2).[32]

O'Toole's conclusion notwithstanding, it is extremely unlikely
that the intermittent chlorination of Cincinnati's water supply for
eight long years produced a "pure and wholesome water supply."

Figure 12-3, based on data compiled by George Warren Fuller,
shows two inflection points in the reduction of Cincinnati's typhoid
fever death rate.[34] The first major decline occurred in connection

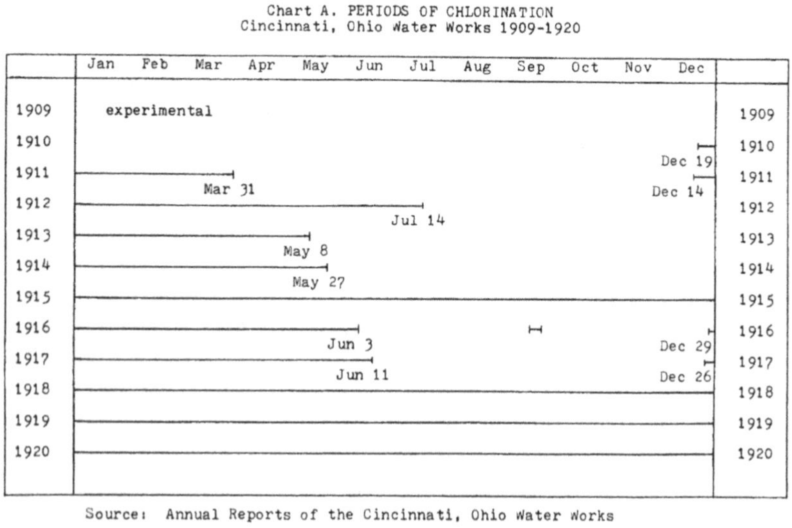

Chart A. PERIODS OF CHLORINATION
Cincinnati, Ohio Water Works 1909-1920

Source: Annual Reports of the Cincinnati, Ohio Water Works

Figure 12-2   Chlorine use at Cincinnati, Ohio, 1909–1920[33]

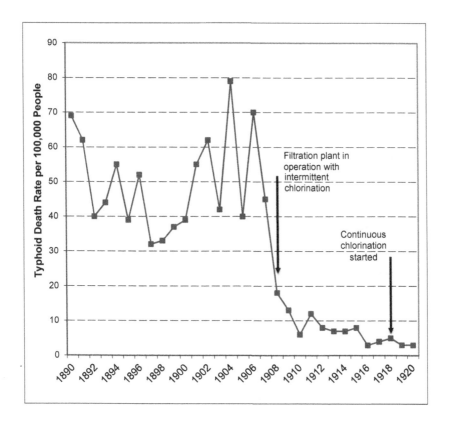

Figure 12-3　Cincinnati's typhoid death rate per 100,000 people, 1890–1920

with the operation of the Cincinnati filtration plant, which started in 1907 and was fully operational in 1908. With intermittent chlorination, the typhoid death rate remained stubbornly high until 1918, when Ellms retired and the new filtration superintendent implemented continuous chlorination. "Ellms'[s] successor, Clarence Bahlman, who had been the Health Department Bacteriologist from 1911 to 1917, began <u>continuously</u> applying chlorine to the water supply in 1918."[35] (underline in original)

There seems to be little justification for Ellms's decision to use intermittent chlorination, given the experiences of many other U.S. cities during this period. The results at Boonton Reservoir showed that continuous application of chlorine in the form of chloride of lime was possible as early as 1908. There is some evidence that Ellms was concerned about the cost of chlorination, even though it was evident to all water treatment practitioners that the cost of chlorination was extremely small. Interestingly, Ellms applied chlorine throughout the year 1915 (Figure 12-2), which corresponded with a new U.S. Treasury Department bacteriological requirement on interstate carriers and the convening of the AWWA conference in Cincinnati that year. In 1916 and 1917, however, Ellms reverted to his usual practice of intermittent chlorination.

O'Toole also speculated about the difficulty of trying to accomplish what Leal did in today's atmosphere of consumer objection to chemicals in drinking water. "Today the initial application of such a chemical to a city's water supply would in all probability engender strong opposition and extended public debate."[36]

Still, O'Toole recognized that the work at Boonton Reservoir blazed a trail for other utilities to follow.

> "At the turn of the century, however, although the chlorination of public water supplies initially met with opposition in some cities, it was relatively mild and short-lived. The rigorous investigations to which the process was subjected during the Boonton Reservoir, New Jersey litigation as well as the demonstration that simple chlorination (i.e., not applied in conjunction with other purification processes) markedly reduced typhoid fever rates, both served to quell the opposition."[37]

In addition, O'Toole noted that the transcript of the second Jersey City trial had revealed the development of an analytical method for estimating the concentration of chlorine remaining in water after the initial dose was applied. Earle B. Phelps discovered that an orthotolidine test could give a qualitative indication of the presence of a chlorine residual[38] (see Chapter 11). Ellms and his assistant, Stephen J. Hauser, subsequently developed a "quantitative" method using orthotolidine in an acid solution,[39] but this method was affected by interfering compounds and chemical conditions, and many years would pass before an accurate method for measuring chlorine residuals would be developed.

OTHER CITIES. Two case studies published in *Engineering News* explained, in part, why chlorination was adopted so quickly in so many places. In Toronto, Canada, and Council Bluffs, Iowa, chloride of lime feed systems were quickly added after typhoid fever outbreaks and the inevitable pressure on political leaders to do something about the epidemics. Even if the feed system was crude, as in Toronto's case, engineers were able to add a treatment step that helped control a waterborne disease within a few days. Then they could follow up later with a better-engineered system.[40]

Table 12-1 shows chlorination data for 10 cities with implementation dates from 1910 to 1913.[41] Death rates from typhoid fever decreased markedly after chlorine was added, but few cities experienced the immediate and spectacular reduction in typhoid deaths achieved by Chicago.

Other cities that installed chlorine in 1911 included Rome, New York; Columbus and Toledo, Ohio; Philadelphia and Pittsburg, Pennsylvania; and Providence, Rhode Island.

Table 12-1 Chlorine implementation in 10 U.S. cities

City	Reference	Source of Suppy	Chlorination Start Date	Typhoid Fever Death Rate Before Chlorination per 100,000 People	Typhoid Fever Death Rate After Chlorination per 100,000 People
Baltimore, Md.	A	Lakes	June 1911	35.4	23.1
Detroit, Mich.	A	Detroit River	March 1913	19.2	15
Waukegan, Ill.	A	Lake Michigan	April 1912	NA	NA
Milwaukee, Wis.	A	Lake Michigan	April 12, 1912*	21.8	8.9
Omaha, Neb.	A	Missouri River	1910	99	25.4
Cleveland, Ohio	B	Lake Erie	September 1911	35.2	10
Des Moines, Iowa	B	Racoon River	December 1910	22.7	13.4
Erie, Pa.	B	Lake Erie	March 1911	38.4	13.5
Evanston, Ill.	B	Lake Michigan	December 1911	26	14.5
Kansas City, Mo.	B	Missouri River	January 1911	42.5	20

*Start of continuous chlorination; used intermittently from June 21, 1910, to that date.

A—Jennings, "Chlorine Compounds in Water Purification," 251.

B—Jennings, "Sterilization of Water Supplies," 317-8.

NA—Not available

Another report showed decreases in the incidence of typhoid fever and other waterborne diseases as a result of improvements in the quality of water supplies. Both filtration and chlorination were responsible for the enhanced quality.[42]

In his book on water purification, Allen Hazen summed up the speed with which chlorine use was adopted after the application at Boonton Reservoir. ". . . as a result of the publicity given by the Jersey City experience, the use of the [chlorination] process extended with **unprecedented rapidity**, until at the present time [1916] the greater part of the water supplied in cities in the United States is treated in this way. . . ."[43] (emphasis added)

Despite the rapid implementation of chlorination, one editorial in 1914 noted that although chemophobia was still an issue with the public, the tide was turning with professionals. ". . . there is a prejudice in the minds of many against a chemical treatment of any kind, which it is difficult to eradicate without evidence from longer experience than is as yet available. Many engineers have joined the procession, however, and chemical methods are popular."[44]

Even seven years later, Joseph W. Ellms from Cincinnati noted that resistance to using chemicals in water was still a problem.

> "Probably there is nothing more difficult to change than human opinions, and especially where such opinions are based upon custom and practice, and where there is possibly an unconscious prejudice against the remedies suggested for effecting a change. Although the opposition to modern methods of water purification, especially where chemical coagulating and sterilizing agents are employed, is by no means as pronounced or as open as formerly, nevertheless it still lingers in the mind of the layman more often than is usually suspected; it is even latent in many sanitarians, who will resort consciously or unconsciously to such measures as storage of surface waters or methods of filtration known to be inadequate for the problem in hand."[45]

Improvements in Chlorine Technology

The 1914 census of utilities using chlorine to disinfect drinking water showed that 80 percent of the responding cities were using chloride of lime.[46] After several years of experience, however, it became clear that chloride of lime had significant drawbacks when applied in large amounts for water treatment. Chloride of lime was great for bleaching laundry, but it tended to form sludge and precipitates that clogged water treatment chemical feeding equipment. Utilities began to replace chloride of lime feeders with liquid chlorine feeders. Philadelphia installed the first permanent liquid chlorine feed system at its Belmont treatment plant in 1913.[47] Wallace & Tiernan was the dominant producer of chlorination equipment in the first decades of the twentieth century. The company's first gas-feed chlorinator, an experimental apparatus, was installed on a tributary of the Rockaway River at Dover, New Jersey, on February 22, 1913.[48] Nonetheless, chloride of lime was crucial to the early adoption of chlorine as a disinfectant for water supplies. Liquid chlorine was not available in 1908, and Leal and Fuller would never have been able to meet the judge's 90-day deadline with anything but chloride of lime. The early experiments at Boonton with electrolytic generation of chlorine were not successful because the equipment was not robust enough to provide continuous, reliable service.

CHLORIDE OF LIME EVOLVED from laundry bleach to deodorizer and putrefaction controller, typhoid fever treatment, miasma manager, filth cleaner, road purifier, diseased patient habitation sterilizer, and drinking water disinfectant/waterborne illness destroyer.

The main chemical form of laundry bleach changed from chloride of lime to sodium hypochlorite in the early twentieth century, and drinking water disinfection soon moved from using chloride of lime to gaseous and liquid chlorine. Roads, filth, and miasmas were no longer considered in need of the disinfecting power of chloride of lime. Since the 1974 discovery of trihalomethanes in drinking water, some water utilities have replaced chlorine with other primary disinfectants (e.g., ozone, C, and ultraviolet light).[49]

Today, chloride of lime is primarily used to disinfect swimming pools when only small amounts are required. Large public pools use gaseous chlorine or have switched to ozone. Water utilities also use calcium hypochlorite for minor disinfection tasks such as disinfecting reservoirs and pipelines after construction. Other forms of chlorine and other disinfectants took over, but in its heyday, chloride of lime was the killer of choice.

WITH THE WIDESPREAD USE OF CHLORINATION, more cities became interested in installing filters to improve the aesthetic characteristics of their water supplies and to ensure multiple-barrier protection from pathogens. In many ways, the expanding use of filtration to treat surface water supplies was due to the overall improvement of water quality and the increasing realization that typhoid disease was completely preventable and should not be tolerated in a modern city.

Figure 12-4 shows that beginning in 1904, the installation of filtration plants began to accelerate. Engineering consultant John W. Alvord's 1917 article in *Journal - American Water Works Association (Journal AWWA)* summarized filtration progress to that date. Alvord catalogued the shift from slow sand filtration to mechanical filtration during the preceding few years. The last

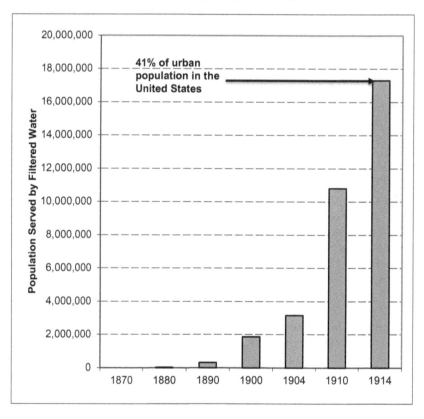

Figure 12-4   U.S. population served by filtered water supplies, 1880–1914[52]

large slow sand filter plants built in the United States served Philadelphia, Pittsburgh, and Washington, D.C.[50]

By 1921, filtration of U.S. water supplies had increased dramatically.

"There are to-day more than eight hundred filter plants in operation in this country and Canada, serving more than twenty millions of people, or more than one-third of the population resident in cities and towns."[51]

In 1917, George Warren Fuller had a contrary point of view and complained that the discovery of chlorine's effectiveness for controlling epidemic disease was limiting the adoption of mechanical filtration.

"Apparently progress in water filtration has been somewhat retarded through efforts to make chlorination serve as a substitute . . . under conditions where it is not entirely adequate. Chlorination has done a great deal of good in the improvement of public water supplies. But it is not a cure-all, and its limitations are far more clearly appreciated with the practical proofs now existing than when academically recited six or eight years ago."[53]

Fuller's former partner, Rudolph Hering, noted that chlorination was not a substitute for filtration.

"Within the past two months, Mr. Rudolph Hering, of New York, reporting on the Buffalo water supply said: 'It is my opinion that sterilization by either method (hypochlorite of lime or liquid chlorine gas) will safeguard the public health at a very much lower cost than the adoption of filtration at this time. As soon as the city is prepared to pay the cost of giving the water at all times an excellent appearance, filtration is the system to be adopted.' It is often a question of whether the results accomplished by filtration of a given water supply are worth the expenditure, provided, of course, that the water as furnished the consumers is free from pathogenic organisms."[54]

In 1922, a committee of public health experts reported how chlorination was being used and misused. Committee members opened their presentation with a startling statement that showed how far the use of chlorine had sunk in the estimation of those responsible for public health. "It is commonly assumed in waterworks practice that chlorination, as a water purification process, is an irregular, haphazard treatment, little understood and poorly controlled."[55]

Part of the reason for the committee's concern was that the early promise of chlorine as a panacea had not been borne out because of complications with applying the process to water supplies of variable quality. Of course, filtration was not a panacea either, nor was sedimentation on its own. The problem for chlorination was the lack of a widely used process to reliably set the dosage by taking into account the chlorine demand of the water and the dosage needed to provide sufficient kill of pathogens.

Other authors were more certain that multiple treatment steps were needed to fully protect the public from waterborne disease.[56] In a 1921 screed against water storage as the sole protective mechanism, George A. Johnson expressed the need for multiple preventive measures in the treatment of surface water supplies.

> "The endeavor has been made to prove that the problem of making surface waters safe for public consumption involves the application of a chain of preventive measures constituting in effect four major lines of defense against water-borne disease, namely:
>
> (a) Maintain the catchment area in as sanitary a condition as practicable; that is, guard against gross pollution entering the streams and lakes which drain the watershed.
>
> (b) Store the water in natural lakes or artificial reservoirs, provided such storage is available or dictated by sound engineering principles.
>
> (c) Coagulate and filter.
>
> (d) Sterilize."[57]

In 1922, Sol Pincus, a sanitary engineer with the U.S. Public Health Service, noted that despite the fact that chlorine could not do everything, it was an effective "barrier" against disease. This was one of the first uses of the word "barrier" in relation to chlorination and its role in protecting the public from waterborne illness.

> "But if chlorination is not to be regarded as the complete treatment process for all kinds of unsafe or polluted waters it is not, on the other hand, to be felt that it has

no large place in the program for water-supply protection. It is not my purpose to convey the impression that the chlorination of water supplies is to be looked at in all cases or the majority of cases with suspicion as to the safety and effective defensive *barrier* that it provides against waterborne infection. Nothing is further from my thoughts. In fact, I fully believe that chlorination has been the greatest single factor to make possible the very high standards that exist to-day for water-supply quality, and in consequence the nearly vanishing typhoid-fever rates now prevalent in our communities."[58] (emphasis added)

Like Johnson, Pincus viewed chlorination as one barrier in the overall treatment process.

"What, then, is the proper place of chlorination in water-supply protection? In general, I believe the consensus of opinion of sanitary engineers would be that chlorination should be considered a supplemental safeguard, after other and more primary defenses have been established. It may be considered as a reserve line of defense, tremendously important, necessarily available for full effectiveness under sudden emergencies, but not the front and sole line of defense, where it would at all times have to be 100 percent perfect under the full attack of the enemy. . . . Though it should as a general rule be considered not the primary and sole protection, its great value and general use as a protecting *barrier,* wherever potential or even remote possibilities exist for unforeseeable contamination reaching the water supply, must be definitely acknowledged and accepted."[59] (emphasis added)

One editorial writer as late as 1914 dragged out the old bromide that it was best to use a source of supply that was not polluted. If cities could have used unpolluted supplies, they would have done so decades before, but they had to use the supplies that were close by and of sufficient quantity even if they were horribly contaminated. The editorial properly cautioned that much could be learned from the new technologies and that making polluted supplies safe depended on people: ". . . [a supply] which is known to be seriously polluted at its source and which must depend upon the human fallibility of the operators and the machinery and the judgments exercised in putting this bad water thru its purifying process."[60]

Even though the multiple-barrier concept of protecting drinking water supplies began to be developed in the first half of the twentieth century, there were still problems with keeping the disinfection component of the barrier operating in an acceptable manner. Early water works practice recognized redundant disinfection equipment as a key part of the multiple-barrier approach, but actual installation of multiple systems was not common.

> "The reliance on one chlorinator, where chlorination is the sole line of defense against a potentially dangerous water supply, has always been considered a serious error, but the shut-down of chlorination under these circumstances [when only chlorine is used] is little short of criminal negligence."[61]

A cartoon commissioned by *American City* magazine illustrated the victory of filtration and chlorination over typhoid fever. In the first frame dated 1890, "typhoid" as the figure of death rules over sufferers with death rates of 80–100 per 100,000 people. In the next frame, a filtration plant delivers the first knockdown blow to typhoid in 1906 by reducing the death rate to 32.1. Finally, in 1918, a cylinder of liquid chlorine appears on the scene to complete the knockout with a death rate reduced to 7.[62]

IN 1912, CHARLES-EDWARD A. WINSLOW noted that since the Jersey City trials, hundreds of cities had adopted the use of chlorine for disinfection. He was also aware that the chlorine dosage had to be carefully controlled or the results would not be satisfactory.[63]

> "It is quite certain that hypochlorite, added in proper amounts, will produce a very high percentage reduction of ordinary water bacteria and of intestinal forms of the *B. coli* group. Its bacterial efficiency, measured in this way, is equal or superior to that of the ordinary water filter of either type [slow sand or mechanical]. In order to effect this result however, the amount of disinfectant must be adjusted to the amount of organic matter in the water, since chlorin [sic] is taken up by some forms of organic matter more promptly than by the bacteria. . . . With sudden variations in organic content there is danger that the process may fail, unless a large excess of chlorin be added at ordinary times, for it is usually impossible to determine such changes until a considerable volume of raw water has passed the purifying works."[64]

Pincus catalogued the many issues that caused variations in the effectiveness of chlorine and made a plea for better control of the process. He noted the following issues affecting chlorine control:

- Excessive doses can result in taste and odor problems and influence the amount of disinfectant that can be applied.
- Suspended matter in the water protects bacteria from the disinfectant.
- Varying chlorine demand of organic materials in the water affect the effectiveness of the process.
- Mixing problems decrease the efficiency of the disinfection process.
- Mechanical breakdowns reduce the protective ability of the disinfectant to zero.
- Full effectiveness depends on vigilant operator attention.

"Chlorination without control—control of dosage, analytical control of quality of the treated water, and control over operation—is practically worthless as a *barrier* against water-borne infection."[65] (emphasis added)

Control of the chlorination process still requires vigilance from water treatment plant operators, but experience and automatic monitors for parameters that have been correlated with increases in organic content (e.g., total organic carbon or surrogates such as conductivity) have mitigated many of these problems. However, it took decades to establish these tools and procedures.

Many improvements over the rudimentary dosages determined by Leal were achieved over the years following 1908. One of the first refinements—combining the concept of chlorine demand and a target residual after a period of time (now referred to as CT, or concentration × contact time)—was introduced in 1919 and revised in 1921.[66] These advances increased the efficiency of chlorination and improved public health protection.

ANYONE WHO USED CHLORIDE OF LIME in the late 1890s and early 1900s knew that high concentrations of chlorine had a strong bleach or chemical smell. Leal knew he had to keep the chlorine dosage at Boonton Reservoir small to avoid taste and odor problems in Jersey City. One publication noted the efficacy of the absorption test developed by Abel Wolman in helping to control taste and odor problems resulting from overdosing.[67] In a paper published in 1914, E. J. Tully, a chemist with the Wisconsin State

Hygiene Laboratory, made one of the first suggestions regarding a chlorine concentration that, when exceeded, might cause taste and odor problems. He posited that a slight chlorine taste and odor was present in water treated with dosages greater than 0.5 ppm.[68] He did not reference any methodology for his determination, but his conclusion was not far from the often-quoted odor threshold concentrations for chlorine in water determined much later in the 1980s.[69] Complicating the issues related to minimizing the amount of chlorine in water was the fact that no robust method for monitoring chlorine residuals was available in the early twentieth century.

After the success at Boonton Reservoir, a limited number of English and continental European treatment plants began using chlorine disinfection.[70] Chloride of lime was dosed in a manner similar to that used in the Boonton process. A 1914 article described two treatment plants in undisclosed locations in England that used chlorine disinfection followed by complete removal of the chlorine to avoid detection of chlorine odors. In both cases, chlorine removal was accomplished by passing the chlorinated water through a bed of "prepared vegetable carbon."[71] The prepared vegetable carbon must have been some form of granular activated carbon, which is well known to remove chlorine from water.[72] The vegetable carbon became exhausted after a period of operation and could be "revivified" by heating it to redness in the absence of air.[73]

The article included a brief description of a small pressure filter plant followed by the dechlorinating vegetable carbon filters. A much longer narrative described treatment with chlorine after slow sand filtration, followed by directing the water through a baffled "contact chamber" for a short time to allow some of the chlorine to dissipate, and finally sending the water through the vegetable carbon filters. The slow sand filters did a poor job of removing B. coli, and a serious typhoid outbreak that occurred in a town upstream of the treatment plant raised serious concerns in the small town. Chloride of lime treatment was started immediately, and the solution and dosing tanks appeared to be similar to those used at Boonton Reservoir. The dosage was 1 ppm of available chlorine.

> "The difficulty was not so much in the chlorination, as in the de-chlorination of the supply. Absolute freedom from taste or smell was insisted upon as any suspicion of flavor in the water would have started a scare immediately."[74]

AFTER THE BENEFITS OF FILTRATION AND CHLORINATION in preventing waterborne disease were well established, a number of engineers and scientists scolded water utilities and others for not installing the best treatment and for not operating that treatment to its fullest advantage.[75]

In 1913, Albert Hooker was technical director of Hooker Electrochemical Company, which produced chloride of lime. He had a definite opinion on the continuing incidence of typhoid fever caused by contaminated water supplies.

> "But woe betide the city that does not take care of its water supply: the consequences are epidemics—wholesale murder—nothing less. *A case of typhoid fever due to a polluted water supply should be good ground for legal redress and recovery of damages as a broken limb due to a defective sidewalk.*"[76] (italics in original)

In 1916, George A. Johnson published a 77-page *Journal AWWA* paper, which took everyone to task for the continued presence of typhoid fever in the population. In dramatic and accusatory language, Johnson laid out the mechanism by which the scourge of typhoid fever occurred as well as the solution to the problem.

> "Since the disease cannot be contracted naturally without taking the specific germ into the mouth, and since the manner in which this act is commonly performed can be said to be associated almost exclusively with the consumption of typhoid infected food and drink, it follows that to eradicate the disease involves only the exercise of really simple measures of precaution. . . ."[77]

Johnson castigated the public and public servants for not spending pennies per person per year to filter drinking water supplies to arrest the continued death toll from typhoid fever. He claimed that to solve the problem, filtration was needed on all surface supplies.[78]

Johnson presented impressive statistics showing that the number of typhoid fever deaths had dropped as a result of the installation of filtration from 1910 to 1913.[79] However, he completely ignored the rapid and dramatic increase in the implementation of chlorination during the same period—a baffling omission of critical information. In fact, nowhere in his 77-page diatribe did he mention the word "chlorine" or advocate the installation of chloride of lime or liquid chlorine feed systems to prevent drinking water supplies from spreading disease. For any reader unfamiliar with the facts or the merits of disinfection, his paper would have been grossly misleading.

Five years later, Johnson noted that in many cases, typhoid fever outbreaks were caused by operators simply turning off the chlorine feed for the weekend or because they did not think it was needed.

> "In all cases the practice [disinfection] should be maintained of applying the sterilizing agent continuously, never periodically. To sterilize only when it 'seems to be necessary' is as pernicious a practice as temporarily suspending the operation of a filter plant or the use of a coagulant when the water looks all right."[80]

Another author related a 1916 incident that showed how critical it was to maintain constant disinfection of a polluted water supply.

> "It is only necessary to refer to the experience in Milwaukee in 1916, where the night operator permitted the chlorinator to be shut down for only 8 hours, and in the next few days, 50,000 to 60,000 cases of enteritis occurred, followed in a fortnight by about 400 to 500 typhoid fever cases with over 40 deaths. . . ."[81]

In 1931, Abel Wolman and Arthur E. Gorman published a slim volume examining the occurrence of typhoid epidemics during the decade 1920–1930. They observed that typhoid fever was far too prevalent, given the knowledge about how to safeguard water supplies and the tremendous investment in filtration and chlorination that had been made over the previous few decades. More than half of the cases of waterborne typhoid fever and dysentery during the decade were caused by "inadequate control over [the] purification method."[82]

The book's Foreword, written by Thomas Parran, the sixth U.S. surgeon general, was unrelenting in its condemnation of lax operations.

> "There may be some excuse when epidemics result from hidden defects in water purification plants, or from cross connections on private property unknown to water works officials; but there is no excuse when these officials fail to operate properly the water purification plants, and fail to secure water of safe sanitary quality at all times. The continued occurrence of typhoid fever

due to neglect in the operation of water purification equipment, therefore is an indictment of the public officials in charge of this municipal equipment. The deplorable negligence of water plant operators has resulted in a loss of confidence on the part of the public. Competent and conscientious officials suffer for the shortcomings of a few incompetents with a resulting harm to the public health movement as a whole."[83]

Wolman and Gorman warned that liability should now be a serious concern for water utilities that did not operate in a competent manner.

"Financial and personal liability for illness resulting from inadequate control of public water supplies [is] being increasingly demanded by the courts in the United States, Canada and abroad. Municipalities, as well as private water companies are now subject to heavy financial damages for neglect of this important public duty."[84]

Conquest of Waterborne Diseases in the United States

The rapid increase in the use of chlorine and the lagging increase in the installation of filtration plants led to the overall decrease in typhoid fever. Figure 12-5 shows the decrease in the typhoid fever death rate in the United States from 1900 to 1956. After the first flush of excitement about chlorination, water suppliers took several decades to deal with the hard realities of implementing the new technology to eliminate the residual incidence of typhoid fever. In 1916, a New York City Health Department staff member commented on the progress in reducing the typhoid fever death rate and made a startling admission: "Until lately all of us felt ashamed of our typhoid death rate."[85]

In 1941, the U.S. Public Health Service estimated that 85 percent of U.S. drinking water supplies were chlorinated.[86] Based on these statistics and other data, the conquest of typhoid fever and other waterborne disease in the United States can probably be fixed at 1941, just before the outbreak of World War II. During this year, the national death rate from typhoid fever dipped below 1 per 100,000. In this context, "conquest" refers to the end of mass killings of adults and children caused by contaminated drinking water. Certainly, the job was not complete because outbreaks of waterborne disease still occur in the United States.[87] However, contemporary outbreaks are far rarer and far less costly in terms of loss of life than the epidemics of the 1800s and early 1900s. We do not live in a sterile world and so long as waterborne diseases exist, there is the potential for outbreaks in the United States. Still, it is worthwhile to recognize chlorine disinfection as a giant step in public health protection and to celebrate the successes achieved so far while continuing to work on the problems still to be solved.

During the period shown on Figure 12-5, advances in medical care and protection of milk supplies contributed to the lower typhoid fever death rate. As early as 1910, an editorial noted that not all of the success with controlling typhoid fever death rates could be attributed to chlorination and filtration.[89] However,

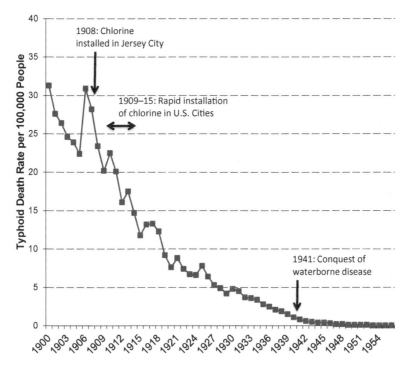

U.S Bureau of the Census. The Statistical History of the United States from Colonial Times to the Present. Vols. 1 & 2 vols. Stamford, Connecticut: Fairfield Publishers, 1965.

Figure 12-5   Typhoid fever death rate in the United States, 1900–1956[88]

plotting a graph of a specific city's typhoid death rate against the installation dates for chlorination and filtration will clearly demonstrate how crucial water treatment was to the victory over typhoid fever.

Life expectancy in the United States increased from about 46.5 years to 70 years during the first half of the twentieth century (Figure 12-6).[90] As defined by the U.S. Census Bureau, life expectancy at birth is ". . . the average number of years that members of a hypothetical [age group] would live if they were subject throughout their lives to the age-specific mortality rates observed at the time of their birth."[91] The huge dip in life expectancy in 1918 reflected the enormous loss of life associated with the influenza

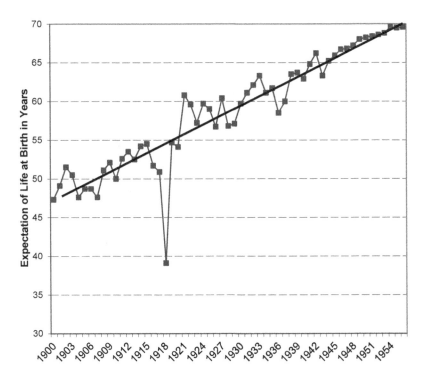

U.S. Bureau of the Cenus. The Statistical History of the United States from Colonial Times to the Present. Vols. 1 & 2 vols. Stamford, Connecticut: Fairfield Publishers, 1965.

Figure 12-6   Increase in life expectancy in the United States, 1900–1956[92]

pandemic that occurred that year. The reduction in waterborne disease from the beginning of the century to its eventual conquest in 1941 constituted no small part of the overall increase in life expectancy.

Minority populations also benefited from improved water supplies. University of Pittsburgh history professor Werner Troesken studied the effect of improved water supplies on death rates in white and black populations in a number of U.S. cities. He charted the huge difference in life expectancies for whites and blacks around 1900 and noted that over the first four decades of the twentieth century, the differences narrowed significantly. He contended that when a city instituted water supply improvements, they appeared to benefit both white and black populations. His final conclusion summarized his findings.

> "During the era of Jim Crow, there was less discrimination and inequality associated with the provision of public water and sewer facilities than one might otherwise expect, especially when one considers the widespread and severe discrimination that occurred in education, public parks, employment, police protection, and other public arenas."[93]

IN 1910, WILLIAM T. SEDGWICK published an article in which he coined the term Mills–Reincke Phenomenon. In this theory, Sedgwick collected the observances of Hiram Mills in Lawrence, Massachusetts, and Dr. J. J. Reincke in Hamburg, Germany, both of whom had noted that with filtration, the overall death rate decreased far beyond what would be expected from the elimination of waterborne disease alone. Sedgwick related these observations to the Hazen Theorem, which posited that for every death avoided by the elimination of waterborne disease, two or three deaths from other causes were also avoided.[94] Others were not so sure that either the phenomenon or the theory was correct. Dr. Charles V. Chapin, health officer for the city of Providence, Rhode Island, was very critical of the theorem, claiming that it was too general and vastly oversimplified the complex factors that were bringing about decreased death rates during this period.[95]

It is not surprising that Sedgwick would come up with a theory such as the Mills–Reincke Phenomenon because as he stated in his article, it was contaminated water sources that reduced the *vital resistance* of people exposed to such untreated supplies.

According to Sedgwick, filtration presumably removed not only the organisms that caused typhoid fever but also unknown substances that weakened a person's immunity or vital resistance.[96]

What the Mills–Reincke Phenomenon may actually have represented was that the diminished overall death rate included decreases in waterborne diseases that had been misdiagnosed (see Edwin O. Jordan's testimony in the Chicago canal case)[97] or deaths that had been attributed to "unknown causes." The prevalence of typhoid fever was probably much higher than evidenced by the official statistics, and once the quality of a city's water supply improved, the decrease in its overall death rate likely reflected the elimination of undiagnosed cases of typhoid fever and diarrheal diseases that had resulted in death.

THE DECLINE IN INFANT DEATH RATES in the United States from 1915 to 1956 was dramatic (Figure 12-7). For these data, an infant is defined as a child less than one year of age. Correlating the decrease in infant mortality solely with improved water supplies is difficult, but chlorination and filtration appear to be two of the major factors that contributed to reducing infant deaths.

The excessive death rate among young children was decried by many in the early twentieth century. A number of published studies have reported that advances in drinking water treatment and improvements to the milk supply were important factors in reducing child mortality. [99]

A study published in 2005 linked a significant decrease in child mortality with improvements in drinking water quality during the first few decades of the twentieth century. "Our results also suggest that clean water was responsible for three quarters of the decline in infant mortality and nearly two thirds of the decline in child mortality." Even with some concerns about the model results presented in this paper, the findings on reduced child mortality are noteworthy.[100]

Striking decreases in childhood mortality indeed occurred during the first 20 to 30 years of the twentieth century. During this time, the quality of water supplies improved dramatically, and the milk supply was cleaned up significantly. No doubt enhanced safety of the milk supply was a major reason for the decrease in childhood mortality during these three decades, but the chlorination and filtration of water supplies also contributed. Contaminated water was implicated in many of the typhoid fever epidemics caused by contaminated milk.[101] Washing equipment for preparing

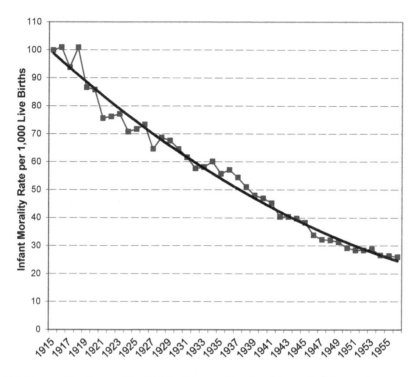

U.S. Bureau of the Census. The Statistical History of the United States from Colonial Times to the Present. Vols. 1 & 2 vols. Stamford, Connecticut: Fairfield Publishers, 1965.

Figure 12-7   U.S. infant mortality rate per 1,000 live births, 1915–1956[98]

and transporting milk with contaminated water was not uncommon. The milk-borne typhoid fever epidemic that occurred in Paterson, New Jersey, in 1899 was typical of these incidents.[102]

Another way to consider the milk supply versus the water supply as vehicles for spreading diarrheal diseases is to ask the question: Would the milk supply have improved as quickly as it did without the parallel improvement in water supplies? In other words, how could a milk supply, even with universal pasteurization, be kept pure until an infant consumed it if sewage-contaminated water was being delivered to households? In the home, milk was often diluted with water, and washing bottles and related materials with contaminated water would have perpetuated a high toll of infant deaths. Wholesale watering of milk by unscrupulous retailers was also known to occur in cities such as Chicago, and this practice would have caused a direct connection between sewage-contaminated water and a contaminated milk supply.[103]

Increased infant survival would have been long delayed without the rapid implementation of chlorination in most municipal water supplies during 1909 to 1914 (Figure 12-1). Saving children from an early death and sparing families the pain of young lives lost are victories for which sanitary engineers can be well proud.

OUR UNDERSTANDING OF HOW WATER SUPPLIES can be protected has continually improved since 1941. The multiple-barrier concept, which was well established by the middle of the twentieth century, put many safeguards between contamination and water consumption by the public. Since passage of the Safe Drinking Water Act in 1974, the federal and state regulations that have been promulgated have strengthened the barriers to waterborne disease. However, well-documented instances of waterborne disease have occurred since the 1941 milestone of the conquest of waterborne disease. Water supplies can still become contaminated. Illness and death can still occur as a result of improper operation of water treatment facilities.

Large outbreaks of gastrointestinal illness caused by contaminated drinking water are unusual in the United States; their separation from the norm makes them remarkable. Each such event over the past several decades generated the proper amount of concern and investigation as to the causes and the measures required to prevent a recurrence. One of the most dramatic examples of such an outbreak occurred when approximately 400,000 people were infected with gastrointestinal disease caused by the organism *Cryptosporidium parvum* in Milwaukee, Wisconsin, during April 1993. Promulgation of a federal regulation called the Long Term 2 Enhanced Surface Water Treatment Rule was, in part, a direct result of that outbreak. Microbiological barriers to disease in Milwaukee and hundreds of cities across the United States have been strengthened as a result.

CHLORINE DISINFECTION EFFECTIVELY BROKE the Sewer Pipe–Water Pipe Death Spiral in the United States. Many water practitioners and historians have praised the use of chlorination in U.S. water supplies.

"Certainly the hypochlorite treatment has proved a blessing."[104]

"There can be no doubt that the development of a practical method of water disinfection during the last two years marks an epoch in the art of water purification."[105]

In a 2007 *Scientific American* article by Peter Jedicke, chlorine was rightly recognized as one of the great inventions of the twentieth century. "Millions of lives around the world have been saved since."[106] In 1997, *Life* magazine stated, "The filtration of drinking water (plus the use of chlorine) is probably the most significant public health advance of the millennium."[107] In Gunther Craun's book on waterborne disease, he stated that ". . . chlorine should be noted as perhaps saving more lives throughout the world than any other chemical. . . ."[108]

The conquest of waterborne disease in the United States was assured by a courageous physician and the greatest sanitary engineer of his time. Dr. John L. Leal and George Warren Fuller were two of the greatest public health heroes of the twentieth century.

Others have denigrated the contribution of chlorine to the protection of public health by focusing on consequences that can be mitigated or outlandish claims that require no discussion here.[109]

DESPITE THE TRIUMPH OVER WATERBORNE DISEASE in the developed world, waterborne illness is still a major killer in developing countries.

A modern cholera pandemic illustrated the importance of public health vigilance and the maintenance of proper drinking water chlorination systems.

> "Beginning in 1961, *Vibrio cholerae* O1, El Tor biotype spread from Indonesia through most of Asia into eastern Europe and Africa. From North Africa, it spread to the Iberian Peninsula and into Italy in 1973. In the late 1970s, small outbreaks occurred in Japan and in the South Pacific.
>
> In January 1991, epidemic cholera appeared in Perú and spread rapidly through most of Latin America, causing more than a million cases by 1994. This was the first time in 100 years that a cholera pandemic had reached the New World."[110]

The original cause of the outbreak in Perú has been much debated, but water supplies contaminated by sewage containing *V. cholerae* and degraded chlorination systems were key to its spread initially. The death toll in Perú was more than 2,800, and an additional 1,000 people died when the disease spread to other South American countries.[111]

"Officials say that Perú's poor water supply and over-crowding of the shanty towns that surround the coastal cities helped to spread the disease. The rapid population growth in Lima and other coastal cities of Perú during recent decades exceeded the infrastructure available to deal with fecal contamination of water supplies. Chlorination is not kept at proper levels and the water pressure is not maintained for twenty-four hours a day, so waste water can flow into pipes that are cracked."[112]

Some sources claimed (though without any hard evidence) that improper chlorination of drinking water in some Peruvian cities occurred because of misplaced concern about the production of disinfection by-products.[113] However, others have shown, convincingly, that other factors, including poor source water quality and severe problems with water treatment infrastructure, caused the epidemic in Perú.[114]

In Haiti, the cholera epidemic that began in October 2010 was one of the outcomes of the devastating earthquake that occurred there on January 10, 2010. "To date [October 19, 2011], over 470,000 cases of cholera have been reported in Haiti with 6,631 attributable deaths."[115] At the time this book was published, there was no evidence that the epidemic had spread far beyond Haiti, but cholera continues to be stubbornly present in that poverty-stricken country and the disease has spilled over into neighboring Dominican Republic.

Cholera is still endemic in Bangladesh and the Indian subcontinent. A recent story in *The New York Times* described efforts to reduce the exposure of families to *V. cholerae* by training women who bring water from polluted streams to filter it through four folded layers of their traditional garment, the sari. Although *V. cholerae* are obviously too small to be filtered out by cloth, the organisms are generally attached to plankton, which can be filtered out by freshly laundered, folded fabric. More than 99 percent removal of the cholera-producing organism can be achieved by this inexpensive change in personal habits.[116]

---

1   Longley, "Disinfection of Water Supplies," 679.

2   Per capita use factor for converting billions of gallons of water served to millions of people was determined to be 90 gallons per capita per day—the median of data from more than 150 cities; *American City*, Water Consumption, 41–2.

3   U.S. Bureau of the Census, *Statistical History*, 240.

4   Longley, "Disinfection of Water Supplies," 680; U.S. Bureau of the Census, *Statistical History*, 240.

5   Race, *Chlorination of Water*, 12.

6   Pincus, "Chlorination Insuring Safety of Drinking Water," 30.

7   Bunker, "The Use of Chlorine in Water Purification," 3–4.

8   Between Jersey City and Water Company, February 4, 1909, 4972.

9   Baker, *Quest*, 153; Between Jersey City and Water Company, May 21, 1909, 6513–4.

10   Between Jersey City and Water Company, May 21, 1909, 6519.

11   Jennings, "Chlorine Compounds in Water Purification," 251.

12   Ericson, "Chlorination of Chicago's Water Supply," 772.

13   Ibid., 773.

14   Brewster, *Engineering News*, 763; from Proceedings of the Annual Convention of the Indiana Sanitary & Water Supply Association, February 25, 1910

15   Ibid.

16   Gelston, "Hypochlorite of Lime as an Adjunct," 394–5.

17   Ibid.

18   Palmer, "Hypochlorite Treatment," 430.

19   "Compulsory Order," *Engineering News*," 712.

20   "Lawsuit to Compel Water Purification," *Engineering News*, 723.

21   Palmer, "Hypochlorite Treatment," 419–20; *Trenton Evening Times*, December 6, 1911.

22   *Trenton Evening Times*, December 6, 1911.

23   Palmer, "Hypochlorite Treatment," 420–1.

24   *Trenton Evening Times*, December 21, 1911.

25   Daggett, "Trenton, N.J., Filtration Plant," 148.

26   Schevitz, "Typhoid Fever in 1920," 42 Adv.

27   Reyer, "Use of Sulphate of Alumina and Hypochlorite of Lime," 390–1.

28   Jensen, "A 20,000,000-Gal. Hypochlorite Water-Disinfecting Plant," 391–2.

29   O'Toole, *Search for Purity*.

30   Ellms, *Water Purification*.

31   O'Toole, *Search for Purity*, 131–2.

32   Ibid., after 124.

33   Ibid.

34   Fuller and McClintock, *Solving Sewage Problems*, 106–7.

35   O'Toole, *Search for Purity*, 136–7.

36   Ibid., 137.

37   Ibid.

38   Ibid., 55.

39   Ellms and Hauser, "Orthotolidine as a Reagent," 915–1030.

40   "Hypochlorite of Lime," *Engineering News*, 416; "Hypochlorite or Bleaching Powder," *Engineering News*, 575.

41   Jennings, "Chlorine Compounds in Water Purification," 251.

42 "Mortality Rates in Relation to the Water Supply," 50.

43 Hazen, *Clean Water and How to Get It*, 102.

44 Editorial, *Municipal Engineering*, 542.

45 Ellms, "Discussion," 301.

46 Longley, "Disinfection of Water Supplies," 680.

47 Baker, *Quest*, 342.

48 Tiernan, "Controlling the Green Goddess," 1045; Wallace & Tiernan's Fiftieth Anniversary, 10. There were many connections between the early days of Wallace & Tiernan and the Jersey City water supply. William Griffin, superintendent of the Jersey City water department, hired Charles F. Wallace and Martin F. Tiernan to disinfect the polluted stream near Dover that was contaminating the Rockaway River as it flowed into Boonton Reservoir. Two of the expert witness in the Jersey City trials, Charles E. North and Earle B. Phelps, hired the two men in the very beginning of their careers to help install disinfection systems in cities as part of North and Phelps's consulting practice. Tiernan actually ran the chloride of lime feed system at Boonton Reservoir in the early fall of 1912 when the chemist was on vacation.

49 McGuire, "Eight Revolutions," 123–49.

50 Alvord, "Recent Progress and Tendencies in Municipal Water Supply."

51 Fuller, "Influence of Sanitary Engineering," 17.

52 Alvord, "Recent Progress and Tendencies in Municipal Water Supply," 283.

53 Fuller, "Relations Between Sewage Disposal and Water Supply," 11.

54 Jennings, "Sterilization of Water Supplies," 314–5.

55 "Report of the Committee on Limitation and Control of Chlorination," 19.

56 Clark and Gage, "Disinfection as an Adjunct," 319; see also the discussion comments appended to the paper.

57 Johnson, "Romance of Water Storage," 299.

58 Pincus, "Chlorination Insuring Safety of Drinking Water," 31.

59 Ibid.

60 Editorial, *Municipal Engineering*, 542.

61 Wolman and Gorman, *Significance of Waterborne Typhoid Fever Outbreaks*, 14.

62 Zim, "Is Your City in the Vanguard Fighting Water-Borne Typhoid?" 247.

63 Winslow, "Field for Water Disinfection," 1.

64 Ibid., 4.

65 Pincus, "Chlorination Insuring Safety of Drinking Water," 31.

66 Wolman and Enslow, "Chlorine Absorption," 209; Wolman, "Water Chlorination Control," 639–41.

67 Sperry and Billings. "Tastes and Odors from Chlorination," 603.

68 Tully, "Calcium Hypochlorite," 137.

69 Krasner and Barrett, "Aroma and Flavor Characteristics," 384.

70 Race, "Forty Years of Chlorination," 479–80.

71 "Water Sterilization Plants," *Municipal Engineering*, 11.

72 McGuire, Suffet, and Radziul, "Assessment of Unit Processes," 569.

73 "Water Sterilization Plants," *Municipal Engineering*, 11.

74 Ibid.; presumably the "scare" would be that chlorine was being added to the water.

75 Lest readers assume that all was well in Europe, a short article in *Engineering News* 63:16 (April 21, 1910), 476, summarized the state of water treatment in France. Only about half of the cities using river supplies had filtration plants, and the much-vaunted ozone disinfection process was used in only 5 of the 434 French cities surveyed. No mention of chlorine disinfection appeared in the article.

76 Hooker, *Chloride of Lime*, 12.

77 Johnson, "The Typhoid Toll," 249.

78 Ibid., 260.

79 Ibid., 273–5.

80 Johnson, "Romance of Water Storage," 300.

81 Pincus, "Chlorination Insuring Safety of Drinking Water," 31.

82 Wolman and Gorman, *Significance of Waterborne Typhoid Fever Outbreaks*, 13.

83 Ibid., vi.

84 Ibid., 50.

85 Park, "Discussion—The Typhoid Toll," 812.

86 Crittenden, et al., *Water Treatment: Principles and Design*, 1037.

87 Rochelle and Clancy, "Evolution of Microbiology in the Drinking Water Industry," 171.

88 U.S. Bureau of the Census, *Statistical History*, 26.

89 Editorial, *Engineering News*, February 17, 1910, 201.

90 The variability in life expectancy statistics is much greater between 1900 and 1944 compared with after 1944.

91 U.S. Bureau of the Census, *Statistical History*, 19.

92 Ibid., 25.

93 Troesken, *Water, Race and Disease*, 201.

94 Sedgwick and MacNutt, "Mills–Reincke Phenomenon."

95 Cassedy, *Charles V. Chapin*, 59.

96 Sedgwick and MacNutt, "Mills–Reincke Phenomenon."

97 Leighton, "Pollution of Illinois and Mississippi Rivers," 241–3.

98 U.S. Bureau of the Census, *Statistical History*, 25.

99 Johnson, "The Typhoid Toll," 285; Ellms, "Discussion," 301; Wolf, *Don't Kill Your Baby*; Condran, Williams, and Cheney, "Decline in Mortality in Philadelphia from 1870 to 1930," 463; Chicago Department of Health, "Bulletin: Chicago School of Sanitary Instruction," Vol. 16, 4; Chicago Department of Health, "Bulletin: Chicago School of Sanitary Instruction," Vol. 4, 4; Condran and Lentzner, "Early Death," 352; Van Poppel and van der Heijden, "Effects of Water Supply on Infant and Childhood Mortality," 142.

100 Cutler and Miller, "Role of Public Health Improvements," 3.

101 Hazen, *Filtration of Public Water-Supplies*, 228 and 230.

102 Leal, "Annual Report of Board of Health 1898," 242–8.

103 Wolf, *Don't Kill Your Baby*, 50.

104 Jennings, "Sterilization of Water Supplies," 318.

105 Winslow, "Field for Water Disinfection," 1.

106 Jedicke, *Scientific American: Great Inventions*, 18. Bizarrely, the author gives credit for the chlorination of drinking water to Wolman and Enslow on the basis of a chlorination article they wrote in 1919.

107 *Life*, "The 100 Events," 80.

108 Craun, *Waterborne Disease in the United States.*

109 The health effects of chlorine have been fully documented in the *Federal Register* in connection with publication of the proposed and final rules that established Maximum Residual Disinfectant Levels. U.S. Environmental Protection Agency (USEPA), "Disinfectants/Disinfection By-Products. Proposed Rule;" USEPA, "Disinfectants and Disinfection By-Products. Final Rule." The health effects of disinfection by-products have been published in a number of *Federal Register* documents, including those establishing Maximum Contaminant Levels for specific chlorine disinfection by-products and for surrogates that reflect the occurrence of a variety of other disinfection by-products. USEPA, "Control of Trihalomethanes;" USEPA, "Disinfectants/Disinfection By-Products. Proposed Rule;" USEPA, "Disinfectants and Disinfection By-Products. Final Rule;" USEPA, "Stage 2 Disinfectants and Disinfection By-Products Rule."

110 Dowell and Levitt, "Protecting the Nation's Health," 13.

111 Suarez and Bradford, "The Economic Impact of the Cholera Epidemic in Perú," 4.

112 Arbona and Crum, "Medical Geography and Cholera in Perú."

113 Conko and Miller, "Precaution Without Principle;" Anderson, "Cholera Epidemic Traced to Risk Miscalculation, 255.

114 Tickner and Gouveia-Vigeant, "1991 Cholera Epidemic in Perú," 495.

115 "Haiti Cholera Outbreak," Centers for Disease Control and Prevention.

116 Zuger, *New York Times*, September 27, 2011.

13

Epilogue—1908 and Beyond

". . . the use of the bleaching powder at the Boonton
supply works of Jersey City, N.J., was done under the
direction of Dr. John L. Leal. . . ."
— Wilson, Letters to the Editor, 28

The year 1908 signaled the beginning of the drinking water disinfection revolution in the United States, and many other significant events occurred that year as well. In his book *America 1908*, Jim Rasenberger chronicled the technological, exploratory, political, and sociological milestones the country experienced that year. On the first page of the book, Rasenberger stated succinctly the thrills attendant to the year: ". . . 1908, by whatever quirk of history or cosmology, was one hell of a ride around the sun." During these 366 days, the Wright brothers amazed the world with extended flights of heavier-than-air machines, the Model T went into production, two explorers set out for the North Pole, a new President was elected, the national pastime captured the country's attention with a strange pennant race, the U.S. Navy's Great White Fleet started its round-the-world cruise, and deadly race riots and other violence scarred the national conscience.

In summing up the Wright Brothers' accomplishments, Wilbur's obituary placed their achievements in the context of their time and the history of applying scientific principles.

"The death of Wilbur Wright has brought intense personal sorrow to all who were in any way associated with him. . . . The science of aviation has lost its greatest student, and in time to come the name of Wilbur Wright will be recorded in the annals of invention with the names of such pioneers as Robert Fulton, Stephenson [inventor of the steam locomotive engine], Bell, and others who have given to the world the value of *practical experiments and successful achievements*."[1] (emphasis added)

Wilbur Wright was not the first person to gaze at a bird and wonder how humans could fly. Nor was he the first person to build an airplane and try to lift it off the ground. He and his brother Orville were the first to actually accomplish powered flight, but, more importantly, they demonstrated in a practical manner how to control that flight. Technological progress is achieved by those who make an idea work.[2]

Wilbur Wright died of typhoid fever on May 30, 1912. He was just 45 years old. Four years earlier, he had astonished the world with his extended flights near Paris. What might he have achieved in continued partnership with his brother had he not been struck down so early? The chlorine revolution did not spread fast enough to save the life of this world-renowned inventor, but chlorination traveled fast enough and far enough to save the lives of hundreds of thousands of future inventors, scientists, and engineers whose creativity transformed the United States and the world.

What did the author of *America 1908* make of the disinfection revolution that was launched during this seminal year a few miles from the center of his story, New York City? He did not mention the chlorination project at Boonton Reservoir. Only passing mention is made of public health and water supply: millions dying from infectious diseases, a cholera epidemic in Manila when the Great White Fleet visited, and William Mulholland's construction of a new water supply for Los Angeles. The cause of Wilbur Wright's typhoid fever has been described as coming from various sources. His obituary mentioned bad fish in a Boston restaurant, but the author of the obituary had no particular reason for believing that was the source.[3] The only thing that appears to be certain is that he contracted the disease on a business trip.

Although chlorination had been instituted in hundreds of U.S. cities by 1912, the death rate from typhoid fever remained high. Boston had a typhoid fever death rate of 8 people per 100,000 in 1912. In that same year, Washington, D.C., and Baltimore, Maryland, had typhoid death rates of 22 and 24 per 100,000, respectively.[4] Wilbur Wright more likely died from contaminated water than bad fish.

Immunology

Understanding how the human body resists disease began with seminal work in the 1880s and was recognized around the world in the same year that drinking water disinfection began to conquer epidemic waterborne disease. In 1908, Elie Metchnikoff won the Nobel Prize for Physiology or Medicine, sharing the prize with Paul Ehrlich "in recognition for their work on immunity."[5] Metchnikoff laid the foundation for the field of immunology by describing the mechanism of phagocytosis—the process in which specialized white blood cells, or phagocytes, surround and destroy harmful microorganisms that have invaded the human body.

Decades later, in the early 1980s, two research teams showed definitively that white blood cells (phagocytic leucocytes) kill microbiological invaders of the human body through a process involving the production of hypochlorous acid and chloramines at the cellular level. Both of these chemicals are toxic to invading organisms.[6] Online videos demonstrating the process of phagocytosis are helpful in understanding the mechanisms.[7] A figure and the accompanying text in a recently published book on immunology illustrate the reaction mechanisms that produce hypochlorous acid and chloramines.[8]

Leal would have had an easier time convincing the New Jersey Chancery Court that adding chlorine to drinking water was an excellent tool for killing the typhoid bacillus if he had known that cells in the human body use the same chemical as part of an innate mechanism for defense against pathogens. The information would also have been of great help to engineers and city leaders who later added chlorine and chloramines to drinking water in the face of continuing public chemophobia.

The Death of John L. Leal

John Leal's death was page-one news in Paterson, New Jersey, on March 14, 1914. The *Paterson Morning Call* published a picture of "The Late Dr. John L. Leal" and his obituary on the front page.[9] The cause of death was reported as diabetes. His obituary identified him as ". . . one of the foremost physicians and bacteriologists. . . . He was especially recognized as a bacteria expert and was often called to cities where an epidemic of typhoid fever was prevalent. His record as a health officer and later as a member of the board of health will go down in the history of this department."[10] He was also recognized for his contribution to drinking water treatment. "Those who are qualified to judge have pronounced John L. Leal one of the very best experts on potable water in the world. . . . [he] was a recognized authority on the detection of pollution in watersheds, and the methods of securing potable water for municipalities."[11]

Leal had that rare combination of talents—the technical knowledge of a scientist/physician and the vision to convert that knowledge into practical applications. Even though he was not widely celebrated during his life, he was able to achieve dramatic technological progress by adapting theoretical understanding and scientific experimentation to the benefit of mankind.

The name "Leal" is associated with an old Scottish term— "land of the leal"—which refers to heaven, and, in particular, a heaven for simple peasant folk. John L. Leal was born to rural life and was a product of an urban upbringing. He focused his entire career on improving the harsh realities of urban life and achieving something closer to a "land of the leal" on Earth by treating the sick and improving public health.

Leal was buried in Paterson's Cedar Lawn Cemetery in the family plot on a hill overlooking the Passaic River, which figured so prominently in his career. Along with his grave are those of his father and mother, his brother Charles, and his mother's sister, Anna L. Laing. An 1876 publication accurately described the cemetery's beautiful setting as viewed from the Leal family plot.

"In the foreground is Dundee Lake, the charming sheet of water formed by the expansion of the Passaic river for three or four miles. . . . From this gentle vale there rises a hill with not too steep an ascent, to a height of perhaps one hundred feet above the lawn at its base. . . . This hillside is dotted all over with graves, while its summit has been regarded as a peculiarly choice situation. And no wonder. It is illumined by the dawn's first blush, emblem of the Eternal Day, which all who there repose are waiting to rise and greet. . . . There be few who find those sublime heights of intellectual serenity in this existence, but on this hilltop at Cedar Lawn there is literally repose, and the weary are at rest. . . . There is something surpassingly lovely in the view from the higher parts of the Cemetery—the meadow and the lake below, the placid fields and blue hills of Bergen County, rising into the Palisades far away in the east. . . ."[12]

About 100 feet southwest of the Leal family plot is a plot for the Scott family. A large obelisk marks a collection of graves identified with tombstones for members of the Scott and Shaw families.

Headstones identify the resting places for John L. Leal's son, Graham Leal; daughter-in-law, Marjorie Shaw Leal; and granddaughter, Marjorie Leal Van Benschoten.

Until October 2012, the ground around John L. Leal's grave was bare—no marker, no headstone, no monument. Thus, one of the world's greatest public health heroes was buried in obscurity. However, a memorial now heralds his accomplishments. A group of people from the drinking water community in New Jersey and elsewhere plus Leal's descendants came together and erected a grave marker declaring John L. Leal a "Hero of Public Health."[13]

The Death of George Warren Fuller

After a long and distinguished career, George Warren Fuller died at 10:30 P.M. on June 15, 1934. He was at his home in New York City and was attended by his physician. The cause of death listed on his death certificate was "acute myocarditis terminal to cirrhosis of the liver."[14] Apparently, Fuller had recovered from a serious illness several years prior to his death.[15]

The tributes were long and well deserved.

"In almost every era of real progress in civilization a few individuals are born who carry on the major activities which distinguish that era. Such an individual, during the advance of sanitary engineering practice in the past 50 years was George W. Fuller. His death . . . marks the passing of one who had probably more to do with the technical development and the public stimulation in the installation of municipal sanitary devices than almost any man in this country."[16]

"Many, however, are born at the right time who are either ill equipped or are lacking in sufficient vision to make the most of that good fortune. In Fuller's case, heredity and environmental influences, coupled with remarkable energy, all contributed to the development of a practitioner of outstanding stature. He will be remembered long in the future, as much for his distinctive personal characteristics as for his long list of contributions in sanitary science and practice."[17]

"*Municipal Sanitation* records this last tribute to one of the important figures who have made this country safe to live in. His name will be among the first in public health achievements of this century."[18]

A memorial written by his staff and published in one of his obituaries emphasized his positive qualities.

"Mr. Fuller's success in business life was quite largely due to his tremendous energy. . . . In whatever he undertook he was absolutely tireless. Having once decided on a course of action, he made quick and accurate decisions, he was not turned aside but drove everything rapidly through to a final and satisfactory conclusion."[19]

Another reminiscence, written by James C. Harding, a former employee and colleague, was published 27 years after Fuller's death and was, perhaps, a little more honest in depicting the positive and negative aspects of Fuller's personality.

"Fuller had a pleasant personality and an almost overwhelming one. He was genial, courteous, and very impressive. He could turn on the charm whenever he wished and vary its type with the person or audience. He was a good story teller and a brilliant conversationalist. He was normally dignified yet warm, but he could be very much otherwise when he felt so inclined. . . . [Fuller's] principal interests were wine, women and real estate. I don't know about the real estate, but he certainly liked the wine or rather the stronger alcoholic beverages, and the women. . . . His appetite for food was as great as that for liquor; he could consume it in astounding quantities if it happened to be something that he liked."[20]

Abel Wolman was acutely aware of Fuller's commitment to making things work. Ivory tower ruminations were far from Fuller's mind. He wanted to create practical solutions, and that is what he did at Boonton.

"He spent a lifetime in converting theoretical principles into practicable services to man. How to do this, with economy and common sense, he conceived to be his permanent and undeviating duty."[21]

George Warren Fuller was buried in the Fuller family plot in West Medway, Massachusetts. His headstone is simple, listing his name, date of death, and his age, "Aged 65 years, 5 mos." The marker for the Fuller family plot is a grand obelisk that rises above the surrounding monuments. On the sides of the obelisk are the names of Fuller's father, mother, and several other relatives who are interred nearby. Fuller's grave marker and burial plot are not about telling the world what he had accomplished in life; they are about family. Next to his headstone is a marker that reads, "Lucy, Wife of George W. Fuller, Died Mar. 19, 1895, Aged 25 yrs. 5 mos." Little information exists about Fuller's first wife, but she is buried right next to him.

If the engineering of the Jersey City chloride of lime system had been a failure, the chlorination of drinking water would have been delayed for many years or possibly even decades. Fuller is remembered primarily for his development of rapid sand filtration and for his service as president of both the American Water Works Association (AWWA) and the American Public Health Association (APHA). His legend should be expanded to include his central role as a hero of public health protection through the revolutionary implementation of chlorine as a drinking water disinfectant.

Undeserved Credit for Chlorine Use

The majority of published summaries of what occurred at Boonton Reservoir in 1908 have given primary credit for the addition of chlorine to the Jersey City water supply to George A. Johnson. It is hard to believe this could have happened, given the historical record in the transcripts of court proceedings when witnesses were required to testify under oath. There is no doubt that Dr. John L. Leal came up with the idea to add chloride of lime to the Jersey City water supply (on May 1, 1908) and that he was responsible for its implementation by hiring the engineering firm of Hering and Fuller (on June 19, 1908) to design and supervise construction of the new technology.

Unfortunately, Leal was not around to remind everyone of his legacy. To paraphrase the famous but anonymous expression: History is written by the survivors.[22] Leal died soon after the New Jersey Supreme Court's decision affirming the lower court's ruling, and he wrote only one article about the Boonton disinfection plant after his landmark 1909 paper was published in the AWWA conference proceedings. Both Johnson and Fuller lived an additional 20 years after Leal's death, and they were able to shape the story behind the chlorine revolution at Boonton.

George A. Johnson was a laboratory analyst (who conducted bacterial and chemical tests) and a facility operator with the firm of

Hering and Fuller. He provided operational oversight of the Boonton sterilization plant from September 26 to December 31, 1908. After that date, Leal managed operation of the Boonton plant.

At no time during his testimony in the Jersey City trials did Johnson state that adding chlorine to the Jersey City water supply was his idea. In a recitation of his experience, he described his role as a treatment plant operator and laboratory technician for the firm of Hering and Fuller. He acknowledged in his testimony that he held no academic degree but developed his skills through "practical experience."[23]

During the second trial, on February 8, 1909, William H. Corbin questioned George A. Johnson:

> "Q. State, Mr. Johnson, what you have had to do with and when you began to have to do with, the works which have been built at the Boonton dam for the purification of the water.
>
> A. I was present at the conference had between Doctor Leal, the sanitary adviser of the Jersey City Water Supply Company, and members of our firm, in the summer of 1908 [June 19, 1908]. It was at that time decided that upon the completion of construction of the plant proposed that I was to be given the actual charge of the operation of this plant. The works were completed in September, and for the week prior to September 26, the date on which the plant went into operation, I was on the ground making preparations for the official operation of the plant. From September 26th until January 1st, 1909, I was in actual charge of the operation of those works. During this entire period I was assisted by three men acting directly under my orders. . . . I made it a practice to be there on an average of twice a week. Other duties outside of New York sometimes made it necessary for me to absent myself from the plant as long as a week."[24]

Subsequent histories of the early use of chlorine made much of Johnson's association with the addition of chlorine at the Bubbly Creek treatment plant. This application of chloride of lime was also introduced in 1908. On October 5, 1909, the thirty-fourth day of testimony during the second trial, Johnson was questioned by William D. Edwards, an attorney with the law firm of Collins & Corbin. During this testimony, Johnson described his role at the Bubbly Creek plant in 1908.

"Q. Did you have charge of the first plant at Bubbly Creek in the stock yards in Chicago, and the purification of waters therefrom?

A. I was appointed *referee* in a *test* which was run on the stock yards filter plant in the months of April and September 1908, and the plant was, during that period, *virtually* under my direction."[25] (emphasis added)

Under oath, Johnson chose his words carefully when answering that question. It is evident that his work for the Bubbly Creek plant was a test, not a full-scale demonstration of chlorine disinfection technology. Nor was it anything approaching the continuous use of chlorine to disinfect a water supply for human consumption. Treatment at the Bubbly Creek plant consisted of coagulation with lime and iron (or alum) followed by sedimentation for 3 hours. The settled water was treated with chloride of lime at a chlorine dose of more than 1 part per million followed by filtration through a sand filter. Johnson's testimony then described the Bubbly Creek plant's source water.

"Q. What is the character of the water treated so far as pollution is concerned?

A. The raw water pumps take out of probably the foulest estuary in the world—

Q. The Chicago river?

A. It is a branch of the Chicago river. It has been notorious for a great many years and was given the name of Bubbly Creek by reason of the fact that gas is rising over the entire surface all of the time. The water which flows through this creek is merely the drainage from seventeen thousand acres of southeastern Chicago and on this area there is resident about 200 thousand people. The sewage from this area and the street washings also are discharged in this creek.

"Q. What is done with this water after it is treated?

A. It is used for watering the stock in the Union Stock Yard in Chicago."[26]

At the end of his testimony on October 5, 1909, Johnson stated:

> "I recommended the use of hypochloride [sic] of lime in connection with the Bubbly Creek filter plant in May, 1908, believing that was the only chemical that would make possible the achievement of satisfactory results. I had had no actual experience of much moment with this chemical but had gained a good deal of knowledge regarding its sterilizing powers from conversations with various scientists in this country and in Europe and from reading various documents descriptive of tests that had been made with it. The plant at Bubbly Creek was a failure until this sterilization agent was used."[27]

Johnson's testimony also included statements acknowledging Leal's role at Boonton: ". . . the process installed by Dr. Leal . . ." and ". . . the result of purification in Dr. Leal's system of purification at Boonton. . . ."[28]

A two-page paper published obscurely in 1909 could have set the record straight as to what actually happened at the Bubbly Creek plant if anyone had read it. The paper was written by Adolph Gehrmann, who was part of the two-man team conducting tests for the Bubbly Creek plant.

> "I desire to bring to your attention some of the data relating to a purification plant *now under test* at the Union Stock Yards, Chicago. . . . Mr. George A. Johnson and myself were selected to conduct *tests* during operation as a basis for determining the various elements of efficiency as required under the contract."[29] (emphasis added)

> "During the *fourth period of test* it had been determined, on the suggestion of Mr. Johnson, to introduce chloride of lime as an oxidizing and germicidal agent in place of copper sulphate."[30] (emphasis added)

Because the "fourth period of test" at the Union Stock Yards was carried out September 3–17, 1908, and Johnson had been working with Fuller on the full-scale Boonton plant since July 19, 1908, it is not hard to figure out where Johnson got the idea to test chloride of lime. It is highly unlikely that in May 1908 he figured out all on his own that chlorine should be added at the Bubbly Creek plant.

Apparently, the tests at the Union Stock Yards satisfied the researchers, although B. coli were still found in the effluent. Nonetheless, the water was deemed good enough, despite some "taste" to the water, to be put in the hog and cattle pens. ". . . it was drunk by the stock very readily."[31]

A 1910 article in *Engineering News* reviewed all of the issues associated with construction of the Bubbly Creek treatment plant and the use of water produced by the plant. In 1909, the city of Chicago had filed a lawsuit demanding that the Union Stock Yards discontinue use of any water from the Bubbly Creek plant for any purposes related to watering cattle. In the trial, the city claimed that water from the plant was being consumed by people, in direct conflict with any imaginable grain of good sense. A key point in the case was that cattle producers believed their cattle put on less weight when they were given water from the Bubbly Creek plant as opposed to city water. The company denied the contention but agreed as a smart business practice to stop using the water. The treatment plant was subsequently shut down.[32]

Three weeks later, *Engineering News* published a 5,000-word letter from Johnson defending the Union Stock Yards and himself.[33] Despite his creative use of the calendar and a few facts, his defense was not persuasive. In 1913, Johnson again tried to make the case that treated water from the foul Bubbly Creek was as good as other treated water from polluted sources.[34] His arguments were not convincing.

In summary, Johnson's testimony and Gehrmann's paper showed that for a few days in September 1908, Johnson was a *referee* for a *test* of chloride of lime at a Chicago plant treating raw sewage to be used as a water supply for cattle.

Why Johnson has been given any credit for the first use of chlorine in drinking water is somewhat of a mystery until one examines the written record after 1909. One reason Johnson is given primary credit is that he published papers giving the impression that he was responsible for it all. One of his papers described the Boonton chlorination process without mentioning Leal, Fuller, or anyone else involved in the project. The paper, published in a reputable journal in 1911, would have been widely circulated among water treatment engineers and would have influenced many readers.[35]

Moses N. Baker, author of the landmark book *The Quest for Pure Water*, quoted a paper written by Johnson as an example of how the credit for chlorination at Boonton Reservoir was misdirected.[36]

Johnson's paper was published in 1911 in the *Journal of the American Public Health Association,* one of the preeminent journals for disseminating new information on disinfection efficacy at the time. A key passage shows that Johnson intended to leave little doubt in the reader's mind as to who was responsible for the chlorination at Boonton Reservoir and how that application of chlorine was the result of his work at the Union Stock Yards.

> "The first demonstration in this country in a practical way of the usefulness of hypochlorites in connection with water purification was made in the fall of 1908 at the filter plant of the Chicago Stock Yards, on the recommendation and under the direction of the writer. Following directly on the heels of the *spectacular results* obtained at Chicago, came the adoption of this process for the sterilization at Boonton, New Jersey, of the impounded and unfiltered water supply of Jersey City, with which work the writer was also connected."[37] (emphasis added)

Virtually the identical paper with the same misattribution had been published the previous year in *Engineering—Contracting*.[38] In addition, Johnson presented and published another 1911 paper that, by omission, did not give credit to anyone else but himself.[39]

Johnson's participation in a 1919 discussion of a chlorination paper shows that he let no opportunity to promote himself go by.

> "An almost unparalleled opportunity for a *spectacular* exhibit of the advantages of chlorination was offered at the Chicago stock-yards where a very unique water problem was presented, and the immediate advantages of this treatment [chloride of lime] became apparent."[40] (emphasis added)

He made no mention of Leal or the Boonton water supply during this discussion.

An examination of early historical references to Johnson's role gives some indication of how his role was inflated and Leal's was largely ignored. "Jersey City has the distinction of being the first to employ the safeguarding of its water supply by chlorination. The system was first installed by Mr. George Johnson and has since been copied by nearly every large city in the country."[41]

A trade journal for industry and finance repeated the misinformation in 1939. "Jersey City has the distinction of being the first to safeguard its water supply by chlorination. The system was installed in September 1908 by the late George R. [sic] Johnson at the Boonton Reservoir."[42]

A brochure produced by Jersey City regarding the potential purchase of the Boonton water supply listed experts for the Jersey City Water Supply Company. "The Water Supply Company . . . [has] engaged a staff of fifteen or more experts including: Messrs. Hering, Fuller and Johnson. . . ." The long list of experts did not include Dr. John L. Leal.[43]

A 1910 paper by Charles-Edward A. Winslow got it completely wrong. Included in a compilation of articles published by the Massachusetts Institute of Technology (MIT), Winslow's paper stated: "The treatment of water supplies by bleaching powder was a different problem however; and the credit for this great advance rests with Mr. G. A. Johnson and his associates. . . ." Leal is mentioned a few paragraphs later but only as a partner with Johnson, who was credited with providing direction to the project.[44]

In a discussion of one of Johnson's papers,[45] Winslow made an astounding statement that bordered on something akin to hero worship.

> "The disinfection of water marks an even greater revolution in engineering, and this whole matter of water disinfection on a practical scale is the work of Mr. Johnson, and of Mr. Johnson alone. It was he who had the courage to face the situation as he saw it. Not only did he have the courage, but he also had the skill to carry out the work, and he did it by such well-proved methods that the one demonstration of the Boonton year's experiment has practically convinced every one as to the importance of this process of water treatment."[46]

Winslow's errors are all the more inexplicable because he was an expert witness for the plaintiffs in the second Jersey City trial, which dealt with the specifics of chloride of lime use at Boonton. Winslow was already a well-respected public health expert in 1910, and he later became president of APHA, editor of the *American Journal of Public Health*, and a giant in the public health field.

Published in the same compilation, a paper by Earle B. Phelps gave proper credit to Leal. "This plant [Boonton], which has been in operation since September, 1908, has been so fully described in various papers by Dr. Leal, who is primarily responsible for its installation. . . ." Phelps was also one of the plaintiffs' expert witnesses.[47]

It makes little sense that in the same volume, MIT published two different accounts from two of its professors, both of whom were involved on the same side in the second Jersey City trial.

Historical accounts that built on the original misinformation have given credit to George A. Johnson as the person responsible for the first use of chlorine in drinking water. An editorial in a prestigious medical journal stated: "The sanitary engineers of this country are responsible for the introduction of chlorine disinfection of water supplies. The first *experiments* were conducted with calcium hypochlorite by George A. Johnson at the Union Stock Yards in Chicago. This was then followed by the treatment of the Jersey City water supply at Boonton, N.J."[48] (emphasis added)

Dr. Leal's role in the use of chlorine at Boonton Reservoir was not mentioned in the editorial. Although the editorial correctly described chlorine use at the Union Stock Yards in Chicago as experimental, the statement implied that this use of chlorine was to treat a water supply for human consumption. The record clearly shows that the treatment plant at the Union Stock Yards treated a disgusting brew of raw sewage and slaughterhouse wastes and produced water for consumption only by cattle and pigs.

In his 1912 book *Sewage Disposal*, Fuller further confused the issue by taking credit for the Boonton chlorination idea for himself and Rudolph Hering. Johnson is mentioned for the Bubbly Creek application. "In connection with the treatment of public water supplies the use of hypochlorites has been the most widely followed in America. In 1908 this treatment was recommended by Messrs. Hering & Fuller for the impounded water supply of Jersey City at the Boonton reservoir, and by Mr. George A. Johnson for the effluent of the mechanical filter plant for treating the water of Bubbly creek, at the Chicago Union Stock Yards."[49] Leal is mentioned later only as one of the authors of the three papers presented at AWWA's annual meeting in 1909.

In 1921, Fuller baldly stated that chlorination was all Johnson's idea. "About a dozen years ago the so-called sterilization or chlorination or disinfection of water supplies was proposed by Col. George A. Johnson, as a means of removing objectionable bacteria."[50]

In some cases, joint credit was given to Leal and the engineers for developing the idea of chlorination at Boonton Reservoir. "The company, through its sanitary officer, Dr. John L. Leal, and its consulting sanitary engineers, Messrs. Rudolph Hering, George W. Fuller and George A. Johnson, commenced a series of experiments to find a means [chloride of lime] of insuring the absolute purity of the water at all times."[51]

The fiction of Johnson's involvement at Boonton Reservoir persisted and was included in Johnson's 1934 obituary, which was actually written by Fuller. "His work on the purification of waters from the Chicago Drainage Canal is well known and there in 1908 as a partner in the firm of Hering and Fuller he employed bleaching powder to sterilize the water supplied to the stock yards. The same year he introduced the process to the Jersey City supply—marking the first American usage of chlorination as a means of purifying water for human consumption."[52]

The fiction continued in Fuller's obituary published in the same journal later that year. "Johnson, who was responsible for the first American installations for the chlorination of water, was at that time an associate with Mr. Fuller who was consultant on the projects."[53]

If anything, George Warren Fuller's role in the Boonton chlorination project has been unfairly minimized. He was responsible for the critically important design of the chloride of lime feed system. If the engineering component of the Boonton chloride of lime plant had not been foolproof, the application of chlorine to drinking water would have been a failure. At this point in his career, Fuller was one of the most famous sanitary engineers in the world. It may be that he did not seek any recognition for his contribution and instead emphasized the role of his protégé, Johnson, who was first associated with him as a laboratory technician in Louisville in 1895.

In AWWA's first compilation of the state of knowledge of water quality and treatment in 1940, the confusion about Johnson's role at Boonton was repeated.

> "The first commercially successful attempt to chlorinate was in America, made by George A. Johnson in 1908. The raw water supply of the Bubbly Creek (Chicago stock yards) filter plant was sterilized with a solution of bleaching powder. The extraordinary results secured in elimination of bacteria without detriment to the quality of the plant effluent caused Johnson *in cooperation with Professor Leal*, to apply the treatment to the entire water supply (40,000,000 gallons daily) of Jersey City, N. J."[54]
> (emphasis added)

In the 1970s, the George A. Johnson myth persisted even in *Journal AWWA*:

"In 1908 a softening plant was built at Columbus, Ohio, and one of the most important steps in public health was taken when bleaching powder was added to the Boonton water supply of Jersey City, N.J., under the direction of George Johnson following a practice begun in Middlekerke, Belgium, in 1902."[55]

In a letter to the editor, Percy S. Wilson, former secretary of AWWA's New Jersey Section, set the record straight in July 1972, a few months after the mistake was published.

"... the use of the bleaching powder at the Boonton supply works of Jersey City, N.J., was done under the direction of Dr. John L. Leal, not Goerge [sic] Johnson. . . . These facts are on record in the papers presented by Dr. Leal, George W. Fuller, George Johnson, and others at the 1909 Convention of the AWWA, at Milwaukee, and published in the Proceedings for that year, starting on page 100."[56]

Wilson was even familiar with the transcript of the trials. "Further verification may be had from other sources, notably the very extensive court records of a trial held in 1908-09. . . ."[57]

Only two years later, another *Journal AWWA* article continued to inflate Johnson's role at Boonton. "In the western hemisphere, the first commercial application of chlorine in water treatment was made in 1908 by G.A. Johnson, who used chloride of lime to chlorinate the Bubbly Creek water supply of the Union Stock Yards in Chicago. In the same year, Johnson and J. L. Leal applied chlorination to the 40-mgd [million gallons per day] Boonton reservoir of Jersey City, N. J.—the first continuous, municipal application of chlorine for water disinfection in the US."[58]

A recent book got the story so wrong that one has to wonder if the author did any research on the Boonton events. "As Johnson supervised the construction of the [Bubbly Creek] plant, he got a call from Jersey City. The beleaguered contractor for the Jersey City Reservoir had heard about his project. If Johnson's system was good enough for cows and pigs in Chicago, he wondered, why wouldn't it work for the people of New Jersey?"[59]

Over the years, some authors got the story right. George E. Symons published a paper on the history of water treatment in

1981, and it was reprinted in 2006. In that paper he gave Leal full credit for the work at Boonton Reservoir.

"The first large-scale chlorination of a municipal water supply took place in spite of the efforts of the city fathers of Jersey City, N.J. . . . Dr. John L. Leal, the company's sanitary adviser, conducted a series of experiments with chlorine liquid and gas and convinced the company that chlorine purification was the way to go. . . . Dr. Leal's perseverance lead to further litigation, after which the court held that chlorine was an effective disinfectant."[60]

Perhaps Moses N. Baker published the clearest statement on the issue of who was responsible for chlorination at Boonton Reservoir and whether Boonton was the first place to use chlorine. "Careful study of available data shows that although the Bubbly Creek plant was put in operation a few days before the one at the Boonton Reservoir, the decision to use chlorination at Boonton was made first, and preliminary tests were made there first as well. The Boonton plant was the conception of Dr. John L. Leal."[61] Unfortunately, Baker's excellent analysis was buried in his 500-page tome, *The Quest for Pure Water*, which was published many years after the early errors were widely known and quoted.

Despite the many attempts to rewrite history, the record is clear. Dr. John L. Leal was responsible for the idea and the execution of the idea to add chlorine to drinking water permanently. Johnson played a minor, supporting role in the Boonton Reservoir project. Regrettably, Johnson's role has been grossly inflated, and this error deserves to be corrected in any future discussions of this historical event.[62]

IN A REWRITTEN AND LATER EDITION of Sedgwick's famous book on sanitary science and public health, Prescott and Horwood published a curiously inaccurate account of Fuller testing chlorine at Louisville. "In 1896 George W. Fuller applied chloride of lime to the raw water flowing on to his experimental rapid sand filters at Louisville with favorable results."[63] As discussed in Chapter 7 of this book, Fuller and the staff at Louisville tested the electrolytic generation of chlorine gas and applied the gas for a few days to the water under test. No bacteriological data demonstrating effective bacteria kill at Louisville has ever been described or published.

Much later, George Clifford White published a strange account of early chlorine use at the Broad Street pump in London. "One of the first known uses of chlorine for disinfection was in the form of hypochlorite, known as chloride of lime. Snow used it in 1850 [sic] in an attempt to disinfect the Broad Street Pump water supply in London after an outbreak of cholera caused by sewage contamination."[64] There is no evidence that this ever happened. The only known use of chloride of lime in connection with the London cholera outbreak was in the streets surrounding the Broad Street pump as a miasma-based sanitary measure to disinfect and deodorize the filth from animals and humans.[65] Unfortunately, many other authors have repeated the fiction published in White's chlorination handbook.[66]

Jersey City Embraces Chlorination

After the long trials, verdicts, and appeals, Jersey City might have been expected to rip out the chlorination system and build a filtration plant or do something else to erase the cause of its courtroom defeat. Not so. In 1910, Jersey City assumed control of the Boonton water supply after paying the Jersey City Water Supply Company the money specified by the court. With all of the positive press and wide adaptation of the Boonton chlorination system, Jersey City became quite proud of the fact that the chlorine revolution had started there, and the city took plenty of opportunities to brag about it.

> "Jersey City has the distinction of being the first to employ the safeguarding of its water supply by chlorination. . . . It is now also being extensively used abroad. It has been a tremendous boon in the safeguarding of public health all over the world and is probably the most important and efficient sanitary measure of protection ever introduced."[67]

In June 1978, Jersey City began operating an 80-mgd conventional water treatment plant on the grounds of the Boonton Reservoir site. The primary purpose of the new treatment plant was to enable the city to meet current and future U.S. Environmental Protection Agency (USEPA) drinking water regulations.[68] It took 70 years, but Jersey City was finally able to provide its customers with multiple-barrier protection from microbial contamination.

Water Industry Hall of Fame

On October 19, 1972, George Warren Fuller's 1971 induction into AWWA's Water Industry Hall of Fame was celebrated with speeches and a reception at the Little Falls Water Purification Plant in Totowa, New Jersey. The Water Industry Hall of Fame honors ". . . those who have made the most significant contributions to the field of public water supply." Fuller was among the first five inductees into this group, which befitted his long career and outstanding accomplishments for the drinking water community. Allen Hazen was also inducted that year.

The festivities surrounding Fuller's induction were recorded in a scrapbook created by Wendell R. Inhoffer, then general superintendent of the water utility and treatment plant on the original site of Fuller's Little Falls treatment plant. More than 100 people attended the celebration, which was billed as a special meeting of the New Jersey Section of AWWA. Notables attending included George E. Symons, vice president of AWWA at the time; officers of the New Jersey Section, including Arnold S. Giannetti, section chairman; and Thaddeus A. Barsh, president of the Passaic Valley Water Commission.

Several former associates and colleagues of Fuller also attended: Percy S. "Syd" Wilson, Ernest Whitlock, James C. Harding, G. F. Wertz, David H. Carmichael, and Rose Barkum Miller. Rose Miller was Fuller's secretary for many years.[69] James Harding gave a speech recounting his memories of working for Fuller in his consulting firms. Harding's talk was modeled after his 1961 *Journal AWWA* paper about Fuller. Pictures of the event show Harding puffing on a cigar and enjoying the hullabaloo about his old boss.

George Symons presented the Water Industry Hall of Fame plaque to Kemp Goodloe Fuller, George Warren Fuller's son. Fuller's grandson, Kemp Goodloe Fuller Jr. was also present. Kemp Fuller Sr. then presented the plaque to Thaddeus Barsh, so it could be displayed in an appropriate location at the Little Falls Water Purification Plant.[70]

On December 20, 1974, a multiple-event celebration was held at Boonton Reservoir. Jersey City Mayor Paul T. Jordan broke ground for the new $30 million water treatment plant. At the same event, AWWA President Chester A. Ring III designated the site of the first chlorination treatment as a "historic landmark" and presented the mayor with a plaque honoring the induction of Dr. John L. Leal into the Water Industry Hall of Fame.[71]

Percy Sidney Wilson, former secretary of AWWA's New Jersey Section,[72] described Dr. John L. Leal as a "truly forgotten man."[73] Since then, the only recognition of Leal's induction into the Water Industry Hall of Fame was his inclusion in a list of New Jersey Hall of Fame inductees in a 1999 New Jersey Section newsletter. Unfortunately, the author of the paragraphs on Leal confused John L. Leal with his father, John R. Leal, and several biographical references in the newsletter were incorrect.[74]

It is not known who nominated Leal for the Water Industry Hall of Fame. Given Wilson's impassioned defense of Leal in *Journal AWWA*, it seems likely that he had a role in pushing for Leal's inclusion among those heroes of public health. Wilson was stating a fact when he wrote in his letter to the *Journal* that "His part in the chlorination at Boonton was soon forgotten, and he died prematurely not long after."[75]

All that is left at the AWWA headquarters today to remind visitors that John L. Leal accomplished something great is a small plaque bearing his name next to the names of other heroes of drinking water protection. The Water Industry Hall of Fame plaque says: "Honoring the Memory of Those Who Have Made Great Contributions Toward Quality Water Service to the Public." Leal's partners on the plaque include those with whom he worked and argued: George Warren Fuller, Allen Hazen, George C. Whipple, Rudolph Hering, and Moses N. Baker. Ironically, the Whipple and Leal plaques are right next to each other—united in honor if not in opinion.

Physician/Engineer Partnership Status

In the twenty-first century, there is typically no direct partnership between engineers and physicians. Environmental engineers who are responsible for providing safe drinking water to the public generally team up with officials from state Departments of Public Health or the USEPA. The officials who work in these agencies are usually environmental engineers just like the folks who work for water utilities. In fact, some individuals move freely among the worlds of water utilities, consulting engineering firms, and state regulatory agencies. USEPA personnel also move among these worlds.

The people who are responsible for protecting public health by controlling waterborne disease have done amazing work, an assertion borne out by the continuing successes in controlling waterborne disease. But instead of working with physicians, today's engineers generally interact with toxicologists, epidemiologists, and other specialists in the fields of chemistry and microbiology. In partnership with these professionals, engineers are responsible for modernizing treatment and distribution systems to ensure that safe drinking water is provided to consumers. There is no need to turn back the clock to a time when physicians were in charge of improving public health in the United States, but there may be an opportunity to bring physicians back into the conversation about what is needed to make drinking water safe.

Currently, there is no organized effort to reach out to physicians and include them in the dialogue about drinking water's role in public health. A few years ago, AWWA developed a Water–Health Workgroup, which led to efforts to improve the association's relationship with the Centers for Disease Control and Prevention (CDC). Guidelines were developed with the CDC and other stakeholders for improving communications with the public, especially when water utilities have to notify the public of contamination problems.[76]

In considering human health, physicians have an ability to focus on what is really important and to disregard the unimportant.[77] Members of the drinking water community have emphasized that we need to keep the legacy of controlling waterborne disease foremost in our minds as we try to understand new contaminants and their potential health effects.[78]

Politicians and policy-makers need to provide funding to both physicians and engineers so that sorting through the competing needs for public investment—including the enhancement of drinking water quality—can be accomplished efficiently. Citizens need to make sure that billions of dollars are not wasted chasing the last chemical molecule and the final statistical anomaly when the most important goal is to protect public health.

Eight Revolutions in Drinking Water Disinfection

As mentioned in the Preface, this book describes only the first revolution in drinking water disinfection outlined in the author's 2006 *Journal AWWA* article—the project that sparked his interest in the history of this technology.[79] The eight disinfection revolutions described in the article were:

1. Jersey City launches full-scale disinfection (1908).

2. Coliforms take center stage: The first federal regulation of bacteria in drinking water was issued (1914).

3. Ammonia–chlorine paves the way for other processes: To control tastes and odors, a new disinfection process was developed (1917).

4. Trihalomethanes are identified: Chlorination of natural organic matter generated a new category of trace organic compounds that were found in drinking water (1974).

5. CT (concentration × contact time) defines disinfection requirements: Federal regulations elevated a theoretical concept to a requirement that had to be met to control viruses and protozoa (1989).

6. Coliforms again take the spotlight: The Total Coliform Rule was adopted in 1989, placing stringent regulatory requirements on this indicator of bacterial contamination.

7. The focus shifts to disinfection by-products: From 1992 to 2000, regulations were negotiated to place stricter controls on organic and inorganic by-products that result from use of any disinfectant.

8. *Cryptosporidium* concerns influence disinfection practices: The severe outbreak of cryptosporidiosis in Milwaukee, Wisconsin, in April 1993 led to more stringent regulations controlling microbial contaminants and an increase in the use of ultraviolet light for inactivation of pathogens.

From a historical perspective, the chlorination of the Boonton water supply is perhaps the most compelling disinfection revolution.

Centennial Celebration of Jersey City's Chlorine Use

On September 24, 2008, a celebration marking the one hundredth anniversary of the use of chlorine at Boonton Reservoir was held at the Liberty House Restaurant in Liberty State Park on the Jersey City waterfront.[80] Jersey City, the American Chemistry Council, and United Water—the private water company currently providing drinking water services to Jersey City—sponsored the event. About 200 people attended. Among the officials who spoke at the reception were Michael Deane, associate administrator of USEPA's Office of Water; Jerramiah T. Healy, the mayor of Jersey City; and Jeffrey Sloan, senior director of the Chlorine Chemistry Division of the American Chemistry Council. Mr. Sloan read a letter of congratulations to Jersey City from President George W. Bush, and a commemorative proclamation from New Jersey's governor was presented. Representatives from the Jersey City Municipal Utilities Authority, the New Jersey Section of AWWA, and United Water also participated in the event.

The Water Quality and Health Council of the Chlorine Chemistry Division of the American Chemistry Council created a Web page and a YouTube video commemorating the anniversary.[81] U.S. Congressman Donald M. Payne also commemorated the event in the *Congressional Record*.[82] A number of other Web pages and publications also noted the anniversary.[83]

If truth prevails, the contributions of a courageous physician and a brilliant engineer to the conquest of waterborne disease will still be remembered in another hundred years.

1   *New York Times,* May 31, 1912.

2   Thomas Edison, interviewed for *The New York Times* on October 11, 1908, stated: "There is a great distinction, however, between the scientific experiment that accomplishes its end and the practical adaptation of it to humanity at large. We read of wonderful things being done experimentally, but whether they can be accomplished practically is another matter."

3   *New York Times,* May 31, 1912.

4   Fuller, "Influence of Sanitary Engineering," 18.

5   Tauber and Chernyak, *Metchnikoff and the Origins of Immunology,* 168.

6   Weis et al., "Chlorination of Taurine by Human Neutrophils," 598; Weis et al., "Oxidative Autoactivation of Latent Collagenase by Human Neutrophils," 747; Grisham et al., "Chlorination of Endogenous Amines by Isolated Neutrophils," 10404; Grisham, et al., "Role of Monochloramine in the Oxidation of Erythrocyte Hemoglobin by Stimulated Neutrophils," 6776.

7   Phagocytosis, YouTube Video; Phagocytosis [HQ], YouTube Video; White Blood Cell Chases Bacteria, YouTube Video.

8   Delves et al., *Roitt's Essential Immunology,* 9.

9   *Paterson Morning Call,* March 14, 1914.

10   Ibid.

11   "Deaths Leal," 208–9.

12   Nelson, *Cedar Lawn Cemetery*, 47.

13   In the spring of 2013, the grave monument was dedicated involving the participation of John L. Leal's great grandchildren.

14   "George W. Fuller," Standard Certificate of Death, June 15, 1934.

15   "George W. Fuller," *Sewage Works Journal*, 807–8.

16   "Sad Milestone," *American Journal of Public Health,* 895.

17   "George W. Fuller," *Transactions*, 1658.

18   "George W. Fuller," *Transactions*, 1659; "George W. Fuller: Industry Pioneer," *Water Engineering Management*, 11.

19   "George W. Fuller," *Transactions*, 1659.

20   Harding, "Personal Reminiscences of George Warren Fuller," 1524.

21   Wolman, "George Warren Fuller," 7.

22   The usual expression is: History is written by the victors.

23   Between Jersey City and Water Company, February 8, 1909, 5126–5128.

24   Ibid., 5128–5129.

25   Between Jersey City and Water Company, October 5, 1909, 6668–6670.

26   Ibid.

27   Between Jersey City and Water Company, October 5, 1909, 6674.

28   Ibid., 6672–6673.

29   Gehrmann, "Experiment in Chemical Purification," 120.

30   Ibid., 121; Baker, in *Quest,* page 339, stated that some testing with hypochlorite was done during the third testing period from July 27 to August 2; however, Gehrmann's article makes it clear that no hypochlorite was tested during that period.

31   Gehrmann, "Experiment in Chemical Purification," 121.

32   "Water Purification Plant of the Chicago Stock Yards," 245.

33   Johnson, "Chicago Stock Yards Water Purification Litigation."

34   Johnson, "Sanitary Significance Common Constituents," 67.

35   Johnson, "Sterilization."

36   Baker, *Quest*, 336.

37   Johnson, "Adaptability and Limitations of Hypochlorite Treatment," 316–9.

38   Ibid.

39   Johnson, "Sterilization."

40   Johnson, "Discussion," 514.

41   See "Scraps of Information"; quote is from a report by D. D. Jackson of Columbia University. Jackson should have known better because he was a witness for the plaintiffs in the second Jersey City trial.

42   Ohland, "The Jersey City Water Supply," 34.

43   Jersey City Board, "Water Question to the Taxpayers."

44   Winslow, "Field for Water Disinfection," 2–3.

45   Johnson, "Sterilization."

46   Winslow, "Discussion," 36.

47   Phelps, "Disinfection of Water and Sewage," 10–11.

48   Editorial, *Journal American Medical Association.*

49   Fuller, *Sewage Disposal,* 730.

50   Fuller, "Influence of Sanitary Engineering," 16–23.

51   Harrison, "Public Water Supplies of Hudson Co."

52   "George Arthur Johnson," *Water Works and Sewerage,* 114.

53   "George W. Fuller," *Water Works and Sewerage,* 235–6.

54   AWWA, *Water Quality and Treatment,* 1940, 133.

55   Singley and Black, "Water Quality and Treatment," 7.

56   Wilson, Letters to the Editor, 28.

57   Ibid.

58   Sletten, "Halogens," 690.

59   Morris, *Blue Death,* 160.

60   Symons, "Water Treatment," 87–98.

61   Baker, *Quest,* 337.

62   There are many other examples of Johnson receiving inflated credit for the chlorination at Boonton, including: Race, *Chlorination of Water,* 11; Prescott and Horwood, *Sedgwick's Principles of Sanitary Science,* 182; Fair and Geyer, *Water Supply and Waste Water Disposal,* 801; Fair, Geyer, and Okun, *Water and Wastewater Engineering,* Vol. 2, 31–14; Hooker, *Chloride of Lime,* 14, 21; AWWA, *Water Works Practice,* 174; James M. Montgomery, *Water Treatment Principles and Design,* 262 (however, Johnson was not mentioned in the second edition—Crittenden, et al., *Water Treatment: Principles and Design,* 1037); "Development in Chlorination Equipment," *Water Works and Sewerage,* 195; Kienle, "Relation of the Chemical Industry to Water Works," 507; Winslow, "Water-Pollution and Water-Purification," 11; Wolman, Donaldson, and Enslow, "Recent Progress in the Art of Water Treatment," 1166.

63   Prescott and Horwood, *Sedgwick's Principles of Sanitary Science,* 182.

64   Black & Veatch, *White's Handbook,* 452.

65   Vinten-Johansen, et al., *Cholera,* 296.

66   On August 14, 2012, I corrected the Wikipedia page that was probably responsible for most of the related errors on other Web sites: Chlorine, http://en.wikipedia.org/wiki/Chloro; Shrestha, "Chlorination"; Bengtson, "Water Chlorination History." Bengtson also claimed that chloride of lime was used on a plant-scale basis to treat drinking water in Hamburg, Germany, in 1893. In fact, chloride of lime was used in Hamburg to treat the sewage discharge from the city.

67   See "Scraps of Information;" quote is from a report by D. D. Jackson of Columbia University.

68   "Jersey City Constructs 80-mgd Treatment Plant," News of the Field, 50–51. Paradoxically, New Jersey's pre-eminence in the field of public health with regard to drinking water treatment did not extend to adoption of other important public health measures. Currently, New Jersey is ranked 49th out of 50 states in the extension of water supply fluoridation. Only 14 percent of the state's citizens are currently provided drinking water containing fluoride; *The New York Times,* March 3, 2012. Many states and cities have struggled with the continuing burden of chemophobia.

69   At the meeting, Rose Miller told one of the guests that she traveled with George Warren Fuller on all of his business trips.

70 It currently hangs in the museum located on the grounds of the treatment plant.

71 Waggoner, *New York Times*, December 21, 1974.

72 Sidney Wilson and Percy S. Wilson are the same man. He signed into the celebration of George Warren Fuller's Hall of Fame induction as Syd Wilson.

73 Waggoner, *New York Times*, December 21, 1974.

74 Hroneich, "New Jersey Hall of Fame," 10.

75 Wilson, Letters to the Editor, 28.

76 Centers for Disease Control and Prevention, "Drinking Water Advisory Communication Toolbox."

77 For example, see the Web page of a family doctor for his opinion on water chlorination: http://www.familydoctormag.com/blog/2008/09/us-celebrates-100-year-anniversary-of-chlorine-in-drinking-water-a-doctors-take-on-the-controversy/

78 Hoffbuhr, "Legacy," 6.

79 McGuire, "Eight Revolutions," 123–49.

80 "Jersey City to Celebrate 100 Years of Safer Drinking Water," September 23, 2008; *Jersey Journal*, December 10, 2008.

81 Water Quality & Health Council, "A Public Health Giant Step: Chlorination of U.S. Drinking Water;" "100 Years of Safer Lives."

82 Payne, "Commemorating the 100th Anniversary of Safe Drinking Water Through Chlorination."

83 Healy, "Jersey City Municipal Water—A Legacy of Innovation, Public Health Protection," http://www.icis.com/Articles/2008/08/25/9150211/chlorine-cleans-water-for-millions-but-comes-under-attack-after-health.html; http://www.dailygotham.com/forum/dr_patrick_moore/100th_anniversary_of_water_chlorination

14

Bibliography

"100 Years of Safer Lives." http://www.youtube.com/
watch?v=aQcAOjFLGoE (accessed October 4, 2012).

Abplanalp, Glen H. 1972 . "Little Falls Treatment Plant: 1902 to
Present." presented at the induction of George W. Fuller into
the AWWA Water Utility Hall of Fame, Little Falls Treatment
Plant, Totowa, N.J., October 19.

Adams, Julius W. 1894. *Sewers and Drains for Populous Districts:
With Rules and Formulae for the Determination of Their
Dimensions.* New York City, N.Y.: Van Nostrand.

Alumni file for John L. Leal. 2007. Princeton College. Sent from
Seeley G. Mudd Manuscript Library. Princeton University.

Alvord, John W. 1917. "Recent Progress and Tendencies in
Municipal Water Supply in the United States." *Journal
AWWA.* 4:3(September): 278–99.

American City. 1912. Water Consumption of Cities 7:41–2.

American Public Health Association (APHA). 1895. *Public Health
Papers and Reports.* Vol. 20, Columbus, Ohio: American Public
Health Association (APHA).

———. 1900. *Public Health Papers and Reports.* Vol. 25, Columbus,
Ohio: APHA.

American Water Works Association Convention. 1906. *Fire and
Water Engineering.* 40:6(July 28): 410–3.

*Appletons' Annual Cyclopaedia and Register of Important Events of
the Year 1893.* 1894. Vol. 8. New York City, N.Y.: D. Appleton
and Co.

Arbona, Sonia and Shannon Crum. "Medical Geography and
Cholera in Perú." http://www.colorado.edu/geography/gcraft/
warmup/cholera/cholera_f.html (accessed October 4, 2012).

"Arrangement and Working of Filter Beds." 1891. *The Engineering
Record.* 23:5 (January 3): 3.

Atkinson, W.B. 1878. *Physicians and Surgeons of the United States.*
n.p. : Robson.

Averill, Chester. 1832. *Facts Regarding the Disinfecting Powers
of Chlorine: With an Explanation of the Mode in Which it
Operates and with Directions How it Should be Applied for*

Disinfecting Purposes. Letter to John I. DeGraff, Mayor of Schenectady. Private printing.

American Water Works Association (AWWA). 1926. *Water Works Practice: A Manual*. Baltimore, Md.: Williams & Wilkins.

AWWA. 1940. *Water Quality and Treatment*. Denver, Co.: AWWA.

AWWA. 1971. *Water Quality and Treatment*. Denver, Co.: AWWA.

Baker, Moses N. 1981. *The Quest for Pure Water: the History of Water Purification from the Earliest Records to the Twentieth Century*. 2nd Edition. Vol. 1. Denver, Co.: American Water Works Association.

———. 1893. *Sewage Purification in America: A Description of the Municipal Sewage Purification Plants in the United States and Canada*. New York City, N.Y.: Engineering Publishing Co.

———. ed. 1897. *The Manual of American Water-Works*. New York City, N.Y.: Engineering News Publishing Co.

Barnes, David S. 2006. *The Great Stink of Paris and the Nineteenth-Century Struggle Against Filth and Germs*. Baltimore, Md.: Johns Hopkins University.

Barry, John M. 2004. *The Great Influenza: The Epic Story of the Deadliest Plague in History*. New York City, N.Y.: Penguin Books.

Bartlett, J.R. 1888. "Outline of Plans for Furnishing an Abundant Supply of Water to the City of New York from a Source Independent of the Croton Watershed Delivered into the Lower Part of the City Under Pressure Sufficient for Domestic, Sanitary, Commercial and Manufacturing Purposes, and for the Extinguishment of Fires with Legal and Engineering and other Papers." Unpublished report.

Bengtson, Harlan. "Water Chlorination History – The mid-1800s through the early 1900s." http://www.brighthub.com/engineering/civil/articles/77511. aspx (accessed October 4, 2012).

Bergen, J.J. and William J. Magie. "Order of Injunction and Modifying Order." In Chancery of New Jersey: Between the Mayor and Aldermen of Jersey City, Complainant, and Patrick H. Flynn and the Jersey City Water Supply Co., Defendants. (August 29, 1905 and October 1, 1907): 4043–6.

Between the Mayor and Aldermen of Jersey City, Complainant, and Patrick H. Flynn and Jersey City Water Supply Company, Defendants: On Bill, etc. (In Chancery of New Jersey) 12 vol. n.p.: privately printed. 1908–10, 1–6987.

Between the Mayor and Aldermen of Jersey City, Complainant, and Patrick H. Flynn and Jersey City Water Supply Company,

Defendants: On Bill, etc. (In Chancery of New Jersey) Exception to Master's Report, 3-45-314, Filed May 20, 1910, 1–4.

Beveridge, T. 1891. "Mayor's Message," *Annual Reports of the City Officers of the City of Paterson, N.J.* (year ending March 20 1891), City of Paterson: City of Paterson.

Bischof, Gustav. 1886. "Notes on Dr. Koch's Water Test." *The Journal of the Society of Chemical Industry.* 5(March 29): 114–21.

Black & Veatch Corporation. 2010. *White's Handbook of Chlorination and Alternative Disinfectants.* New York City, N.Y.: John Wiley & Sons.

Blanck, M.L., "Cholera in New York Bay 1892." http://www. maggieblanck.com/Cholera1892.html, (accessed October 4, 2012).

Board of Health. 1899. *Statement of Mortality.* Paterson, N.J.: City of Paterson. November.

Board of Health. 1899. *Statement of Mortality.* Paterson, N.J.: City of Paterson. December.

Boby, W. 1905. "Discussion on Water Filtration." *Journal of the Royal Sanitary Institute—Transactions,* 26: 682.

Boonton Times. 1899. "Water Company Organized." May 4, 1899. In "Old Boonton and the Jersey City Reservoir." Compiled by Arline Fowler Dempsey.

———. 1899. "The Water Company's Big Mortgage." December 14, 1899. In "Old Boonton and the Jersey City Reservoir." Compiled by Arline Fowler Dempsey.

———. 1901. "Work on Water Works." May 16, 1901. In "Old Boonton and the Jersey City Reservoir." Compiled by Arline Fowler Dempsey, August 9, 1982.

———. 1901. "P.H. Flynn Will Sewer Boonton." July 11, 1901. In "Old Boonton and the Jersey City Reservoir." Compiled by Arline Fowler Dempsey.

———. 1904. "The Visit of the Engineers." January 28, 1904. In "Old Boonton and the Jersey City Reservoir." Compiled by Arline Fowler Dempsey

———. 1906. "Waterworks Suit Begins." February 23, 1906. In "Old Boonton and the Jersey City Reservoir." Compiled by Arline Fowler Dempsey.

Brannan, John W. 1899. "Typhoid Fever: Symptomatology and Treatment." In *Twentieth Century Practice: An International Encyclopedia of Modern Medical Science by Leading Authorities of Europe and America.* Edited by Thomas L. Stedman, Vol. 16, 613–759. New York City, N.Y.: William Wood and Co.

Brewster, J.H. 1910. *Engineering News*. 63:26(June 30): 763; from Proceedings of the Annual Convention of the Indiana Sanitary & Water Supply Association.

Brock, Thomas D. 1999. *Robert Koch: A Life in Medicine and Bacteriology*. Washington, D.C.: ASM Press.

Brown, John J. 1890. "Paterson's Water Supply." In *Paterson, New Jersey, Its Advantages for Manufacturing and Residence: Its Industries, Prominent Men, Banks, Schools, Churches, etc.* edited by Charles A. Shriner. Paterson, N.J.: Paterson Board of Trade, 73–81.

Budd, William. 1861. "Observations on Typhoid or Intestinal Fever: The Pythogenic Theory." *British Medical Journal*. 2:45(November 9): 485–7.

———. 1918. "Typhoid Fever: Its Nature, Mode of Spreading, and Prevention." originally published in 1873, *American Journal of Public Health*. 8(8): 610–12.

"Buffalo". 1894. *The Pharmaceutical Era*. 11:(May 1): 431.

Bunker, George C. 1929. "The Use of Chlorine in Water Purification." *Journal of the American Medical Association*. 92:(January 5): 1–6.

Bureau of Labor Statistics. CPI Inflation Calculator. http://www.bls.gov/data/inflation_calculator.htm (accessed October 4, 2012).

Butler, S.W. 1874. *The Medical Register and Directory of the United States*. Office of the Medical and Surgical Reporter.

Cassedy, James H. 1962. *Charles V. Chapin and the Public Health Movement*. Cambridge, Ma.: Harvard University Press.

Cedar Lawn Cemetery. Lot and Grave Index. Section 4, Lot No. 66. Leal. Selection Date: August 30, 1882.

Census Record. 1910. George W. Fuller. http://search.ancestry.com/cgi-bin/sse.dll?h=16809496&db=1910USCenIndex&indiv=try.

Census Record. 1880. Lucy Hunter. http://search.ancestry.com/cgi-bin/sse.dll?h=21468735&db=1880usfedcen&indiv=try.

Centers for Disease Control and Prevention. "Drinking Water Advisory Communication Toolbox." http://www.cdc.gov/healthywater/emergency/drinking_water_advisory/index.html (accessed October 4, 2012).

Chapin, Charles V. 1895. "The Filtration of Water." *The Medical News*. 66(January 5): 11–4.

———. 1934. "Disinfection in American Cities." *The Medical Officer (London)*. 30(November 17): 232–3, In *Papers of Charles V. Chapin, M.D.* Clarence L. Scamman ed., New York City, N.Y.: Oxford University Press.

Chicago Department of Health. 1907. *City of Chicago Bulletin of the Department of Health*. 10:5(February 2):1–4.

———. 1922. *City of Chicago Bulletin: Chicago School of Sanitary Instruction*. 16:1(January 7): 1–4.

———. 1910. *City of Chicago Bulletin: Chicago School of Sanitary Instruction*. 4:1(January 1): 1–4.

Chick, Harriette. 1908. "An Investigation of the Laws of Disinfection." *The Journal of Hygiene*. 8:1 92–158.

Chlorine. http://en.wikipedia.org/wiki/Chloro. (Accessed November 9, 2012).

Clark, Harry W. 1902. "Purification of Water." In *State Board of Health of Massachusetts, Thirty-Third Annual Report*, 313–21. Boston, Ma.: State of Massachusetts.

———. and Stephen De M. Gage. 1909. "Disinfection as an Adjunct to Water Purification." 23:3 302–19.

———. 1920. "A Study of Massachusetts Water Supplies and the Typhoid Rate: Innocence or Repentance in Drinking Waters." *Journal New England Water Works Association*. 34(9): 203–16.

Clayton, W.W. and William Nelson. 1882. *History of Bergen and Passaic Counties, New Jersey*. Philadelphia, Pa. : Everts & Peck.

Colby, Frank M. and Harry T. Peck eds. 1900. *The International Year Book—A Compendium of the World's Progress During the Year 1899*. n.p.: Dodd, Mead, and Co.

Colwell, Rita R. 1996. "Global Climate and Infectious Disease: The Cholera Paradigm." *Science*. 274:5295(December 20): 2025–31.

Condran, Gretchen A. and Harold R. Lentzner. 2004. "Early Death: Mortality among Young Children in New York, Chicago, and New Orleans." *Journal of Interdisciplinary History* 34:3(Winter): 315–54.

———. Henry Williams and Rose A. Cheney. 1978. "The Decline in Mortality in Philadelphia from 1870 to 1930: The Role of Municipal Services." In *Sickness & Health in America*, 452–66. Edited by Judith W. Leavitt and Ronald L. Numbers. Madison, Wisc.: University of Wisconsin.

Conko, Gregory and Henry I. Miller. 2001. "Precaution (Of a Sort) Without Principle." *Priorities for Health*. 13:3(October 31). http://cei.org/op-eds-and-articles/precaution-sort-without-principle (accessed October 4, 2012).

Corfield, W.H. 1902. *The Etiology of Typhoid Fever and Its Prevention*. London, U.K.: H.K. Lewis.

Craun, Gunther. 1986. *Waterborne Disease in the United States*. New York City, N.Y.: CRC Press.

Crittenden, John C., R. Rhodes Trussell, David W. Hand, Kerry J. Howe and George Tchobanoglous. 2005. *Water Treatment: Principles and Design.* 2nd ed. Hoboken, N.J.: John Wiley & Sons.

"Compulsory Order to Purify the Water-Supply." 1910. *Engineering News.* 63:14(June 16): 712.

Curts, Robert M. and Allen Hazen. 1906. *Joint Committee on Sewage Disposal of the City of Paterson.* Paterson, N.J.: Chronicle Print.

Cutler, David and Grant Miller. 2005. "The Role of Public Health Improvements in Health Advances: The Twentieth-Century United States." *Demography.* 42:1 (February): 1–22.

Daggett, F.W. 1919. "The Trenton, N.J., Filtration Plant." *Journal AWWA.* 6:2(June): 147–56.

De Kruif, Paul. 1996. *Microbe Hunters.* New York City, N.Y.: Harcourt.

"Deaths Leal." 1914. *Jour. Medical Society of New Jersey* 11:4 (April): 208–9.

Debre, Patrice. 1998. *Louis Pasteur.* Baltimore, Md.: Johns Hopkins University.

Delaware County, N.Y. Genealogy and History Site. Marriages from *Bloomville Mirror*—1851–1855, http://www.dcnyhistory. org/oldnewsidx/vrmmalz4.html (accessed October 4, 2012).

Delves, Peter J., Seamus J. Martin, Dennis R. Burton and Ivan M. Riott. 2006. *Roitt's Essential Immunology.* Malden, Mass.: Blackwell Publishing.

Department of Public Health. 1899. City of Newark, N.J. Annual Report.

"Development in Chlorination Equipment." 1934. *Water Works and Sewerage.* 81(June): 195.

Development Proposal for a New Jersey Historic Urban Industrial Park. 1974. The Celebration of a City: Paterson as the First Planned Industrial Center in the Nation. Unpublished Manuscript, Great Falls Development, Inc. Paterson, New Jersey. Presented to Governor Brendan T. Byrne (October 8).

"Discussion." 1895. In *Public Health Papers and Reports.* American Public Health Association. 21, Columbus, OH: APHA, 178.

"Discussion" of three sterilization plant papers by Leal, Fuller and Johnson. 1909. *Proceedings AWWA.* 147-62. Denver, Colo.: American Water Works Association.

Disinfection Committee AWWA. 1978. "Committee Report." *Journal AWWA,* 70:4 219–22.

Dowell, Scott F. and Alexandra M. Levitt. 2002. "Protecting the Nation's Health in an Era of Globalization: CDC's Global Infectious Disease Strategy." Centers for Disease Control and Prevention. Atlanta, Ga.: CDC.

Draft Registration Card, Myron E. Fuller. Philadelphia, Pa. 1918. http://search.ancestry.com/iexec?htx=View&r=an&dbid=64 82&iid=pa-1907635-3239&fn=Myron+E&ln=Fuller&st=r&ss rc=pt_t4622183_p6001826876_kpidz0q3d6001826876z0q26 pgz0q3d32768z0q26pgplz0q3dpid&pid=34018014 (accessed November 19, 2012).

Drown, Thomas M. 1894. "The Electrical Purification of Water." *Journal NEWWA*. 8: 183–7.

Duffy, John. 1990. *The Sanitarians: A History of American Public Health*. Chicago, Ill.: University of Illinois.

Dziejman, Michelle, Balon, Emmy; Boyd, Dana; Fraser, Clare M.; Heidelberg, John F.; and Mekalanos, John J. 2002. "Comparative Genomic Analysis of *Vibrio cholerae*: Genes that Correlate with Cholera Endemic and Pandemic Disease." *Proceedings National Academy of Science*. 99:3(February 5): 1556–61.

Eagle-Tribune. 2007. "Lawrence Cited for Violations of Brand-New Water Treatment Plant." (May 4).

———. 2007. "State Slams City for Putting Drinking Water at Risk: Outside Firm Takes Over Plant at a Cost of $400K." (November 18).

Editorial. 1908. *Engineering News*, 60:(September 3): 257.

Editorial. 1910. *Engineering News*, 63:(February 17): 201.

Editorial. 1914. *Municipal Engineering*. 46:(June): 542.

Editorial. 1919. *Journal American Medical Association.*, (December 27), as quoted in *American Journal of Public Health*. 1920: 276.

Ellms, Joseph W. and S.J. Hauser. 1913. "Orthotolidine as a Reagent for the Colorimetric Estimation of Small Quantities of Free Chlorine." *Journal of Industrial and Engineering Chemistry* 5: 915–1030.

———. *Water Purification*. 1917. New York City, N.Y.: McGraw-Hill.

———. 1921. "Discussion—Romance of Water Storage." *Journal AWWA*. 8:4(July): 300–1.

Ensign Thomas Fuller, Ancestry.com, Story, http://trees. ancestry.com/tree/4622183/person/-136295325/ story/1?pg=32817&pgpl=pid (accessed October 4, 2012).

Ericson, John. 1918. "Chlorination of Chicago's Water Supply." American Journal of Public Health. 8:10 772–5.

Escherich, Theodor. 1886. *Enterobacteria of Infants and Their Relation to Digestion Physiology*. Munich, Germany: n.p.

Evening Journal. 1904. "Water Experts Make Report to the City." (November 19).

———. 1905. "Record has $500,000 Water Bond." (April 22).

———. 1907. "Lawyer Jabs the Water Experts." (April 27).

———. 1909. "In 1998." Editorial Cartoon. (April 30).

———. 1909. "City Won't Stand for More Delays in Water Battle" (May 8).

———. 1909. "Boonton Water Looked Like Ink When Bottled." (May 12).

———. 1909. "More Hard Raps for Jersey City Water." (May 13).

———. 1909. "More Samples of Jersey City Water." (May 20).

———. 1909. "Water Board to Watch Dr. Leal." (June 2).

———. 1909. "Dr. Leal on Boonton Water Process." (June 8).

Evening Times (Trenton). 1898. "Municipal Hygiene." (February 16).

Fair, Gordon M., and John C. Geyer. 1954. *Water Supply and Wastewater Disposal*. New York City, N.Y.: John Wiley & Sons, Inc.

———., and Daniel A. Okun. 1968. *Water and Wastewater Engineering*. Vol. 2. New York City, N.Y.: John Wiley & Sons. Inc.

Frankland, Percy F. 1884. "New Aspects of Filtration and Other Methods of Water Treatment; The Gelatine Process of Water Examination." *Journal of the Society of Chemical Industry*. 4(December 29): 698–709.

Frankland, E. 1896. "The Past, Present, and Future Water Supply of London." *Nature*. 53:1383 (April 30): 619–22.

Frerichs, Ralph R. "John Snow." http://www.ph.ucla.edu/epi/snow.html (accessed October 4, 2012).

Fresenius, C. Reigius. 1871. *A System of Instruction in Quantitative Chemical Analysis*. New York: Wiley.

Fuller, Francis H. 1893. "Descendants of Ensign Thomas Fuller of Dedham,." *The Dedham Historical Register*. Vol. 4: 156–162. Dedham, Ma.:Dedham Historical Society.

Fuller, George W. 1891. "The Specific Organism of Typhoid Fever." *Technology Quarterly—MIT*. 4: 130–46.

———. 1892. "Special Biological Work." In *State Board of Health of Massachusetts, Twenty-Third Annual Report*, 620–33. Boston, Ma.: State of Massachusetts.

———. 1892. "The Differentiation of the Bacillus of Typhoid Fever." In *State Board of Health of Massachusetts, Twenty-Third Annual Report*, 637–44. Boston, Ma.: State of Massachusetts.

———. 1893. "Progress in Sanitary Science." *Science.* 22:549(August 11): 73–4.

———. 1893. "Experiments at the Lawrence Experiment Station Upon the Purification of Water by Sand Filtration." In *State Board of Health of Massachusetts, Twenty-Fourth Annual Report,* 449–538. Boston, Ma.: State of Massachusetts.

———. 1894. "Experiments Upon the Purification of Sewage and Water at the Lawrence Experiment Station." In *State Board of Health of Massachusetts, Twenty-Fourth Annual Report,* 401–61. Boston, Ma.: State of Massachusetts.

———. 1895. "Sand Filtration of Water, with Special Reference to Results Obtained, at Lawrence, Massachusetts." In *American Public Health Association, Public Health Papers and Reports.* Vol. 20, 64–71. Columbus, Ohio: APHA.

———. 1898. *Report on the Investigations into the Purification of the Ohio River Water at Louisville, Kentucky: Made to the President and Directors of the Louisville Water Company.* New York City, N.Y.: Van Nostrand.

———. 1899. *Report on the Investigations into the Purification of the Ohio River Water for the Improved Water Supply of the City of Cincinnati.* Cincinnati, Ohio: City of Cincinnati.

———. 1903. "The Filtration Works of the East Jersey Water Company, at Little Falls, New Jersey." *Transactions of the ASCE.* 29(February): 153–202.

———. 1909. "Description of the Process and Plant of the Jersey City Water Supply Company for the Sterilization of the Water of the Boonton Reservoir." *Proceedings AWWA,* 110–34. Denver, Colo.: American Water Works Association.

———. 1912. *Sewage Disposal.* New York City, N.Y.: McGraw-Hill.

———. 1917. "Relations Between Sewage Disposal and Water Supply are Changing." *Engineering News-Record.* 78:1(April 5): 11–2.

———. 1917. "The Influence of Sanitary Engineering on Public Health." *American Journal of Public Health.* 12:(January): 16–23.

———. 1933. "Progress in Water Purification." *Journal AWWA.* 25(11): 1566–76.

———. and George A. Johnson. 1900. "On the Question of Standard Methods for the Determination of the Numbers of Bacteria in Waters." In *Proceedings APHA—1899.* 25: 574–9.

———. and James R. McClintock. 1926. *Solving Sewage Problems.* New York City, N.Y.: McGraw-Hill.

Garraty, John A. and Mark C. Carnes eds. 1999. *George W. Fuller: American National Biography.* Vol.10, 145–7. New York City, N.Y.: Oxford University Press.

Gehrmann, Adolph. 1909. "An Experiment in Chemical Purification of Water." In *First Report of the Lake Michigan Water Commission,* 120–4. Urbana, Ill.: Lake Michigan Water Commission.

Geison, Gerald L. 1995. *The Private Science of Louis Pasteur.* Princeton, N.J.: Princeton University.

Gelston, W.R. 1910. "Hypochlorite of Lime as an Adjunct to Mechanical Water Filtration at Quincy, Ill." *Engineering News.* 63:14 (April 7): 394–5.

General Catalogue of Princeton University 1746–1906, 260. 1908. Princeton, N.J.: Princeton University.

"George Arthur Johnson." 1934. *Water Works and Sewerage.* 81:(April): 114.

"George W. Fuller." 1934. *Journal AWWA.* 26:7 950–4.

"George Warren Fuller: A Distinguished Expert in Sanitary Science, and an Authority on Waterworks." 1903. *Successful American.* 7:2 100–1.

"George Warren Fuller." 1999. *American National Biography.* John A. Garraty and Mark C. Carnes, eds. Vol.10. New York: Oxford. 145–7.

"George W. Fuller." 1934. *Sewage Works Journal.* 6:4 (July): 807–8.

"George W. Fuller." 1934. *Water Works and Sewerage.* 81:(July): 235–6.

"George W. Fuller." 1935. *Transactions of the American Society of Civil Engineers.* 100: 1653–60.

"George Warren Fuller Award." 1949. *Journal AWWA.* 96:3 284–8.

"George W. Fuller: Industry Pioneer." 2003. *Water Engineering Management.* 23:(May): 10–11.

"George W. Fuller." 1934. Standard Certificate of Death, Department of the City of New York, State of New York, A-75985, Registered Number 14596, June 15.

Gibson, Campbell. 1998. "Population of the 100 Largest Cities and Other Urban Places in The United States: 1790 to 1990." Washington, D.C: U.S. Census Bureau. http://www.census.gov/population/www/documentation/twps0027/twps0027.html (accessed October 4, 2012).

Godlee, Rickman J. 1918. *Lord Lister.* Second edition, London, U.K.: MacMillan Publishers.

Grisham, Matthew B., et al. 1984. "Role of Monochloramine in the Oxidation of Erythrocyte Hemoglobin by Stimulated Neutrophils." *Journal of Biological Chemistry*. 259:11 (June 10): 6766–72.

———. 1984. "Chlorination of Endogenous Amines by Isolated Neutrophils." *Journal of Biological Chemistry*. 259:16 (August 25): 10404–13.

Haines, Michael R. 2001. "The Urban Mortality Transition in the United States, 1800–1940." Historical Paper 134. Cambridge, Ma.: National Bureau of Economic Research.

"Haiti Cholera Outbreak." Atlanta, Ga.: Centers for Disease Control and Prevention. http://www.cdc.gov/haiticholera/ (accessed October 4, 2012).

Halliday, Stephen. 2001. *The Great Stink of London: Sir Joseph Bazalgette and the Cleansing of the Victorian Metropolis*. London, U.K.: History Press.

Harding, James C. 1961. "Personal Reminiscences of George Warren Fuller." *Journal AWWA*. 53(12): 1523–7.

Harrison, Edlow Wingate. 1909. "The Public Water Supplies of Hudson Co., N.J. Particularly with Reference to the Jersey City Supply," read before The Historical Society of Hudson County, November 18, 1909. In *The Historical Society of Hudson County*.

Hartford Courant. "Improvement in Water Department." March 29, 1910.

Hayward, T.E. 1899. "The Causation of Typhoid Fever." *Public Health: The Journal of the Incorporated Society of Medical Officers (England)*. 11(8): 740–4.

Hazen, Allen. 1895. *The Filtration of Public Water-Supplies*. New York City, N.Y.: John Wiley & Sons.

———. Discussion of "Purification of Water for Domestic Use," *Transactions ASCE*. 54:Part D (1905): 247-52.

———. 1916. *Clean Water and How to Get It*. New York City, N.Y.: John Wiley & Sons.

Healy, Jerramiah T. 2008. "Jersey City Municipal Water – A Legacy of Innovation, Public Health Protection." *U.S. Mayor*. April 21.

Hemphill, Sandra. 2007. *The Strange Case of the Broad Street Pump: John Snow and the Mystery of Cholera*. Los Angeles, Ca.: University of California.

Hendricks, David W. 2006. *Water Treatment Unit Processes*. New York: CRC Press.

Herschel, Clemens. Discussion of "The Electrical Purification of Water." *Journal NEWWA*. 8: 186-7.

Hill, Henry W. 1920. "Water Supply: For Municipal, Domestic and Potable Purposes, Including Its Sources, Conservation, Purification and Distribution." In *The Encyclopedia Americana*, 39–65.

Hill, John W. 1893. "Is Our Drinking Water Dangerous," *Proceedings AWWA*, 131–2. Denver, Colo.: American Water Works Association.

Hill, Libby. 2000. *The Chicago River: A Natural and Unnatural History*. Chicago, Ill.: Lake Claremont Press.

Hoffbuhr, Jack W. 1999. "Legacy." *Journal AWWA*. 91(2): 6.

Hooker, Albert D. 1913. *Chloride of Lime in Sanitation*. New York City, N.Y.: John Wiley & Sons.

Houston, Alexander C. 1913. *Studies in Water Supply*. London, U.K. : McMillan and Co.

———. 1921. "B. Welchii, Gastro-Enteritis and Water Supply." *Engineering News-Record*. 87:12(September 22): 484–7.

Howatson, Andrew. 1905. Discussion of "Purification of Water for Domestic Use." *Transactions ASCE*. 54(Part D): 191-2.

Hroneich, J.A. 1999. "New Jersey Hall of Fame Tradition." *New Jersey Pipeline*. New Jersey Section AWWA. 10:.

Hudson v. New Jersey. Supreme Court of United States April 6, 1908. 209 U. S. 349, 62 L. ed., 28 Sup. Ct. Rep: 629.

"Hypochlorite of Lime." 1910. *Engineering News*. 63:14(April 7): 416.

"Hypochlorite or Bleaching Powder." 1910. *Engineering News*. 63:19(May 12): 575.

Jedicke, Peter. 2007. *Scientific American: Great Inventions of the 20th Century*. New York City, N.Y.: Infobase Publishing.

Jennings, C.A. 1914. "Sterilization of Water Supplies: Results of Hypochlorite Process," *Municipal Engineering*. 46(4): 314–8.

———. 1918. "Some Results Secured by Chlorine Compounds in Water Purification and Sewage Treatment." *Municipal Engineering*. 54(6): 249–51.

———. 1948. "The Significance of the Bubbly Creek Experiment." *Journal AWWA*. 40:10 1037-41.

Jensen, J.A. 1910. "A 20,000,000-Gal. Hypochlorite Water-Disinfecting Plant at Minneapolis, Minn." *Engineering News*. 63:14(April 7): 391–2.

"Jersey City's Water Supply." 1898. *City Government*. 4(3): 99–101.

Jersey City Board of Street and Water Commissioners. 1909. "The Water Question to the Taxpayers of Jersey City." Jersey City Library, N352.6, J48Wt, 3–12. Jersey City, N.J. March 1.

"Jersey City's Water Supply Improved." 1925. *Aquafax.* (Feb.–March): 3–8.

"Jersey City Constructs 80-mgd Treatment Plant." 1979. News of the Field. *Journal AWWA.* 71(10): 50–51.

"Jersey City Past and Present." New Jersey City University. http://www.njcu.edu/programs/jchistory/home2.htm (accessed October 11, 2011).

"Jersey City to Celebrate 100 Years of Safer Drinking Water." 2008. Press Release. City of Jersey. Released September 23.

James M. Montgomery, Consulting Engineers. 1985. *Water Treatment Principles and Design.* New York City, N.Y.: John Wiley & Sons.

"John Laing Leal." 1913. *Who's Who in America.* vol. 7, Chicago: A.N. Marquis.

"John L. Leal, M.D." 1916. *Journal of the Medical Society of New Jersey* 13: (August): 427-28.

Johnson, George A. 1909. "Description of Methods of Operation of the Sterilization Plant of the Jersey City Water Supply Company at Boonton, N.J., and Discussion of Results of Analyses of Raw and Treated Water, With Notes on the Cost of Treatment." *Proceedings AWWA,* 135–47. Denver, Colo.: American Water Works Association.

———. 1910. "The Chicago Stock Yards Water Purification Litigation." *Engineering News.* 64:13(September 29): 342–3.

———. 1910. "Adaptability and Limitations of Hypochlorite Treatment of Public Water Supplies." *Engineering—Contracting,* 34:15(October 12): 316–19.

———. 1911. "Sterilization of Public Water Supplies." *Jour. Assoc. Engr. Societies.* 46(1): 12–24.

———. 1913. "Sanitary Significance of the More Common Constituents of Water." *Municipal Engineering.* 44(7): 66–70.

———. 1916. "The Typhoid Toll." *Journal AWWA.* 3:2 249–326.

———. 1919. "Discussion." *Journal AWWA.* 6:3 514–5.

———. 1921. "The Romance of Water Storage." *Journal AWWA.* 8:4 291–300.

Johnson, Steven. 2006. *The Ghost Map: The Story of London's Most Terrifying Epidemic and How It Changed Science, Cities and the Modern World,* New York City, N.Y.: Riverhead Books.

Jordan, Edwin O., George C. Whipple, and C-E.A. Winslow. 1924. *A Pioneer of Public Health: William Thompson Sedgwick*. New Haven, Conn.: Yale University Press.

Keen, William W. 1917. *Medical Research and Human Welfare: A Record of Personal Experiences and Observations During a Professional Life of Fifty-Seven Years*. New York City, N.Y.: Houghton Mifflin.

————. 1923. "A Surgeon's Answer." California State Board of Health *Weekly Bulletin*. 2:8 (April 7): 1.

Kienle, John A. 1919. "The Relation of the Chemical Industry of Niagara Falls to the Water Works." *Journal AWWA*. 6:3 496–523.

Kinnicutt, L.P. 1911. Discussion of "Sterilization of Public Water Supplies." *Jour. Assoc. Engr. Societies*. 46(1): 44–5.

Kirkwood, James P. 1869. *Report on the Filtration of River Waters, for the Supply of Cities, as practised in Europe, made to the Board of Water Commissioners of the City of St. Louis*. Reprint. New York City, N.Y.: D. Van Nostrand.

Kemna, A. 1904. "Purification of Water for Domestic Use— European Practice," *Transactions ASCE*. 61: 167.

Kennedy, Michael. 2004. *A Brief History of Disease, Science and Medicine: From the Ice Age to the Genome Project*. Mission Viejo, Ca.: Asklepiad Press.

Krasner, Stuart W. and Sylvia E. Barrett. 1984. "Aroma and Flavor Characteristics of Free Chlorine and Chloramines," 381–98. Proceedings AWWA Water Quality Technology Conference, Denver Colorado.

Kuhn, Thomas S. 1970. *The Structure of Scientific Revolutions*. Second Edition, Chicago, Ill.: University of Chicago.

Kyriakodis, Harry. "Not Your Ordinary Fire Plugs." http://hiddencityphila.org/2012/03/not-your-ordinary-fire-plugs/ Accessed September 7, 2012.

"Laboratory Tests of the Effect of Storage on the Vitality of the Typhoid Bacillus in the London Water Supply." 1908. *Engineering News*. 60: (September 3): 247–8; summary of report by A.C. Houston, "The Vitality of the Typhoid Bacillus in Artificially Infected Samples of Raw Thames, Lee and New River Water, with Special Reference to the Question of Storage."

Lanier, J. Michael. 1976. "Historical Development of Municipal Water Systems in the United States 1776–1976." *Journal AWWA*. 68(4): 173–80.

"Lawsuit to Compel Water Purification." 1910. *Engineering News.* 63:15(June 23): 723.

Leal, John L. 1887. "Report of the City Physician," *Annual Reports of the City Officers of the City of Paterson, N.J.*, 94–96. Paterson, N.J.: City of Paterson.

————. 1891. "Annual Report of the Board of Health." *Annual Reports of the City Officers of the City of Paterson, N.J.* (the year ending March 20, 1891,) City of Paterson, N.J.: City of Paterson.

————. 1892. "Annual Report of the Board of Health." *Annual Reports of the City Officers of the City of Paterson, N.J.* (the year ending March 20, 1892,) City of Paterson, N.J.: City of Paterson.

————. 1893. "Annual Report of the Board of Health." *Annual Reports of the City Officers of the City of Paterson, N.J.* (the year ending March 20, 1893,) City of Paterson, N.J.: City of Paterson.

————. 1894. "Annual Report of the Board of Health." *Annual Reports of the City Officers of the City of Paterson, N.J.* (the year ending March 20, 1894) City of Paterson, N.J.: City of Paterson.

————. 1895. "Annual Report of the Board of Health." *Annual Reports of the City Officers of the City of Paterson, N.J.* (the year ending March 20, 1895) City of Paterson, N.J.: City of Paterson.

————. 1896. "Annual Report of the Board of Health." *Annual Reports of the City Officers of the City of Paterson, N.J.* (the year ending March 20, 1896) City of Paterson, N.J.: City of Paterson.

————. 1896. "Isolation Hospitals." *Public Health Pap Rep.* 22: 200–206.

————. 1897. "Annual Report of the Board of Health." *Annual Reports of the City Officers of the City of Paterson, N.J.* (the year ending March 20, 1897) City of Paterson, N.J.: City of Paterson.

————. 1898. "Annual Report of the Board of Health." *Annual Reports of the City Officers of the City of Paterson, N.J.* (the year ending March 20, 1898) City of Paterson, N.J.: City of Paterson.

————. 1898. "House Sanitation with Reference to Drainage, Plumbing, and Ventilation." In *American Public Health*

Association, Public Health Papers and Reports. Vol. 23: 401–8. Columbus, Ohio: APHA.

———. 1899. "Annual Report of the Board of Health." *Annual Reports of the City Officers of the City of Paterson, N.J.* (the year ending March 20, 1899) City of Paterson, N.J.: City of Paterson.

———. 1899. "Board of Health, Statement of Mortality— November 1899." City of Paterson, N.J.: City of Paterson.

———. 1899. "Board of Health, Statement of Mortality— December 1899." City of Paterson, N.J.: City of Paterson.

———. 1900. "An Epidemic of Typhoid Fever Due to an Infected Public Water Supply," *American Public Health Association, Public Health Papers and Reports.* Vol. 25: 166–71. Columbus, Ohio: APHA.

———. 1902. "Facts vs. Fallacies of Sanitary Science," *Eleventh Biennial Report of the Board of Health of the State of Iowa for the Period Ending June 30, 1901.* Des Moines, Iowa, 129–40, from *The Christian Advocate.* New York.

———. 1909. "The Sterilization Plant of the Jersey City Water Supply Company at Boonton, N.J." *Proceedings AWWA.* 100–9. Denver, Colo.: American Water Works Association.

———. 1909. "Steralization' (sic) of a Potable Water Supply by Means of 'Bleach,'" *Journal of the Engineers Society of Pennsylvania.* 1(8): 382–91.

———. 1911. "Purification of Drinking Water." *Proceedings AWWA,* 299–314. Denver, Colo.: American Water Works Association.

LeFevre, Edwin. 1912. "Ptomaines and Ptomaine Poisoning." *The Popular Science Monthly.* 80(4): 400–4.

Leighton, Marshall O. 1907. "Pollution of Illinois and Mississippi Rivers by Chicago Sewage: A Digest of the Testimony Taken in the Case of the State of Missouri v. the State of Illinois and the Sanitary District of Chicago." U.S. Geological Survey, Water Supply and Irrigation Paper No. 194, Series L, Quality of Water, 20, Department of the Interior, Washington, D.C.: U.S. Government Printing Office.

Leonard, John W. 1922. *Who's Who in Engineering.* New York City, N.Y.: John W. Leonard Corp.

Life. 1997. "The 100 Events (That Changed the World)." (Fall): 14–134.

"Lincoln Water-Supply." 1898. The Lancet. (Janurary 15): 172.

"Local Government Board Inquiry into the Maidstone Epidemic." 1898. *The Lancet.* (February 5): 391.

"Local Government Board Report on the Epidemic of Typhoid Fever at Maidstone." 1898. *Public Health: The Journal of the Incorporated Society of Medical Officers (England).* 11(10): 50–8.

Longley, Francis F. 1915. "Present Status of Disinfection of Water Supplies." *Journal AWWA.* 2: 679–92.

Lucy Fuller. 1895. Deaths Registered in the City of Boston for the year 1895. Commonwealth of Massachusetts. Volume 456.

Macnamara, C. 1876. *A History of Asiatic Cholera,* London: McMillan.

Magie, William J. 1910. In Chancery of New Jersey: Between the Mayor and Aldermen of Jersey City, Complainant, and the Jersey City Water Supply Co., Defendant. Report for Hon. W.J. Magie, special master on cost of sewers, etc., and on efficiency of sterilization plant at Boonton. (Case Number 27/475-Z-45-314): 1–15. Jersey City, N.J.: Press Chronicle Co.

"Maidstone Epidemic." 1898. *The Sanitary Record and Journal of Sanitary and Municipal Engineering.* 22(September 2): 238.

Maignen, P.A. 1906. "Discussion of Disinfection as a Means of Water Purification." *Proceedings AWWA,* 285–6. Denver, Colo.: American Water Works Association.

Malone, Patrick M. 2009. *Waterpower in Lowell: Engineering and Industry in Nineteenth-Century America.* Baltimore, Md.: Johns Hopkins University.

Manual of the Board of Public Works for the Year 1883–84. 1884. Jersey City, N.J.: Board of Public Works.

Marquis, Albert N. 1913. *Who's Who in America.* Vol. 7. Chicago, Ill.: A.N. Marquis.

Mason, William P. 1906. "Discussion of Disinfection as a Means of Water Purification." *Proceedings AWWA,* 282–3. Denver, Colo.: American Water Works Association.

———. 1907. *Water-Supply: Considered Principally from a Sanitary Standpoint.* 3rd Edition. New York City, N.Y.: John Wiley & Sons.

"Mayor & Aldermen of Jersey City v. Jersey City Water Supply Co." Court of Errors and Appeals. November 21, 1911. In *Atlantic Reporter.* 81 (1912): 1134.

McGuire, Michael J., Irwin H. Suffet and Joseph V. Radziul. 1978. "Assessment of Unit Processes for the Removal of Trace Organic Compounds from Drinking Water." *Journal AWWA,* 70(10): 565-72.

———. Stuart W. Krasner, Cordelia J. Hwang and George Izaguirre. 1983. "An Early Warning System for Detecting

Earthy-Musty Odors in Reservoirs." *Water Science and Technology*, 15:6/7: 267-77.

———. 2006. "Eight Revolutions in the History of U.S. Drinking Water Disinfection," *Journal AWWA*. 98(3): 123–49.

McKee, J.H. 1903. *Back in War Times: History of the 144th Regiment, New York Volunteer Infantry*. n.p.: Horace E. Bailey.

Means, Edward G., Lynn Hanami, Harry F. Ridgway and Betty H. Olson. 1981. "Evaluating Mediums and Plating Techniques for Enumerating Bacteria in Water Distribution Systems." *Journal AWWA*. 73(11): 585–90.

Merriam-Webster. "Revolution—Definition." http://www.merriam-webster.com/dictionary/revolution (accessed October 4, 2012)

Miller, G. Wade and Rip G. Rice. 1978. "European Water Treatment Practices—Their Experience with Ozone." *Civil Engineering*. 48: 76–7.

Milligan, Robert E. 1911. "The Mechanical Filtration Plant at Newport, R.I." *Journal NEWWA*. 25:1 60–5.

Mitchell, H. 1903. *Twenty-Seventh Annual Report of the Board of Health of the State of New Jersey*. October 31.

Moore, Charles. 1901. *Purification of the Washington Water Supply: An Inquiry Held by Direction of the United States Senate Committee on the District of Columbia*. Washington, D.C.: Government Printing Office.

Morris, R.D. 2007. *The Blue Death*. New York City, N.Y.: Harper Collins.

"Mortality Rates in Relation to the Water Supply." 1914. *Municipal Engineering*. 46(1): 49–50.

Nelson, William. 1876. *History and Description of Cedar Lawn Cemetery at Paterson, New Jersey*. Paterson, N.J.: Press Book; Location: New Jersey Historical Society, Newark, N.J..

———. and Charles A. Shriner. 1920. *History of Paterson and Its Environs*. Vol. 2, New York City, N.Y.: Lewis Historical Publishing Company.

Nesfield, V.B. 1903. "A Chemical Method of Sterilizing Water Without Affecting its Potability." *Public Health*. 15(7): 601–3.

———. 1905. "A Simple Chemical Process of Sterilizing Water for Drinking Purposes for Use in the Field and at Home." *The Journal of Preventive Medicine*. 8: 623-32.

New York Passenger List. 1890. George Fuller. http://search.ancestry.com/cgi-bin/sse.dll?h=8245927&db=nypl&indiv=try. *State of Nebraska*. November 5, 1890.

New York Times. 1880. "Princeton's Class of '80." June 24.

————. 1882. "Dr. John R. Leal." August 29.

————. 1889. "To Give Newark Water." August 2.

————. 1892. "Typhoid Epidemic in Chicago." February 14.

————. 1892. "Wiegmann Had Visited Them." September 16.

————. 1895. "Expert in Bacteriology: Convention of Scientists Wise in the Ways of Good and Evil Germs." June 23.

————. 1899. "Jersey City's Water Supply." October 25.

————. 1899. "Paterson Health Inspector Resigns." December 19.

————. 1900. "Died, Arrowsmith." May 8.

————. 1903. "Died, Leal." June 3.

————. 1908. "The Future's Possibilities Outlined by Edison," October 11.

————. 1912. "Wilbur Wright Dies of Typhoid Fever." May 31.

————. 1914. "Dr. John Laing Leal." March 14.

————. 1918. "Mrs. Fuller Gets Divorce." July 19.

————. 1923. "Jersey Water Merger," October 31.

————. 1934. "George W. Fuller, Engineer, Dies, 65." June 16.

————. Photo on Front Page. October 31, 2009.

————. 2012. "In New Jersey, a Battle Over Fluoridation, and the Facts." March 3.

News from the Classes. 1899. *The Technology Review.* Vol 1. Boston, Ma.: Massachusetts Institute of Technology.

North, Charles E. 1921. "Milk and Its Relation to Public Health." In *A Half Century of Public Health: Jubilee Historical Volume of the American Public Health Association.* Mazyck P. Ravenel, ed. New York City, N.Y.: American Public Health Association.

Notes. 1885. *The American Monthly Microscopical Journal* 6:7 138.

Notes. 1885. *Knowledge* 8:198 (August 14) 147.

Ogborn, L. and D.C. Miller. 1990. *Index to Deaths & Marriages in the Delaware Gazette 1819–1879, Delaware County, New York,* 35–6. Derwood, Md.: Muse Productions.

Ohland, H.M. 1939. "The Jersey City Water Supply." *Journal of Industry & Finance.* (September): 34–5.

O'Toole, C.K. 1990. *The Search for Purity: A Retrospective Policy Analysis of the Decision to Chlorinate Cincinnati's Public Water Supply, 1890–1920.* New York City, N.Y.: Garland Publishing.

Palin, A. T. 1957. "The Determination of Free and Combined Chlorine in Water by the Use of Diethyl-p-phenylene Diamine." *Journal AWWA.* 49(7): 873.

————. 1975. "Current DPD Methods for Residual Halogen Compounds and Ozone in Water." *Journal AWWA.* 67(1): 32–3.

Palmer, George T. 1913. "Hypochlorite Treatment of the Water Supply at Trenton, New Jersey." In *Board of Health of the State of New Jersey, Thirty-Sixth Annual Report 1912*, 419–32. Trenton, N.J.: State of New Jersey.

Park, William H. 1916. "Discussion—The Typhoid Toll." *Journal AWWA.* 3:3 812–3.

Passport Application. 1890. George Warren Fuller. No. 14532. Issued May 2. http://search.ancestry.com/iexec?htx=View&r=an&dbid =1174&iid=USM1372_348-0156&fn=George+Warren&ln=Fuller &st=r&ssrc=&pid=1178167 (accessed November 19, 2012).

Pasteur, Louis and Joseph Lister. 1996. *Germ Theory and its Application to Medicine and On the Antiseptic Principle of the Practice of Surgery.* Great Minds Series. Amherst, N.Y.: Prometheus Books.

Paterson Morning Call. 1914. "Dr. John L. Leal has Passed Away." March 14.

Payment, Pierre, D.P. Sartory and D.J. Reasoner. 2003. "The History and Use of HPC in Drinking-Water Quality Management." In *Heterotrophic Plant Counts and Drinking-Water Safety*, 20–48. J. Bartram, J. Cotruvo, M. Exner, C. Fricker and A. Glasmacher eds. London, U.K.: IWA Publishing.

Payne, Donald M. 2008. "Commemorating the 100th Anniversary of Safe Drinking Water Through Chlorination." *Congressional Record—Extension of Remarks,* E2105. Washington D.C: USEPA September 28.

Peck, Ermon M. 1910. "Experiments with Hypochlorite of Lime as a Water Disinfectant at Hartford, Conn." *Engineering News.* 63:14 (April 7): 395.

Phagocytosis, YouTube Video, http://www.youtube.com/watch?v=a1 xPpsxvhVA&feature=related (accessed November 19, 2012).

Phagocytosis [HQ], YouTube Video, http://www.youtube.com/ watch?v=LHuqfN7_QNg&feature=related (accessed November 19, 2012).

Phelps, Earle B. 1910. "The Disinfection of Water and Sewage." In *Contributions from the Sanitary Research Laboratory and Sewage Experiment Station*, 1–17. Boston, Ma.: Massachusetts Institute of Technology.

———. 1910. "Disinfection of Sewage and Sewage Effluents." Transactions American Society for Municipal Improvements. Reprinted in *Contributions from the Sanitary Research*

Laboratory and Sewage Experiment Station, 1–8. Boston, Ma.: Massachusetts Institute of Technology.

Phisterer, Frederick. 1912. *New York in the War of the Rebellion,* 3rd ed. Albany: J. B. Lyon Company.

Pincus, Sol. 1922. "Chlorination as a Means of Insuring the Safety of Drinking Water." In *Public Health Bulletin No. 128,* 29–32. Treasury Department, United States Public Health Service, Washington, D.C.: Government Printing Office.

Plarr, Victor G. 1899. "Woodhead, German Sims." *Men and Women of the Time: A Dictionary of Contemporaries.* 15th edition. London, U.K.: George Routledge and Sons.

Pollitzer, R. 1959. *Cholera.* WHO Monograph No. 43. Geneva:World Health Organization.

Potts, Clyde. 1925. "Sewage Treatment Plant at Boonton," Public Works. 56(7): 231.

Prescott, Samuel C. and Murray P. Horwood. 1935. *Sedgwick's Principles of Sanitary Science and the Public Health: Rewritten and Enlarged.* New York City, N.Y.: McMillan Publishers.

"Progress in 1906 on the Intercepting Sewer System of Chicago." 1907. *Engineering-Contracting.* 27:16 (April 17): 172–3.

"Public Works of Chicago." 1875. *Engineering News.* 2:(April 15): 42–3; continued 2:(May 15): 55–6.

"Pure Water." 1907. *Journal of Tropical Medicine and Hygiene.* 10(January 15): 30.

"Purification of River-Water for Drinking and Industrial Purposes." 1899. *The Lancet.* Vol. 1, (2): 331–2.

Race, Joseph. 1918. *Chlorination of Water.* New York City, N.Y.: John Wiley & Sons.

———. 1950. "Forty Years of Chlorination: 1910–1949." *Journal Institution of Water Engineers.* 4: 479–505.

Rasenberger, Jim. 2007. *America 1908: The Dawn of Flight, the Race to the Pole, the Invention of the Model T, and the Making of a Modern Nation.* New York City, N.Y.: Scribner Publishing.

Ravenel, Mazyck P. 1921. *A Half Century of Public Health: Jubilee Historical Volume of the American Public Health Association.* New York City, N.Y.: American Public Health Association.

Reece, R.J. 1907. "Report on the Epidemic of Enteric Fever in the City of Lincoln, 1904–5." In *Thirty-Fifth Annual Report of the Local Government Board, 1905–6: Supplement Containing the Report of the Medical Officer for 1905–6.* London, U.K.: Local Government Board.

"Remarkable Letter." 1919. *Journal American Medical Association.* 72 (January 11): 122.

Report of Committee on Standard Methods of Water Analysis to the Laboratory Section of the American Public Health Association. 1905. Chicago, Ill.: American Public Health Association.

"Report of the Committee on Limitation and Control of Chlorination of Water." 1923. In *Public Health Bulletin No. 133,* 19–25. Treasury Department, United States Public Health Service. Washington, D.C.: Government Printing Office.

Report of the Filtration Commission of the City of Pittsburgh, Pennsylvania. 1899. Pittsburgh, Pa.: City of Pittsburgh.

"Report on Water-Supply and Sewerage of the State Board of Health of Massachusetts." 1891. Abstract In *Minutes of Proceedings of the Institution of Civil Engineers; with other Selected and Abstracted Papers.* James Forrest ed. Vol. 106, 380–2.

Reverend James Laing: His Legacy, http://andesgazette.net/2009/12/01/reverend-james-laing-his-legacy/ (accessed October 4, 2012).

Reyer, George. 1910. "The Use of Sulphate of Alumina and Hypochlorite of Lime in the Storage and Distributing Reservoir of the Nashville Water-Works." *Engineering News.* 63:14(April 7): 390–1.

Rich, Burdett A. and Henry P. Farnham eds. 1908. *The Lawyers Reports Annotated.* New Series Book 14, 197–209. Rochester, NY: Lawyers Co-Operative.

Rideal, Samuel. 1906. *Sewage and the Bacterial Purification of Sewage.* New York City, N.Y.: John Wiley & Sons.

Rochelle, Paul A. and Clancy, Jennifer L. 2006. "The Evolution of Microbiology in the Drinking Water Industry." *Journal AWWA.* 98(3): 163–91.

Roechling, Herman Alfred. 1898. *Sewer Gas and its Influence Upon Health: Treatise.* London, U.K.: Biggs and Co.

Rosen, George. 1993. *A History of Public Health.* Expanded Edition, Baltimore, Md.: Johns Hopkins University.

Rosenberg, Charles E. 1987. *The Cholera Years.* Chicago, Ill.: Chicago University Press.

"Rudolph Hering." 1924. *Journal AWWA.* 11(1): 304–6.

Sackett, William E., ed. 1917. *Scannell's New Jersey's First Citizens.* Vol. 1, Paterson, NJ: Scannell.

"Sad Milestone in Sanitary Engineering Progress." 1934. *American Journal of Public Health.* 24:8: 895–6.

Sargent, D.A. 1908. "The Physique of Scholars, Athletes and the Average Student." *Popular Science Monthly* 73:13(9): 248–56.

Scannell, J.J., ed. 1920. *Scannell's New Jersey's First Citizens and State Guide,* Vol. 2. Paterson, N.J.: Scannell.

Schevitz, Jules. 1921. "Typhoid Fever in 1920." *The Nation's Health.* 3(7): 42 Adv.

"Scraps of Information." 1929. Department of Streets and Public Improvement. Information Brochure published by Jersey City, New Jersey, located in the Free Public Library of Jersey City.

Sedgwick, William T. 1890. "The Data of Filtration: I. Some Recent Experiments on the Removal of Bacteria from Drinking Water by Continuous Filtration Through Sand." In *Technology Quarterly Massachusetts Institute of Technology,* 69–75. Boston, Ma.: Massachusetts Institute of Technology.

———. 1893. "On Recent Epidemics of Typhoid Fever in the Cities of Lowell and Lawrence Due to Infected Water Supply: With Observations on Typhoid Fever in Other Cities and Towns of the Merrimack Valley, Especially Newburyport." In *State Board of Health of Massachusetts, Twenty-Fourth Annual Report,* 667–704. Boston, Ma.: State of Massachusetts.

———. 1902. *Principles of Sanitary Science and the Public Health: With Special Reference to the Causation and Prevention of Infectious Diseases.* New York City, N.Y.: McMillan Publishers.

———. and J. Scott MacNutt. 1910. "On the Mills-Reincke Phenomenon and Hazen's Theorem Concerning the Decrease in Mortality from Diseases Other Than Typhoid Fever Following the Purification of Public Water-Supplies." *Journal of Infectious Diseases.* 7:4(August 24): 489–564.

Shadwell, A. 1898. "Suicide by Typhoid Fever." *National Review.* 30:179 715–27.

Shrestha, Raju. "Chlorination." http://www.sswm.info/category/implementation-tools/water-purification/hardware/point-use-water-treatment/chlorination (accessed October 4, 2012).

Shriner, Charles A. (ed.) 1890. *Paterson, New Jersey, Its Advantages for Manufacturing and Residence: its Industries, Prominent Men, Banks, Schools, Churches, etc.* Paterson, N.J.: Paterson Board of Trade.

Simard, A. and M.D. Fortier. 1895. "Notes Concerning the Nourishment of Children in their Earliest Infancy, Exclusive of Nursing at the Breast." *Journal APHA.* 20 (11): 367–79.

Sinclair, Upton. 1990. *The Jungle: With an Afterword by Emory Elliott*. New York, City, N.Y.: Signet Classic, original copyright 1905, originally published in 1904.

Singley, Edward G. and A.P. Black. 1972. "Water Quality and Treatment: Past, Present, & Future." *Journal AWWA*. 64(1): 6–10.

"Sir Alexander Houston." 1933. *Nature*. 132 (November 25): 810–1.

Sletten, O. 1974. "Halogens and Their Role in Disinfection." *Journal AWWA*. 66(12): 690.

Smith, H.E. 1894. Discussion of "Electrical Purification of Water." *Journal NEWWA*. 8: 186.

Soper, George A. 1906. "The Role of Public Water Supplies in the Spread of Typhoid Fever as Shown Chiefly by the Greatest Typhoid Epidemics." *Proceedings AWWA*, 79–105. Denver, Colo.: American Water Works Association.

Spalding, Heman and Herman N. Bundesen. 1918. "Control of Typhoid Fever in Chicago." *American Journal of Public Health*. 8: 258–62.

Sperry, Walter A. and Lloyd C. Billings. 1921. "Tastes and Odors from Chlorination." *Journal AWWA*. 8:6 603–15.

Staley, Cady and George S. Pierson. 1899. *The Separate System of Sewerage: Its Theory and Construction*. Third edition, New York City, N.Y.: Van Nostrand Publishing.

Standard Methods for the Examination of Water and Sewage. 1912. Second edition. New York City, N.Y: American Public Health Association.

Standard Methods for the Examination of Water and Wastewater. 1998. APHA, AWWA, WEF. Lenore S. Clesceri, Arnold E. Greenberg and Andrew D. Eaton eds.. Twentieth edition. Washington, D.C.: American Public Health Association.

Stanwell-Smith, R. 1997. "The Maidstone typhoid outbreak of 1897: An Important Centenary." *Euro Surveill*. 1:29(November 13). http://www.eurosurveillance.org/ViewArticle. aspx?ArticleId=1027 (accessed October 4, 2012).

Sternberg, George M. 1888. *Disinfection and Disinfectants: Their Application and Use in the Prevention and Treatment of Disease and in Public and Private Sanitation*. Concord, N.H.: American Public Health Association.

———. 1900. "Disinfection and Individual Prophylaxis Against Infectious Diseases," In *American Public Health Association, Public Health Papers and Reports*. Vol. 25. Columbus, Ohio: American Public Health Association.

Stevens, Frederic W. 1908. In Chancery of New Jersey: Between the Mayor and Aldermen of Jersey City, Complainant, and Patrick H. Flynn and Jersey City Water Supply Co., Defendants; Opinion, May 1, 1908.

———. 1908. In Chancery of New Jersey: Between the Mayor and Aldermen of Jersey City, Complainant, and Patrick H. Flynn and Jersey City Water Supply Co., Defendants; Final Decree, June 4, 1908.

———. 1910. In Chancery of New Jersey: Between the Mayor and Aldermen of Jersey City, Complainant, and Patrick H. Flynn and Jersey City Water Supply Co., Defendants; Decree Confirming Report, etc., Case Number 27/475-Z-45-314, November 15, 1–10.

Stewart, D.D. 1893. *Treatment of Typhoid Fever*. Detroit, Mich.: George S. Davis.

Suarez, Ruben and Bonnie Bradford. 1993. "The Economic Impact of the Cholera Epidemic in Peru: An Application of the Cost of Illness Methodology." Washington Field Report No. 415 for USAID, http://pdf.usaid.gov/pdf_docs/PNABP618.pdf (accessed October 4, 2012).

Sullivan, John L. 1828. *Report, on the Origin and Increase of the Paterson Manufactories and the Intended Diversion of Their Waters by the Morris Canal Company*. Paterson, N.J.: unknown publisher.

Swarts, Gardner T. 1895. "Discussion on the Foregoing Group of Papers From 'The Cart Before the Horse' to 'The Report of the Committee on the Pollution of Water Supplies,' Inclusive." In *American Public Health Association, Public Health Papers and Reports*. Vol. 20, 83–4. Columbus, Ohio: American Public, Health Association.

Symons, George E. 2006. "Water Treatment Through the Ages." *Journal AWWA*. 98(3): 87–98.

Tauber, Alfred I. and Leon Chernyak. 1991. *Metchnikoff and the Origins of Immunology: From Metaphor to Theory*. New York City, N.Y.: Oxford University.

The Times. 1854. "The Cholera Near Golden Square." September 15.

Tickner, Joel and Tami Gouveia-Vigeant. 2005. "The 1991 Cholera Epidemic in Peru: Not a Case of Precaution Gone Awry." *Risk Analysis*. 25:3 495–502.

Tiernan, Martin F. 1948 . "Controlling the Green Goddess." *Journal AWWA*. 40:10 1042-50.

Todd, Robert E. and Frank B. Sanborn. 1912. *The Report of the Lawrence Survey*. Andover, Ma.: Andover Press.

Tomes, Nancy. 1998. *The Gospel of Germs: Men, Women, and the Microbe in American Life*. Cambridge, Ma.: Harvard University.

Transactions of the Medical Society of New Jersey—1894. 1894. Newark, N.J.: L.F. Hardham.

Transactions of the Medical Society of New Jersey—1900. 1900. Newark, N.J.: L.F. Hardham.

Transactions of the Medical Society of New Jersey—1905. 1905–1906. Vol 2. Newark, N.J.: L.F. Hardham.

Trenton Evening Times. 1911. "State Takes Hand in Typhoid Fight." December 6.

———. 1911. "Expert Says Water is Safe and Can be Made Pure by Filter." December 21.

———. 1885. "A Sanitary Tract: Which is Worth Preserving for Future Reference, Disinfection Made Practical, Easy and Effective." June 25.

Troesken, Werner. 2004. *Water, Race, and Disease*. Cambridge, Ma: Massachusetts Institute of Technology Press.

Trowbridge, W.P. 1885. *Reports on the Water Power of the United States, Statistics of Power and Machinery Employed in Manufactures*, Part 1, Census Bureau. Washington D.C.: Government Printing Office.

Turneaure, F.E., and H.L. Russell. 1901. *Public Water-Supplies: Requirements, Resources, and the Construction of Works*. 1st Edition. New York City, N.Y.: John Wiley & Sons, Inc.

Tully, E.J. 1914. "Calcium Hypochlorite as a Water Disinfectant." *Municipal Engineering*. 47:2 137.

Twentieth Annual Report of the Paterson General Hospital Association, Paterson, New Jersey, 1891. Paterson, N.J.:Paterson General Hospital Assocation.

Twenty-Second Report of the Paterson General Hospital Association, Paterson, New Jersey, 1893. Paterson, N.J.:Paterson General Hospital Association

Twenty-Third Report of the Paterson General Hospital Association, Paterson, New Jersey, 1894. Paterson, N.J.:Paterson General Hospital Association

Twenty-Fourth Report of the Paterson General Hospital Association, Paterson, New Jersey, 1895. Paterson, N.J.:Paterson General Hospital Association

Bibliography

Twentieth Annual Report of the State Board of Health, of the State of Rhode Island. 1891. (for the year ending December 31, 1891). Providence, RI:E. L. Freeman Co.

Tyndall, John. 1892. "Louis Pasteur His Life and Labours," in *New Fragments.* Second edition. London, U.K.: Longmans, Green and Co.

"Typhoid Epidemic at Maidstone." 1897. *Journal of the Sanitary Institute.* 18(11): 388.

U.S. Bureau of the Census. 1965. *The Statistical History of the United States from Colonial Times to the Present.* Vols. 1 & 2. Stamford, Conn.: Fairfield Publishers.

U.S. Environmental Protection Agency (USEPA). 1979. "Control of Trihalomethanes in Drinking Water. Final Rule." *Fed. Reg.* 44:231 (November 29) 68624.

USEPA. 1994. "Disinfectants/Disinfection By-Products. Proposed Rule." *Fed. Reg.* 59:145 (July 29): 38668.

USEPA. 1998. "Disinfectants and Disinfection By-Products. Final Rule." *Fed. Reg.* 63:241 (December 16): 69390.

USEPA. 2006. "Stage 2 Disinfectants and Disinfection By-Products Rule." *Fed. Reg.* 71(2) (1): 38.

"U.S. Supreme Court Prohibits NJ Diversions." 1908. *Engineering News.* 59:17(April 23): 454.

Van Poppel, Frans and Cor van der Heijden. 1997. "The Effects of Water Supply on Infant and Childhood Mortality: A review of Historical Evidence." *Health Transition Review.* 7: 113–48.

Vermeule, Cornelius C. *Report on Water-Supply—Water-Power, the Flow of Streams and Attendant Phenomena.* 1894. Vol. 3 of the Final Report of the State Geologist. *Geological Survey of New Jersey.* Trenton, N.J.: John L. Murphy Publishing Co.

Vinten-Johansen, Peter, Howard Brody, Nigel Paneth, Stephen Rachman and Michael Rip. 2003. *Cholera, Chloroform, and the Science of Medicine.* New York City, N.Y.: Oxford University Press.

Waggoner, W.H. 1974. "$30-Milllion Water Plant Is Begun by Jersey City." *New York Times.* December 21.

Walden, A.E. 1909. "Some Results Obtained by the Application of Hypochlorite of Lime on Mechanical and Slow Sand Filters and the Method of Controlling Coagulent (sic) and the Operation of the Orifice Boxes." *Proceedings AWWA,* 26–38. Denver, Colo.: American Water Works Association.

Wallace & Tiernan's Fiftieth Anniversary. 1963. Brochure prepared for the Fiftieth Anniversary of Wallace & Tiernan, Inc.

Washington Post. 1902. "Dr. Wyman at the Head." December 13.

"Water Purification Plant of the Chicago Stock Yards." 1910. *Engineering News.* 64:10 (September 8): 245–6.

Water Quality & Health Council, Chlorine Chemistry Division, American Chemistry Council. "A Public Health Giant Step: Chlorination of U.S. Drinking Water." http://www. waterandhealth.org/drinkingwater/chlorination_history.html (accessed October 4, 2012).

"Water Sterilization Plants." 1914. *Municipal Engineering.* 46(1): 11–3.

"Water Supply of Jersey City." A brochure compiled by the Free Public Library of Jersey City, about 1909.

Watson, B.A. 1883. "Necrology." *Journal of the American Medical Association.* 1 (10): 406.

Weis, Stephen J. et al. 1982. "Chlorination of Taurine by Human Neutrophils." *Journal of Clinical Investigations.* 70(9): 598–607.

———. et al. 1985. "Oxidative Autoactivation of Latent Collagenase by Human Neutrophils." *Science.* 227:4688 747–9.

Welch, William H. 1905. "Introduction to the 1897 Report of the Bacteriological Committee." American Public Health Association. *Public Health Papers and Reports.* Vol. 30, Columbus, OH:APHA.

Weston, Arthur D. 1920. "Epidemic of Gastro-Enteritis in Peabody, Mass., October, 1913." *Journal New England Water Works Association.* 20:3 193–4.

Whipple, George Chandler. 1905. Discussion of "Purification of Water for Domestic Use." *Transactions ASCE.* 54:Part D 192–206.

———. 1906. "Disinfection as a Means of Water Purification." *Proceedings AWWA,* 266–80. Denver, Colo.: American Water Works Association.

———. 1914. *Typhoid Fever.* 1908. New York City, N.Y.: John Wiley & Sons.

———. 1915. "Journal, January 8, 1915." Papers of George Chandler Whipple. Harvard University Archives. Call No. HUG 1876.3005.

———. 1915. "Biographical Memorandum of George W. Fuller," August 13, 1915: 1–4. George Chandler Whipple Papers, Harvard University Archives.

———. 1917. *State Sanitation: A Review of the Work of the Massachusetts State Board of Health.* Vol 1. Cambridge, Ma.: Harvard Press.

Bibliography

White Blood Cell Chases Bacteria, YouTube Video. http://www.
youtube.com/watch?v=e2i3cEar60I (accessed November 20,
2012).

White, George C. 1999. *Handbook of Chlorination and Alternative
Disinfectants.* New York: Wiley. [White, *Handbook of
Chlorination*, 3.]

Wilson, Percy S. Letters to the Editor. 1972. *Journal AWWA.* 64(7):
28.

Winslow, Charles-Edward A. 1910. "Water-Pollution and Water-
Purification at Jersey City, N J." In *Contributions from the
Sanitary Research Laboratory and Sewage Experiment Station.*
Boston, Ma.: Massachusetts Institute of Technology, 1–24.
Reprinted from *Journal of the Western Society of Engineers.*
15:4 (August 1910).

―――. 1910. "The Field for Water Disinfection from a Sanitary
Standpoint." In *Contributions from the Sanitary Research
Laboratory and Sewage Experiment Station.* 1-6. Boston, Ma.:
Massachusetts Institute of Technology.

―――. 1911. "Discussion." *Jour. Assoc. Engr. Societies.* 46:1 36–7.

―――. 1916. Letter to William T. Sedgwick. Charles-Edward
Amory Winslow papers, 1874–1977 (inclusive), 1915–1945
(bulk). : General Correspondence. Yale University Library,
Manuscripts and Archives, May 23, 1916.

―――. and David Greenberg. 1918. "The Effect of Putrefactive
Odors Upon Growth and Upon Resistance to Disease."
American Journal of Public Health. 8 (10): 759–68.

―――. 1953. "They Were Giants in Those Days." *American Journal
of Public Health.* 43(6): 15–9.

Wishnow, Rodney M. and Jesse L. Steinfeld. 1976. "The Conquest
of the Major Infectious Diseases in the United States: A
Bicentennial Retrospective." *Annual Review Microbiology.* 30:
427–50.

Wolf, Harold W. 1972. "The Coliform Count as a Measure of Water
Quality." In *Water Pollution Microbiology,* 333–45. Ralph
Mitchell ed. New York: Wiley-Interscience.

Wolf, Jacqueline H. 2001. *Don't Kill Your Baby: Public Health and
the Decline of Breastfeeding in the Nineteenth and Twentieth
Centuries.* Columbus, Ohio: Ohio State University.

Wolman, Abel and I. H. Enslow. 1919. "Chlorine Absorption and
the Chlorination of Water." *Industrial Engineering Chemistry.*
11:3 209.

————. 1921. "Water Chlorination Control by the Absorption Method." *Engineering News Record.* 86:15 (April 14): 639–41.

————., Wellington Donaldson and Linn H. Enslow. 1930. "Recent Progress in the Art of Water Treatment." *Journal AWWA.* 22(9): 1161–77.

————. and Arthur E. Gorman. 1931. *The Significance of Waterborne Typhoid Fever Outbreaks.* Baltimore, Md.: Williams & Wilkins Co.

————. 1976. "George Warren Fuller: A Reminiscence." Essays in *Public Works History.* Essay Number 2. Washington, D.C.: Public Works Historical Society.

Woodhead, G. Sims and W.J. Ware. 1901. "Disinfection of the Maidstone Water Service Mains." In *Society of Engineers: Transactions for 1900,* 49–58. New York City, N.Y.: Spon and Chamberlain.

Yeo, J. Burney. 1868. "On the Treatment of Typhoid Fever." *The Medical Times and Gazette.* Vol. 1(February 1): 117–8.

Youmans, W.J. 1889. *The Popular Science Monthly.* 35(May to October): 282.

Zim. 1919. "Is Your City in the Vanguard Fighting Water-Borne Typhoid?" Cartoon. *American City.* 21:3 247.

Zuger, Abigail. 2011. "Folded Saris: Cholera Prevention Minimal Cost (Used Saris are Better Than New)." *New York Times.* September 27.

About the Author

Photo Credit: Aloma Ichinose, 2012

Michael J. McGuire is an environmental engineer who has worked in the drinking water quality field for more than 40 years. He has a B.S. in civil engineering from the University of Pennsylvania (1969), and obtained both M.S. and Ph.D. (1977) degrees in environmental engineering from Drexel University. McGuire has been active in professional associations throughout his career—especially with the American Water Works Association. He is a widely published author of technical articles and the co-editor of four books. He has won numerous awards from professional societies for his research on drinking water quality improvement including one he especially treasures from AWWA—the George Warren Fuller Award in 1994. McGuire was elected to the National Academy of Engineering in 2009.

Index

Index

Index

Index

Index

Index

Index